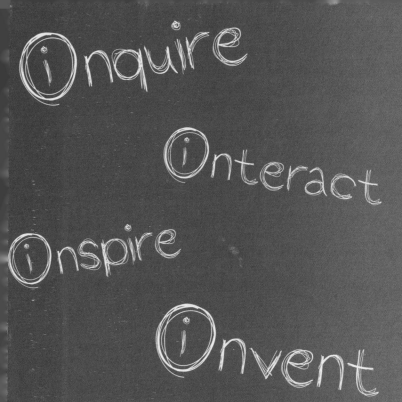

Ⓘnquire

Ⓘnteract

Ⓘnspire

Ⓘnvent

This ⒾScience Interactive Student Textbook Belongs to:

_____

Name

_____

Teacher/Class

Where am I located?

The dot on the map shows where my school is.

McGraw Hill Education

ⒾSCIENCE

Glencoe

Northern Saw-Whet Owl, *Aegolius acadicus*
This small owl is nocturnal and, therefore, seldom is seen. It is only about 17 cm–22 cm in length and has a wingspan of about 50 cm–56 cm. Its habitat includes short conifers and dense thickets across most of the United States and southern Canada.

The McGraw·Hill Companies

 Education

Send all inquiries to:
McGraw-Hill Education
8787 Orion Place
Columbus, OH 43240-4027

ISBN: 978-0-07-660222-3
MHID: 0-07-660222-2

Printed in the United States of America.

8 9 10 11 12 13 QVS 19 18 17 16 15

Florida Teacher Advisory Board

The Florida Teacher Advisory Board provided valuable input in the development of the © 2012 Florida student textbooks.

Ray Amil
Union Park Middle School
Orlando, FL

Maria Swain Kearns
Venice Middle School
Venice, FL

Ivette M. Acevedo Santiago, MEd
Resource Teacher
Lake Nona High School
Orlando, FL

Christy Bowman
Montford Middle School
Tallahassee, FL

Susan Leeds
Department Chair
Howard Middle School
Orlando, FL

Rachel Cassandra Scott
Bair Middle School
Sunrise, FL

Authors and Contributors

Authors

American Museum of Natural History
New York, NY

Michelle Anderson, MS
Lecturer
The Ohio State University
Columbus, OH

Juli Berwald, PhD
Science Writer
Austin, TX

John F. Bolzan, PhD
Science Writer
Columbus, OH

Rachel Clark, MS
Science Writer
Moscow, ID

Patricia Craig, MS
Science Writer
Bozeman, MT

Randall Frost, PhD
Science Writer
Pleasanton, CA

Lisa S. Gardiner, PhD
Science Writer
Denver, CO

Jennifer Gonya, PhD
The Ohio State University
Columbus, OH

Mary Ann Grobbel, MD
Science Writer
Grand Rapids, MI

Whitney Crispen Hagins, MA, MAT
Biology Teacher
Lexington High School
Lexington, MA

Carole Holmberg, BS
Planetarium Director
Calusa Nature Center and Planetarium, Inc.
Fort Myers, FL

Tina C. Hopper
Science Writer
Rockwall, TX

Jonathan D. W. Kahl, PhD
Professor of Atmospheric Science
University of Wisconsin-Milwaukee
Milwaukee, WI

Nanette Kalis
Science Writer
Athens, OH

S. Page Keeley, MEd
Maine Mathematics and Science Alliance
Augusta, ME

Cindy Klevickis, PhD
Professor of Integrated Science and Technology
James Madison University
Harrisonburg, VA

Kimberly Fekany Lee, PhD
Science Writer
La Grange, IL

Michael Manga, PhD
Professor
University of California, Berkeley
Berkeley, CA

Devi Ried Mathieu
Science Writer
Sebastopol, CA

Elizabeth A. Nagy-Shadman, PhD
Geology Professor
Pasadena City College
Pasadena, CA

William D. Rogers, DA
Professor of Biology
Ball State University
Muncie, IN

Donna L. Ross, PhD
Associate Professor
San Diego State University
San Diego, CA

Marion B. Sewer, PhD
Assistant Professor
School of Biology
Georgia Institute of Technology
Atlanta, GA

Julia Meyer Sheets, PhD
Lecturer
School of Earth Sciences
The Ohio State University
Columbus, OH

Michael J. Singer, PhD
Professor of Soil Science
Department of Land, Air and Water Resources
University of California
Davis, CA

Karen S. Sottosanti, MA
Science Writer
Pickerington, Ohio

Paul K. Strode, PhD
I.B. Biology Teacher
Fairview High School
Boulder, CO

Jan M. Vermilye, PhD
Research Geologist
Seismo-Tectonic Reservoir Monitoring (STRM)
Boulder, CO

Judith A. Yero, MA
Director
Teacher's Mind Resources
Hamilton, MT

Dinah Zike, MEd
Author, Consultant, Inventor of Foldables
Dinah Zike Academy; Dinah-Might Adventures, LP
San Antonio, TX

Margaret Zorn, MS
Science Writer
Yorktown, VA

Authors and Contributors

Consulting Authors

Alton L. Biggs
Biggs Educational Consulting
Commerce, TX

Ralph M. Feather, Jr., PhD
Assistant Professor
Department of Educational
Studies and Secondary Education
Bloomsburg University
Bloomsburg, PA

Douglas Fisher, PhD
Professor of Teacher Education
San Diego State University
San Diego, CA

Edward P. Ortleb
Science/Safety Consultant
St. Louis, MO

Series Consultants

Science

Solomon Bililign, PhD
Professor
Department of Physics
North Carolina Agricultural and
Technical State University
Greensboro, NC

John Choinski
Professor
Department of Biology
University of Central Arkansas
Conway, AR

Anastasia Chopelas, PhD
Research Professor
Department of Earth and Space
Sciences
UCLA
Los Angeles, CA

David T. Crowther, PhD
Professor of Science Education
University of Nevada, Reno
Reno, NV

A. John Gatz
Professor of Zoology
Ohio Wesleyan University
Delaware, OH

Sarah Gille, PhD
Professor
University of California San
Diego
La Jolla, CA

David G. Haase, PhD
Professor of Physics
North Carolina State University
Raleigh, NC

Janet S. Herman, PhD
Professor
Department of Environmental
Sciences
University of Virginia
Charlottesville, VA

David T. Ho, PhD
Associate Professor
Department of Oceanography
University of Hawaii
Honolulu, HI

Ruth Howes, PhD
Professor of Physics
Marquette University
Milwaukee, WI

Jose Miguel Hurtado, Jr., PhD
Associate Professor
Department of Geological
Sciences
University of Texas at El Paso
El Paso, TX

Monika Kress, PhD
Assistant Professor
San Jose State University
San Jose, CA

Mark E. Lee, PhD
Associate Chair & Assistant
Professor
Department of Biology
Spelman College
Atlanta, GA

Linda Lundgren
Science writer
Lakewood, CO

Keith O. Mann, PhD
Ohio Wesleyan University
Delaware, OH

Charles W. McLaughlin, PhD
Adjunct Professor of Chemistry
Montana State University
Bozeman, MT

Katharina Pahnke, PhD
Research Professor
Department of Geology and
Geophysics
University of Hawaii
Honolulu, HI

Jesús Pando, PhD
Associate Professor
DePaul University
Chicago, IL

Hay-Oak Park, PhD
Associate Professor
Department of Molecular
Genetics
Ohio State University
Columbus, OH

David A. Rubin, PhD
Associate Professor of Physiology
School of Biological Sciences
Illinois State University
Normal, IL

Toni D. Sauncy
Assistant Professor of Physics
Department of Physics
Angelo State University
San Angelo, TX

Series Consultants, continued

Malathi Srivatsan, PhD
Associate Professor of
Neurobiology
College of Sciences and
Mathematics
Arkansas State University
Jonesboro, AR

Cheryl Wistrom, PhD
Associate Professor of Chemistry
Saint Joseph's College
Rensselaer, IN

Reading

ReLeah Cossett Lent
Author/Educational Consultant
Blue Ridge, GA

Math

Vik Hovsepian
Professor of Mathematics
Rio Hondo College
Whittier, CA

Series Reviewers

Thad Boggs
Mandarin High School
Jacksonville, FL

Catherine Butcher
Webster Junior High School
Minden, LA

Erin Darichuk
West Frederick Middle School
Frederick, MD

Joanne Hedrick Davis
Murphy High School
Murphy, NC

Anthony J. DiSipio, Jr.
Octorara Middle School
Atglen, PA

Adrienne Elder
Tulsa Public Schools
Tulsa, OK

Carolyn Elliott
Iredell-Statesville Schools
Statesville, NC

Christine M. Jacobs
Ranger Middle School
Murphy, NC

Jason O. L. Johnson
Thurmont Middle School
Thurmont, MD

Felecia Joiner
Stony Point Ninth Grade Center
Round Rock, TX

Joseph L. Kowalski, MS
Lamar Academy
McAllen, TX

Brian McClain
Amos P. Godby High School
Tallahassee, FL

Von W. Mosser
Thurmont Middle School
Thurmont, MD

Ashlea Peterson
Heritage Intermediate Grade
Center
Coweta, OK

Nicole Lenihan Rhoades
Walkersville Middle School
Walkersvillle, MD

Maria A. Rozenberg
Indian Ridge Middle School
Davie, FL

Barb Seymour
Westridge Middle School
Overland Park, KS

Ginger Shirley
Our Lady of Providence Junior-
Senior High School
Clarksville, IN

Curtis Smith
Elmwood Middle School
Rogers, AR

Sheila Smith
Jackson Public School
Jackson, MS

Sabra Soileau
Moss Bluff Middle School
Lake Charles, LA

Tony Spoores
Switzerland County Middle
School
Vevay, IN

Nancy A. Stearns
Switzerland County Middle
School
Vevay, IN

Kari Vogel
Princeton Middle School
Princeton, MN

Alison Welch
Wm. D. Slider Middle School
El Paso, TX

Linda Workman
Parkway Northeast Middle
School
Creve Coeur, MO

 interact . . .

With your book!

Answer questions, record data, and interact with images directly in your book!

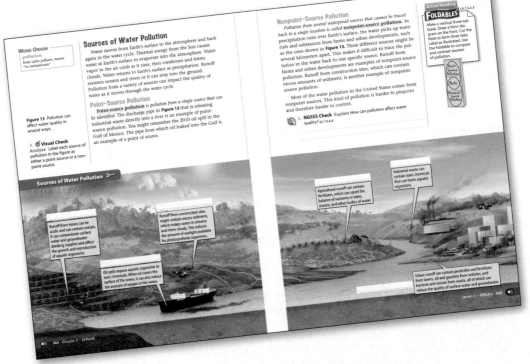

Online!

Log on to **ConnectED** for a digital version of this book that includes

- audio;
- animations;
- virtual labs.

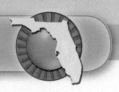

Labs, Labs, Labs

Launch Labs at the beginning of every lesson let you be the scientist! The ⓘLAB Station on **Connect ED** has all the labs for each chapter.

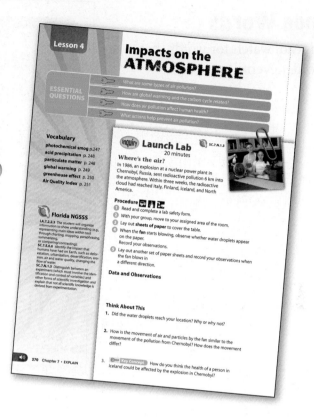

Virtual Labs

Virtual Labs provide a highly interactive lab experience.

. . .with ⓘSCIENCE.

ⓘ Read ⓘ Science

Sequence Words

While you read, watch for words that show the order events happen:

- first
- next
- last
- begins
- second
- later

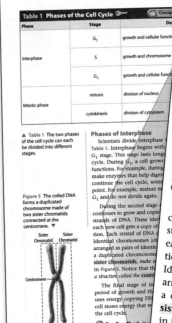

Vocabulary Help

Science terms are highlighted and reviewed to check your understanding.

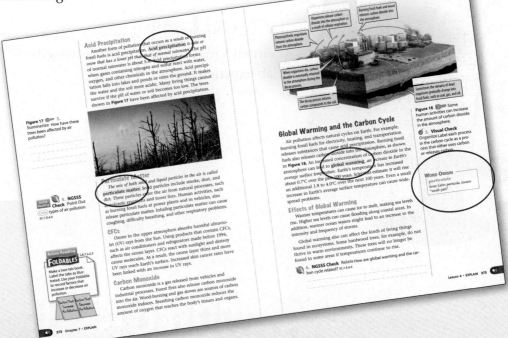

Write the answers to questions right in your book!

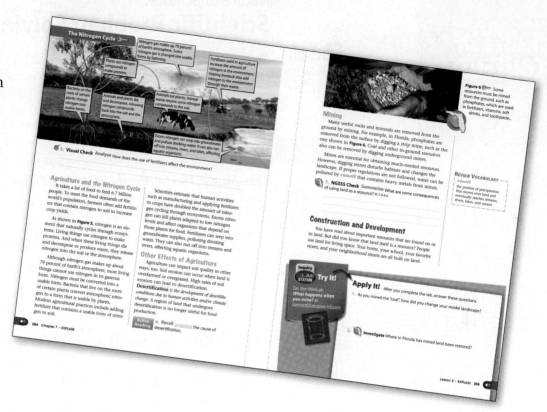

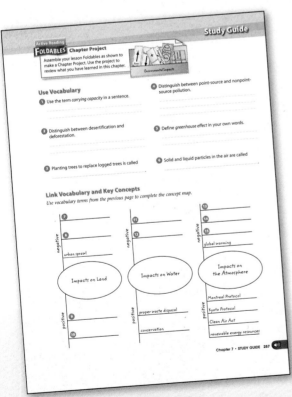

Concept Map

Each chapter's **Concept Map** gives you a place to show all the science connections you've learned.

TABLE OF CONTENTS

Inquiry
iLAB STATION

Skill Lab: How does the strength of geometric shapes differ?

Inquiry Lab: Build and Test a Bridge

Your online portal to everything you need!
Video · Audio · Review · iLab Station · WebQuests · Assessment · Concepts in Motion · Personal Tutors · Virtual Labs

Here are some of the exciting digital activities for this chapter!

Virtual Lab: What strategies are involved in solving a science problem?

BrainPOP: Scientific Methods

Page Keeley: Science Probe

Check It! → ☐ Lesson 1   ☐ Lesson 2   ☐ Lesson 3

☐ **MiniLabs:**   LESSON 1: What keeps Earth in orbit?

LESSON 2: How can the Moon be rotating if the same side of the Moon is always facing Earth?

LESSON 3: What does the Moon's shadow look like?

Try It! then Apply It!

Skill Lab: How does Earth's tilted rotation axis affect the seasons?

Inquiry Lab: Phases of the Moon

Your online portal to everything you need!
Video • Audio • Review • ⓘLab Station • WebQuests • Assessment • Concepts in Motion • Personal Tutors • Virtual Labs

Here are some of the exciting digital activities for this chapter!

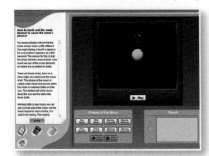

Virtual Lab: How do Earth and the moon interact to cause the moon's phases?

BrainPOP: Tides

Glancing collision of the Earth with a Mars-sized planet

Concepts in Motion: Moon Impact Theory

Inquiry iLAB STATION

☐ **MiniLabs:**   *Try It! then Apply It!*

LESSON 1: How can you model an elliptical orbit?

LESSON 2: How can you model the inner planets?

LESSON 3: How do Saturn's moons affect its rings?

LESSON 4: How do impact craters form?

Skill Practice: What can we learn about planets by graphing their characteristics?

Inquiry Lab: Scaling Down the Solar System

ConnectED   **Your online portal to everything you need!**
Video • Audio • Review • ⓘLab Station • WebQuests • Assessment • Concepts in Motion • Personal Tutors • Virtual Labs

Here are some of the exciting digital activities for this chapter!

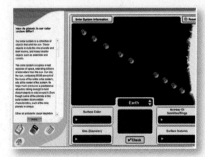

Virtual Lab: How do planets in our solar system differ?

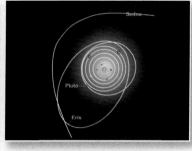

Concepts in Motion: Kuiper Belt

What's Science got to do With It? Home Sweet Home

Check It!   ☐ Lesson 1   ☐ Lesson 2   ☐ Lesson 3   ☐ Lesson 4

FLORIDA SCIENCE COURSE 3

Inquiry LAB STATION

☐ **MiniLabs:**

Try It! then Apply It!

LESSON 1: How does light differ?
LESSON 2: Can you model the Sun's structure?
LESSON 3: How do astronomers detect black holes?
LESSON 4: Can you identify a galaxy?

Skill Practice: How can you use scientific illustrations?
How can graphing data help you understand stars?

Inquiry Lab: Describe a Trip Through Space

ConnectED   **Your online portal to everything you need!**
Video • Audio • Review • ⓘLab Station • WebQuests • Assessment • Concepts in Motion • Personal Tutors • Virtual Labs

Here are some of the exciting digital activities for this chapter!

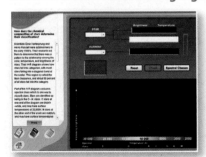

Virtual Lab: How does the chemical composition of stars determine their classification?

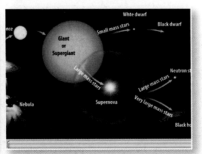

Concepts in Motion: Life Cycle of a Star

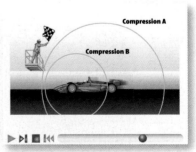

Concepts in Motion: Doppler Effect

Check It! ☐ Lesson 1 ☐ Lesson 2 ☐ Lesson 3

Inquiry
①LAB STATION

Try It! then Apply It!

☐ **MiniLabs:** LESSON 1: What is white light?

LESSON 2: How does lack of friction in space affect simple tasks?

LESSON 3: What conditions are required for life on Earth?

Skill Practice: How can you construct a simple telescope?

Inquiry Lab: Design and Construct a Moon Habitat

ConnectED **Your online portal to everything you need!**

Video • Audio • Review • ①Lab Station • WebQuests • Assessment • Concepts in Motion • Personal Tutors • Virtual Labs

Here are some of the exciting digital activities for this chapter!

Virtual Lab: How does an artificial satellite stay in orbit?

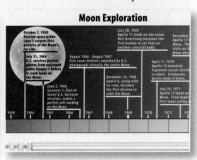

Concepts in Motion: Moon Exploration

Page Keeley: Science Probe

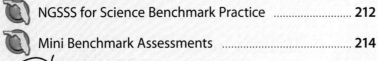

 ☐ **MiniLabs:** LESSON 1: How can you find an object's mass and volume?

Try It! then Apply It!   LESSON 2: Is mass conserved during a chemical reaction?

Skill Practice: How can you calculate density?

Inquiry Lab: Identifying Unknown Materials

ConnectED **Your online portal to everything you need!**

Video • Audio • Review • ①Lab Station • WebQuests • Assessment • Concepts in Motion • Personal Tutors • Virtual Labs

Here are some of the exciting digital activities for this chapter!

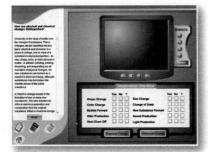
Virtual Lab: How are physical and chemical changes distinguished?

Concepts in Motion: States of Matter

BrainPOP: Property Changes

FLORIDA SCIENCE COURSE 3

TABLE OF CONTENTS

Check It! ☐ Lesson 1 ☐ Lesson 2

Inquiry **iLAB STATION**

☐ **MiniLabs:** LESSON 1: How do elements, compounds, and mixtures differ?

Try It! then Apply It!

LESSON 2: How can you model atoms?

Inquiry Lab: Balloon Molecules

ConnectED

Your online portal to everything you need!
Video • Audio • Review • iLab Station • WebQuests • Assessment • Concepts in Motion • Personal Tutors • Virtual Labs

Here are some of the exciting digital activities for this chapter!

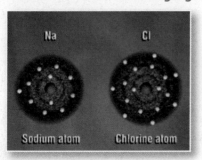

Concepts in Motion: Ionic Bonds

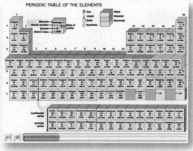

Concepts in Motion: Periodic Table of the Elements

BrainPOP: Atomic Model

Check It! ☐ Lesson 1   ☐ Lesson 2   ☐ Lesson 3

Inquiry ☐LAB STATION

☐ **MiniLabs:** LESSON 1: How does atom size change across a period?

Try It! then Apply It!

LESSON 2: How well do materials conduct thermal energy?

LESSON 3: Which insulates better?

Skill Practice: How is the periodic table arranged?

Inquiry Lab: Alien Insect Periodic Table

ConnectED

Your online portal to everything you need!

Video • Audio • Review • ⓘLab Station • WebQuests • Assessment • Concepts in Motion • Personal Tutors • Virtual Labs

Here are some of the exciting digital activities for this chapter!

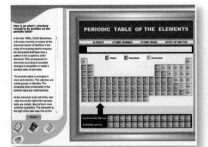

Virtual Lab: How is an atom's structure related to its position on the periodic table?

Concepts in Motion: Atomic Structure

Page Keeley: Science Probe

FLORIDA ⓘSCIENCE COURSE 3

TABLE OF CONTENTS

Check It! ☐ Lesson 1   ☐ Lesson 2   ☐ Lesson 3

 Inquiry ⓘLAB STATION

☐ **MiniLabs:** LESSON 1: How does an electron's energy relate to its position in an atom?

LESSON 2: How do compounds form?

LESSON 3: How many ionic compounds can you make?

Try It! then Apply It!

Skill Practice: How can you model compounds?

Inquiry Lab: Ions in Solution

Connect**ED** **Your online portal to everything you need!**

Video • Audio • Review • ⓘLab Station • WebQuests • Assessment • Concepts in Motion • Personal Tutors • Virtual Labs

Here are some of the exciting digital activities for this chapter!

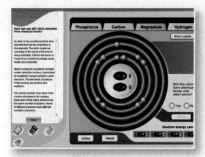

Virtual Lab: How can you tell which elements form chemical bonds?

Concepts in Motion: Na Cl Bonding

Concepts in Motion: Ionic Bonding

FLORIDA
iSCIENCE
COURSE 3

Check It! ☐ Lesson 1   ☐ Lesson 2   ☐ Lesson 3

Inquiry
iLAB STATION

☐ **MiniLabs:** LESSON 1: Which one is the mixture?
LESSON 2: How much is dissolved?
LESSON 3: Is it an acid or a base?

Try It! then Apply It!

Skill Practice: How does a solute affect the conductivity of a solution?

Inquiry Lab: Can the pH of a solution be changed?

ConnectED **Your online portal to everything you need!**
Video • Audio • Review • iLab Station • WebQuests • Assessment • Concepts in Motion • Personal Tutors • Virtual Labs

Here are some of the exciting digital activities for this chapter!

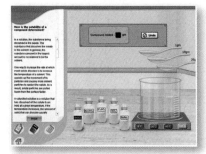

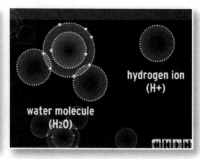

Virtual Lab: How is the solubility of a compound determined?

What's Science got to do With It? Sports Drinks

BrainPOP: Acids and Bases

FLORIDA ⓘSCIENCE Course 3

TABLE OF CONTENTS

Check It! ☐ Lesson 1   ☐ Lesson 2

☐ **MiniLabs:** LESSON 1: How does an equation represent a reaction?

LESSON 2: Can you speed up a reaction?

Try It! then Apply It!

Skill Practice: What can you learn from an experiment?

Inquiry Lab: Design an Experiment to Test Advertising Claims

💻 **ConnectED**

Your online portal to everything you need!
Video • Audio • Review • ⓘLab Station • WebQuests • Assessment • Concepts in Motion • Personal Tutors • Virtual Labs

Here are some of the exciting digital activities for this chapter!

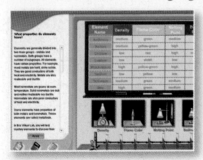

Virtual Lab: What properties do elements have?

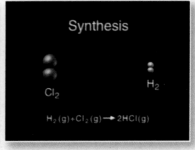

Concepts in Motion: Chemical Reactions

What's Science Got to do With It? Arson Investigation

Inquiry ①LAB STATION

☐ **MiniLabs:** LESSON 1: Can you observe plant processes?
LESSON 2: When will plants flower?

Try It! then Apply It!

Inquiry Lab: Design a Stimulating Environment for Plants

 **Connect ED** **Your online portal to everything you need!**
Video • Audio • Review • ①Lab Station • WebQuests • Assessment • Concepts in Motion • Personal Tutors • Virtual Labs

Here are some of the exciting digital activities for this chapter!

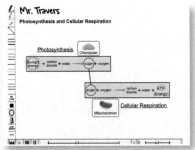

Page Keeley: Science Probe

Personal Tutor: Photosynthesis and Cellular Respiration

BrainPOP: Plant Growth

FLORIDA
iSCIENCE
COURSE 3

TABLE OF
CONTENTS

Check It! ☐ Lesson 1   ☐ Lesson 2   ☐ Lesson 3

Inquiry
iLAB
STATION

☐ **MiniLabs:** LESSON 1: Is your soil rich in nitrogen?
LESSON 2: How can you classify organisms?

Try It! then Apply It!

Skill Practice: How do scientists use variables?
Inquiry Lab: Can you observe part of the carbon cycle?

Your online portal to everything you need!
Video • Audio • Review • iLab Station • WebQuests • Assessment • Concepts in Motion • Personal Tutors • Virtual Labs

Here are some of the exciting digital activities for this chapter!

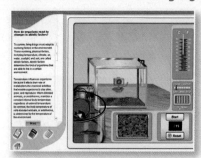

Virtual Lab: How do organisms react to changes in abiotic factors?

Concepts in Motion: The Nitrogen Cycle

BrainPOP: Water Cycle

Prove or Disprove?

Two students discussed scientific methods. They disagreed about why scientists test a hypothesis.

Sharla: I think scientists test a hypothesis to disprove it.

Marcos: I think scientists test a hypothesis to prove it.

(Circle) the student you most agree with. Explain why you agree with that student.

FLORIDA
Nature of Science

Scientific Problem
SOLVING

Nature of Science

This chapter begins your study of the nature of science, but there is even more information about the nature of science in this book. Each unit begins by exploring an important topic that is fundamental to scientific study. As you read these topics, you will learn even more about the nature of science.

FLORIDA BIG IDEAS

1 **The Practice of Science**
2 **The Characteristics of Scientific Knowledge**
3 **The Role of Theories, Laws, Hypotheses, and Models**
4 **Science and Society**
5 **Earth in Space and Time**

Think About It!

What is scientific inquiry?

This might look like a weird spaceship docking in a science-fiction movie. However, it is actually the back of an airplane engine being tested in a huge wind tunnel. An experiment is an important part of scientific investigations.

1. Why do you think an experiment is important?

_____

_____

_____

2. What is scientific inquiry?

_____

_____

_____

Florida NGSSS

LA.8.2.2.3 The student will organize information to show understanding or relationships among facts, ideas, and events (e.g., representing key points within text through charting, mapping, paraphrasing, summarizing, or comparing/contrasting);

MA.6.A.3.6 Construct and analyze tables, graphs, and equations to describe linear functions and other simple relations using both common language and algebraic notation.

SC.8.E.5.10 Assess how technology is essential to science for such purposes as access to outer space and other remote locations, sample collection, measurement, data collection and storage, computation, and communication of information.

SC.8.N.1.2 Design and conduct a study using repeated trials and replication.

SC.8.N.1.3 Use phrases such as "results support" or "fail to support" in science, understanding that science does not offer conclusive 'proof' of a knowledge claim.

SC.8.N.1.4 Explain how hypotheses are valuable if they lead to further investigations, even if they turn out not to be supported by the data.

SC.8.N.1.5 Analyze the methods used to develop a scientific explanation as seen in different fields of science.

SC.8.N.1.6 Understand that scientific investigations involve the collection of relevant empirical evidence, the use of logical reasoning, and the application of imagination in devising hypotheses, predictions, explanations and models to make sense of the collected evidence.

SC.8.N.2.1 Distinguish between scientific and pseudoscientific ideas.

SC.8.N.2.2 Discuss what characterizes science and its methods.

SC.8.N.3.1 Select models useful in relating the results of their own investigations.

SC.8.N.3.2 Explain why theories may be modified but are rarely discarded.

SC.8.N.4.1 Explain that science is one of the processes that can be used to inform decision making at the community, state, national, and international levels.

SC.8.N.4.2 Explain how political, social, and economic concerns can affect science, and vice versa.

There's More Online!
Video • Audio • Review • ⓘLab Station • WebQuest • Assessment • Concepts in Motion • Multilingual eGlossary

Scientific INQUIRY

Vocabulary

science p. NOS 4

observation p. NOS 6

inference p. NOS 6

hypothesis p. NOS 6

prediction p. NOS 6

scientific theory p. NOS 8

scientific law p. NOS 8

technology p. NOS 9

critical thinking p. NOS 10

Understanding Science

In a clear night sky, the stars seem to shine like diamonds scattered on black velvet. Why do stars seem to shine more brightly some nights than others?

When you ask questions, such as the one above, you are practicing science. **Science** *is the investigation and exploration of natural events and of the new information that results from those investigations.* You can help shape the future by accumulating knowledge, developing new technologies, and sharing ideas with others.

Throughout history, people of many different backgrounds, interests, and talents have made scientific contributions. Sometimes they overcame a limited educational background and excelled in science. One example is Marie Curie, shown in **Figure 1**. She was a scientist who won two Nobel prizes in the early 1900s for her work with radioactivity. As a young student, Marie was not allowed to study at the University of Warsaw in Poland because she was a woman. Despite this obstacle, she made significant contributions to science.

Active Reading **1. Reflect** Infer how people with different backgrounds have contributed to science. How have attitudes changed to include everyone who has an interest in science?

Figure 1 Modern medical procedures such as X-rays, radioactive cancer treatments, and nuclear-power generation are some of the technologies made possible because of the pioneering work of Marie Curie and her associates.

Branches of Science

Scientific study is organized into several branches. The three most common branches are physical science, Earth science, and life science. Each branch focuses on a different part of the natural world.

WORD ORIGIN

science

from Latin *scientia*, means "knowledge" or "to know"

Active Reading

2. Question In the boxes below, suggest some possible questions people in different branches of science might ask.

Physical Science

Physical science, or physics and chemistry, is the study of matter and energy. The physicist is using an instrument to measure radiation in space.

Physical Science Questions:

- [blank]
- [blank]
- [blank]

Earth Science

Earth scientists study the many processes that occur on Earth, in space, and deep within Earth. This scientist will study a water sample from Mexico.

Earth Science Questions:

- [blank]
- [blank]
- [blank]

Life Science

Life scientists study all organisms and the many processes that occur in them. This life scientist is studying the avian flu virus.

Life Science Questions:

- [blank]
- [blank]
- [blank]

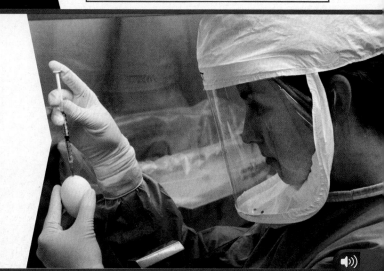

What is Scientific Inquiry?

When scientists conduct investigations, they often want to answer questions about the natural world. They use scientific inquiry—a process that uses a set of skills to answer questions or test ideas. You might have heard these steps called "the scientific method." However, there is no one scientific method. The skills that scientists use to conduct an investigation can be used in any order. One possible sequence is shown in **Figure 2**.

 3. Consider Highlight the term *scientific inquiry* and its definition.

Ask Questions

Imagine warming yourself near a campfire. As you place twigs and logs onto the fire, the fire releases smoke and light. You feel the warmth of the thermal energy being released. These are **observations**—*the results of using one or more of your senses to gather information and taking note of what occurs*. Observations often lead to questions. You ask yourself, "When logs burn, what happens to the wood? Do the logs disappear?"

You might recall that matter can change form, but it cannot be created or destroyed. Therefore, you could infer that the logs do not just disappear. They must undergo some type of change. An **inference** *is a logical explanation of an observation that is drawn from prior knowledge or experience.*

Hypothesize and Predict

You decide to investigate further. You might develop a **hypothesis**—*a possible explanation for an observation that can be tested by scientific investigations*. Your hypothesis about what happens might be: When logs burn, new substances form because matter cannot be destroyed.

When scientists state a hypothesis, they often use it to make predictions. *A* **prediction** *is a statement of what will happen next in a sequence of events.* Predictions based on information might be found when testing the hypothesis. Based on a hypothesis, you might predict that if logs burn, then the substances that make up the logs change into other substances.

Active Reading **4. Analyze** Illustrate the relationship between a hypothesis and a prediction.

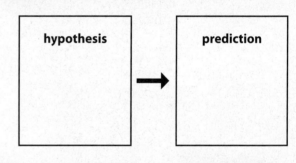

Figure 2 There are many possible steps in the process of scientific inquiry, and they can be performed in a variety of different sequences.

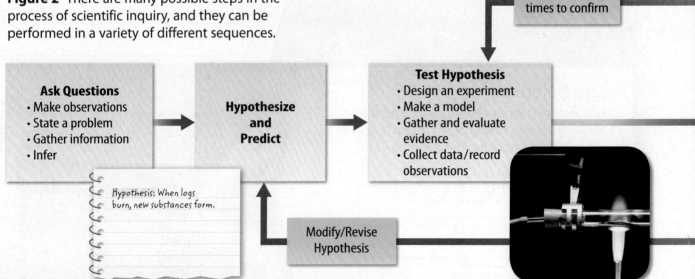

Test Hypothesis and Analyze Results

When you test a hypothesis, you often test your predictions. If a prediction is confirmed, then it supports your hypothesis. If your prediction is not confirmed, you might modify your hypothesis and retest it.

To test your predictions and hypothesis, design an experiment to find out what substances make up wood. Then determine what makes up the ash, the smoke, and other products that form during the burning process. You also could research this topic to find answers to questions.

After doing an experiment or research, analyze your results. You might make additional inferences after reviewing your data. If you find that new substances actually do form when wood burns, your hypothesis is supported. Some methods of testing a hypothesis and analyzing results are shown in **Figure 2**.

Active Reading 5. **Suggest** Explain why results might not support a hypothesis.

_____

_____

Discuss what you might do next if your hypothesis is not supported.

_____

_____

Draw Conclusions

After analyzing your results, draw conclusions about your investigation. A conclusion is a summary of the information gained from testing a hypothesis. Like a scientist does, you should test and retest your hypothesis several times to make sure the results are consistent.

Active Reading 6. **Define** Highlight the term *conclusion* and its definition.

Communicate Results

Sharing the results of a scientific inquiry is an important part of science. By exchanging information, scientists can evaluate and test others' work, apply new knowledge in their own research, and help keep scientific information accurate. As you do research on the Internet or in science books, you use information that someone else communicated. Scientists exchange information in many ways, as shown below in **Figure 2**.

Active Reading 7. **Point out** Underline three reasons scientists exchange experimental results and scientific information.

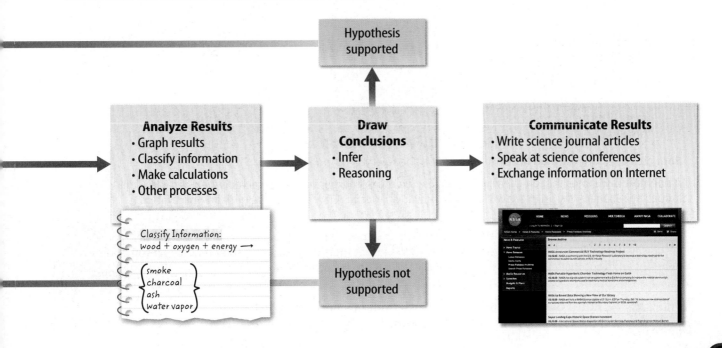

Analyze Results
- Graph results
- Classify information
- Make calculations
- Other processes

Classify Information:
wood + oxygen + energy →
(smoke
charcoal
ash
water vapor)

Draw Conclusions
- Infer
- Reasoning

Hypothesis supported

Hypothesis not supported

Communicate Results
- Write science journal articles
- Speak at science conferences
- Exchange information on Internet

Unsupported or Supported Hypotheses

Is a scientific investigation a failure and a waste of time if the hypothesis is not supported? Absolutely not! Valuable information can still be gained. The hypothesis can be revised and tested again. Each time a hypothesis is tested, scientists learn more about the topic they are studying.

Scientific Theory

When hypotheses (or a group of closely related hypotheses) are supported through many tests, a scientific theory can develop. A **scientific theory** *is an explanation of observations or events that is based on knowledge gained from many observations and investigations.*

A scientific theory does not develop from just one hypothesis, but from many hypotheses that are connected by a common idea. The kinetic molecular theory described below explains the behavior and energy of particles that make up a gas.

Scientific Law

A scientific law is different from a societal law, which is an agreement on a set of behaviors. A **scientific law** *is a rule that describes a repeatable pattern in nature.* A scientific law does not explain why or how the pattern happens, it only states that it will happen. For example, if you drop a ball, it will fall towards the ground every time. This is a repeated pattern that relates to the law of universal gravitation. The law of conservation of energy, described below, is also a scientific law.

 8. Distinguish <u>Underline</u> the terms scientific law and scientific theory and their definitions.

Kinetic Molecular Theory

The kinetic molecular theory explains how particles that make up a gas move in constant, random motions.

The kinetic molecular theory also assumes that the collisions of particles in a gas are elastic collisions. An elastic collision is a collision in which no kinetic energy is lost. Therefore, kinetic energy among gas particles is conserved.

Law of Conservation of Energy

The law of conservation of energy states that in any chemical reaction or physical change, energy is neither created nor destroyed. The total energy of particles before and after collisions is the same.

Scientific Law v. Scientific Theory

Both are based on repeated observations and can be rejected or modified.

A scientific law states that an event *will* occur. For example, energy will be conserved when particles collide. It does not explain why an event will occur or how it will occur. A law stands true until an observation is made that does not follow the law.

A scientific theory is an explanation of *why* or *how* an event occurred. For example, collisions of particles of a gas are elastic collisions. Therefore, no kinetic energy is lost. A theory can be rejected or modified if someone observes an event that disproves the theory. A theory will never become a law.

Active Reading **9. Point out** Highlight the terms scientific law and scientific theory and their descriptions in the box above.

Results of Scientific Inquiry

Why do you and others ask questions and investigate the natural world? Often, the results of a scientific investigation are new materials and technology, the discovery of new objects, or answers to questions.

Active Reading   **10. Analyze** In the spaces provided below, discuss how technology helps scientists develop new materials, discover new objects or events, and answer questions.

New Materials and Technology

Corporations and governments research and design new materials and technologies. **Technology** *is the practical use of scientific knowledge, especially for industrial or commercial use.* Scientists use technology to design new materials that make bicycles and cycling gear lighter, more durable, safer, and more aerodynamic. Using wind tunnels, scientists test these new materials to see whether they improve the cyclist's performance.

Summarize How might safer, more efficient bike equipment be a product of technology and scientific inquiry?

_____

_____

_____

New Objects or Events

Scientific investigations also lead to newly discovered objects or events. For example, NASA's *Hubble Space Telescope* captured this image of two colliding galaxies. They have been nicknamed the mice, because of their long tails. The tails are composed of gases and young, hot blue stars. If computer models are correct, these galaxies will combine in the future and form one large galaxy.

Explain How did technology aid scientists in the discovery of colliding galaxies?

_____

_____

_____

Answers to Questions

Often scientific investigations are launched to answer *who, what, when, where, why,* or *how* questions. This research chemist investigates new substances found in mushrooms and bacteria. New drug treatments for diseases might be found using new substances. Other scientists look for clues about what causes diseases, whether they can be passed from person to person, and when the disease first appeared.

Infer How is technology used by chemists to develop new knowledge and treatments for the medical field?

_____

_____

Create a two-tab book and label it as shown. Use it to discuss the importance of evaluating scientific information.

Why is it important to...

| ...be scientifically literate? | ...use critical thinking? |

Evaluating Scientific Information

Are you able to determine if information that claims to be scientifically proven is actually true and scientific instead of pseudoscientific (information incorrectly represented as scientific)? It is important that you are skeptical, identify facts and opinions, and think critically about information. **Critical thinking** *is comparing what you already know with the information you are given in order to decide whether you agree with it.*

Active Reading **11. Differentiate** (Circle) the terms *skepticism, critical thinking, opinion,* and *misleading information* and their definitions in both the text above and the text boxes below.

Be A Rock Star!
Do you dream of being a rock star?

Sing, dance, and play guitar like a rock star with the new Rocker-rific Spotlight. A new scientific process developed by Rising Star Laboratories allows you to overcome your lack of musical talent and enables you to perform like a real rock star.

This amazing new light actually changes your voice quality and enhances your brain chemistry so that you can sing, dance, and play a guitar like a professional rock star. Now, there is no need to practice or pay for expensive lessons. The Rocker-rific Spotlight does the work for you.

Dr. Sammy Truelove says, "Before lack of talent stopped someone from achieving his or her dreams of being a rockstar. This scientific breakthrough transforms people with absolutely no talent into amazing rock stars in just minutes. Of the many patients that I have tested with this product, no one has failed to achieve his or her dreams."

Disclaimer: This product was tested on laboratory rats and might not work for everyone.

Skepticism
To be skeptical is to doubt the truthfulness of something. A scientifically literate person can recognize when information misrepresents the facts. Science often is self-correcting because someone usually challenges inaccurate information and tests scientific results for accuracy.

Identifying Facts and Misleading Information
Misleading information often is worded to sound like scientific facts. A scientifically literate person can recognize fake claims and quickly determine when information is false.

Critical Thinking
Use critical thinking skills to compare what you know with the new information given to you. If the information does not sound reliable, either research and find more information about the topic or dismiss the information as unreliable.

Identify Opinions
An opinion is a personal view, feeling, or claim about a topic. Opinions cannot be proven true or false. An opinion might contain inaccurate information.

Science cannot answer all questions.

It might seem that scientific inquiry is the best way to answer all questions. But, science cannot answer all questions. Questions that deal with beliefs, values, personal opinions, and feelings cannot be answered scientifically. It is impossible to collect scientific data on these topics.

Figure 3 Always follow safety procedures when doing scientific investigations.

Active Reading

12. Apply Provide two questions that cannot be answered through scientific inquiry.

_____

_____

Safety in Science

Scientists use safe procedures in scientific investigations. During a scientific inquiry, you should always wear protective equipment, as shown in **Figure 3**. You also should learn the meaning of safety symbols, follow your teacher's instructions, and learn to recognize potential hazards.

Lesson Review 1

Use Vocabulary

1 **Define** *technology* in your own words. SC.8.E.5.10

_____

_____

2 **Use the term** *observation* in a sentence to show its scientific meaning.

_____

_____

Understand Key Concepts

3 Which action is NOT a way to test a hypothesis? SC.8.N.1.6

(A) analyze results   (C) make a model

(B) design an experiment   (D) gather and evaluate evidence

4 **Give an example** of a time when you practiced critical thinking.

_____

_____

Interpret Graphics

5 **Complete** the graphic organizer below with some examples of how to communicate the results of scientific inquiry. LA.8.2.2.3

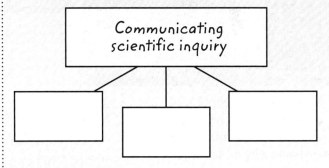

Communicating scientific inquiry

Critical Thinking

6 **Summarize** Your classmate writes the following as a hypothesis:

Red is a beautiful color.

Write a brief explanation to your classmate explaining why this is not a hypothesis. SC.8.N.1.6

_____

_____

Measurement and
SCIENTIFIC TOOLS

- Why did scientists create the International System of Units (SI)?
- Why is scientific notation a useful tool for scientists?
- How can tools, such as graduated cylinders and triple-beam balances, assist physical scientists?

Vocabulary

description p. NOS 12

explanation p. NOS 12

International System of Units (SI) p. NOS 13

scientific notation p. NOS 15

percent error p. NOS 15

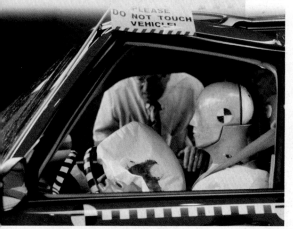

Figure 4 A description of an event details what you observed. An explanation explains why or how the event occurred.

Description and Explanation

Suppose you work for a company to calculate how cars perform during accidents, as shown in **Figure 4**. You measure the acceleration of cars as they crash into other objects.

A **description** *is a spoken or written summary of observations.* Measurements are descriptions of the results of the crash tests. A report discusses the results recorded. An **explanation** *is an interpretation of observations.* You make inferences and explain why the crash damaged the vehicles in specific ways.

A description and an explanation differ. When you describe something, you report your observations. When you explain something, you interpret your observations.

Active Reading

1. Distinguish Describe how a description differs from an explanation. Use an example to support your response.

_____

_____

The International System of Units

Different parts of the world use different systems of measurements. This can cause confusion when people communicate their measurements. In 1960, a new system of measurement was adopted. *The internationally accepted system of measurement is the* **International System of Units (SI).**

Active Reading

2. Apply Why did scientists establish the International System of Units?

_____

_____

SI Base Units

When you take measurements during scientific investigations, you will use the SI system, which consists of measurement base units, as shown in **Table 1**. Other units used in the SI system that are not base units are derived from the base units. For example, the liter, used to measure volume, was derived from the base unit for length.

SI Unit Prefixes

The SI system is based on multiples of ten represented by prefixes, as shown in **Table 2**. For example, the prefix *milli-* means 0.001 or 10-3. So, a milliliter is 0.001 L, or 1/1,000 L. Another way to say this is: 1 L is 1,000 times greater than 1 mL.

Converting Among SI Units

To convert from one SI unit to another, you either multiply or divide by a factor of ten. You also can use proportion calculations to make conversions. An example of how to convert between SI units is shown in **Figure 5**.

 Active Reading 3. **Summarize** Restate in your own words how to convert 400 mL to liters.

_____

_____

Table 1	SI Base Units
Quantity Measured	**Unit (symbol)**
Length	meter (m)
Mass	kilogram (kg)
Time	second (s)
Electric current	ampere (A)
Temperature	kelvin (K)
Substance amount	mole (mol)
Light intensity	candela (cd)

Table 2	Prefixes
Prefix	**Meaning**
Mega- (M)	1,000,000 or (10^6)
Kilo- (k)	1,000 or (10^3)
Hecto- (h)	100 or (10^2)
Deka- (da)	10 or (10^1)
Deci- (d)	0.1 or $\left(\frac{1}{10}\right)$ or (10^{-1})
Centi- (c)	0.01 or $\left(\frac{1}{100}\right)$ or (10^{-2})
Milli- (m)	0.001 or $\left(\frac{1}{1,000}\right)$ or (10^{-3})
Micro- (μ)	0.000001 or $\left(\frac{1}{1,000,000}\right)$ or (10^{-6})

Figure 5 The rock in the photograph has a mass of 17.5 grams. Convert that measurement to kilograms. ▼

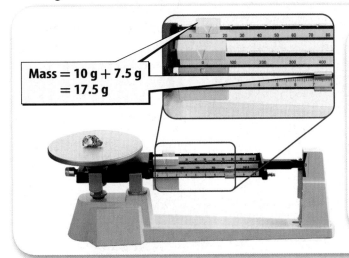

Mass = 10 g + 7.5 g = 17.5 g

1. Determine the correct relationship between grams and kilograms. There are 1,000 g in 1 kg.

$$\frac{1\ kg}{1,000\ g}$$

$$\frac{x}{17.5\ g} = \frac{1\ kg}{1,000\ g}$$

$$x = \frac{(17.5\ g)(1\ kg)}{1,000\ g}; x = 0.0175\ kg$$

2. Check your units. The unit *grams* is canceled out in the equation, so the answer is 0.0175 kg.

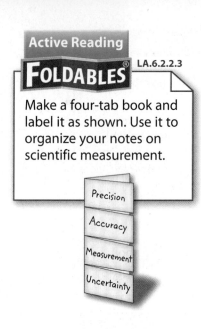
WORD ORIGIN

notation

from Latin *notationem*, means "a marking or explanation"

Figure 6 The graduated cylinder is marked in 1-mL increments. The beaker is marked in 50-mL increments. A graduated cylinder provides greater accuracy.

Table 3 Student Density and Error Data (Accepted value: Density of sodium chloride, 21.7 g/cm³)			
	Student A	**Student B**	**Student C**
	Density	**Density**	**Density**
Trial 1	23.4 g/cm³	18.9 g/cm³	21.9 g/cm³
Trial 2	23.5 g/cm³	27.2 g/cm³	21.4 g/cm³
Trial 3	23.4 g/cm³	29.1 g/cm³	21.3 g/cm³
Mean	23.4 g/cm³	25.1 g/cm³	21.5 g/cm³

Measurement and Uncertainty

The terms *precision* and *accuracy* have specific scientific meanings. Precision is a description of how similar repeated measurements are to each other. Accuracy is a description of how close a measurement is to an accepted value.

The difference between precision and accuracy is illustrated in **Table 3**. Students were asked to find the density of sodium chloride (NaCl). In three trials, students measured the volume and the mass of sodium chloride (NaCl). Then, they calculated the density for each trial and the mean of all three trials. Student A's measurements are the most precise because they are closest to each other. Student C's measurements are the most accurate because they are closest to the scientifically accepted value. Student B's measurements are neither precise nor accurate. They are not close to each other or to the accepted value.

Active Reading **4. Differentiate** Highlight the terms precision and accuracy and their scientific meanings.

Tools and Accuracy

No tool provides a perfect measurement. All measurements have some degree of uncertainty. Some tools produce more accurate measurements than other tools, as shown in **Figure 6**.

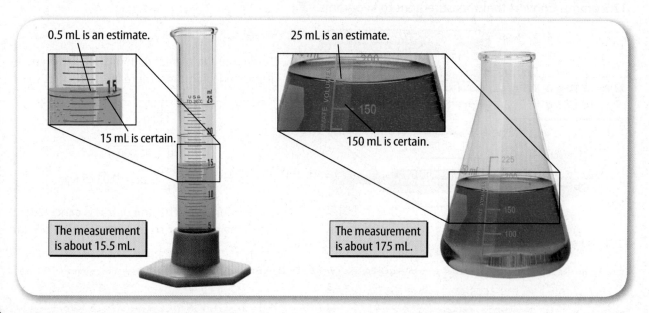

0.5 mL is an estimate.

15 mL is certain.

The measurement is about 15.5 mL.

25 mL is an estimate.

150 mL is certain.

The measurement is about 175 mL.

Scientific Notation

Suppose you are writing a report that includes Earth's distance from the Sun—149,600,000 km—and the density of the Sun's lower atmosphere—0.000000028 g/cm^3. These numerals take up too much space and might be difficult to read, so you use **scientific notation**—*a method of writing or displaying very small or very large values in a short form.* To write numerals in scientific notation, use the steps shown to the right.

 5. Analyze Why is scientific notation a useful tool?

_____

_____

Percent Error

The densities recorded for NaCl are experimental values because they were calculated during an experiment. Each of these values has some error because the accepted value for table salt density is 21.65 g/cm^3. Percent error can help you determine the size of your experimental error. **Percent error** *is the expression of error as a percentage of the accepted value.*

How to Write in Scientific Notation

1 Write the original number.
  A. 149,600,000
  B. 0.000000028

2 Move the decimal point to the right or the left to make the number between 1 and 10. Count the number of decimal places moved and note the direction.
  A. 1.49600000 = 8 places to the left
  B. 00000002.8 = 8 places to the right

3 Rewrite the number deleting all extra zeros to the right or to the left of the decimal point.
  A. 1.496
  B. 2.8

4 Write a multiplication symbol and the number *10* with an exponent. The exponent should equal the number of places that you moved the decimal point in step 2. If you moved the decimal point to the left, the exponent is positive. If you moved the decimal point to the right, the exponent is negative.
  A. 1.496×10^8
  B. 2.8×10^{-8}

Math Skills   MA.6.A.3.6

Solve for Percent Error A student in the laboratory measures the boiling point of water at 97.5°C. If the accepted value for the boiling point of water is 100.0°C, what is the percent error?

1. This is what you know:

   experimental value = 97.5°C
   accepted value = 100.0°C

2. This is what you need to find: percent error

3. Use this formula:

   $$\text{percent error} = \frac{|\text{experimental value} - \text{accepted value}|}{\text{accepted value}} \times 100\%$$

   Note that |experimental value – accepted value| refers to the absolute value of this operation.

4. Substitute the known values into the equation and perform the calculations

   $$\text{percent error} = \frac{|97.5° - 100.0°|}{100.0°} \times 100\% = 2.50\%$$

Practice

6. Calculate the percent error if the experimental value of the density of gold is 18.7 g/cm^3 and the accepted value is 19.3 g/cm^3.

Scientific Tools

As you conduct scientific investigations, you will use tools to take measurements. The tools listed here are some of the tools commonly used in science.

Active Reading 7. **Apply** Infer a safety rule for each scientific tool discussed.

◄ Science Journal

Use a science journal to record observations, write questions and hypotheses, collect data, and analyze the results of scientific inquiry. All scientists record the information they learn while conducting investigations.

Balances ►

A balance is used to measure the mass of an object. Units often used for mass are kilograms (kg), grams (g), and milligrams (mg). Two common types of balances are the electronic balance and the triple-beam balance.

Balance safety rule: _____

_____

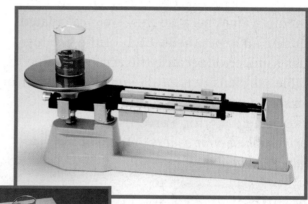

◄ Glassware

Laboratory glassware is used to hold or measure the volume of liquids. Flasks, beakers, test tubes, and graduated cylinders are just some of the different types of glassware available. Volume usually is measured in liters (L) and milliliters (mL).

Glassware safety rule: _____

_____

Thermometers ▶

A thermometer is used to measure the temperature of substances. Although Kelvin is the SI unit of measurement for temperature, in the science classroom, you often measure temperature in degrees Celsius (°C). To avoid breakage, never stir a substance with a thermometer.

Thermometer safety rule: _____

◀ Calculators

A calculator is a scientific tool that you might use in math class. But you also can use it to make quick calculations using your data.

Calculator safety rule: _____

Computers ▼

For today's students and scientists, electronic probes can be attached to computers and handheld calculators to record measurements. There are probes for collecting different kinds of information, such as temperature and the speed of objects. It is difficult to think of a time when computers were not readily available. Scientists can collect, compile, and analyze data more quickly using a computer. Computers are also used to prepare research reports and to share data and ideas with investigators worldwide.

Hardware refers to the physical components of a computer, such as the monitor and the mouse. Computer software refers to the programs that are run on computers, such as word processing, spreadsheet, and presentation programs.

Computer safety rule:

Additional Tools Used by Physical Scientists

Active Reading 8. **Revisit** Briefly recall a time when you used scientific equipment.

_____

_____

pH paper is used to quickly estimate the acidity of a substance. The paper changes color when it comes into contact with an acid or a base.

A hot plate is a small heating device that can be placed on a table or desk to heat substances in the laboratory.

Scientists use stopwatches to measure the time it takes for an event to occur. The SI unit for time is seconds (s).

A spring scale can be used to measure the weight or the amount of force applied to an object. The SI unit for weight is the newton (N).

Lesson Review 2

Use Vocabulary

1. A spoken or written summary of observations is a(n) _____, while a(n) _____ is an interpretation of observations.

Understand Key Concepts 🔑

2. Which type of glassware would you use to measure the volume of a liquid?
   - (A) beaker
   - (B) flask
   - (C) graduated cylinder
   - (D) test tube

3. **Summarize** why recording the diameter of an atom or the distance to the Moon would use scientific notation. MA.6.A.3.6

_____

_____

4. **Explain** why scientists use the International System of Units (SI). SC.8.N.1.5

_____

_____

Interpret Graphics

5. **Identify** List some scientific tools used to collect data. SC.8.N.1.5

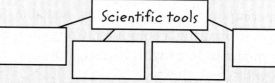

Scientific tools

Critical Thinking

6. **Explain** why precision and accuracy should be part of a scientific investigation.

_____

_____

Math Skills   MA.6.A.3.6

7. **Calculate** the percent error if the experimental value for the density of zinc is 9.95 g/cm³, but the accepted value is 7.13 g/cm³.

The Design Process

Create a Solution

RURAL WATER
RRWA 6

Scientists investigate and explore natural events and then interpret data and share information learned from those investigations. How do engineers differ from scientists? Engineers design, construct, and maintain things that do not occur in nature. Look around. The results of engineering include cell phones, bicycles, contact lenses, roller coasters, computers, cars, and buildings. Science involves the practice of scientific inquiry, but engineering involves The Design Process—a set of methods used to create solutions to problems or needs.

What do a water park and a water tower have in common? Both are designed for a purpose. A water tower is an elevated water container that uses pressure to supply water to buildings. Some water towers even have creative designs. Engineers design water parks to provide entertainment to people. However, engineers must design water parks to meet specific safety regulations.

The Design Process

1. Identify a Problem or Need
- Determine a problem or need
- Document all questions, research, and procedures throughout the process

2. Research and Development
- Brainstorm solutions
- Research any existing solutions that address the problem or need
- Suggest limitations

5. Communicate Results and Redesign
- Communicate the design process and results to others
- Redesign and modify the solution
- Construct the final solution

3. Construct a Prototype
- Develop possible solutions
- Estimate materials, costs, resources, and time to develop the solutions
- Select the best possible solution
- Construct a prototype

4. Test and Evaluate Solutions
- Use models to test the solution
- Use graphs, charts, and tables to evaluate results
- Analyze the solution's strengths and weaknesses

It's Your Turn

SC.8.N.1.1
SC.8.N.1.4

Inquiry
LAB STATION

DESIGN PROCESS LAB Design a Pollution Solution.

Case STUDY

ESSENTIAL QUESTIONS

🔑 Why are evaluation and testing important in the design process?

🔑 How is scientific inquiry used in a real-life scientific investigation?

Vocabulary

variable p. NOS 21

constant p. NOS 21

independent variable p. NOS 21

dependent variable p. NOS 21

experimental group p. NOS 21

control group p. NOS 21

qualitative data p. NOS 24

quantitative data p. NOS 24

The Minneapolis Bridge Failure

On August 1, 2007, the center section of the Interstate-35W (I-35W) bridge in Minneapolis, Minnesota, suddenly collapsed. A major portion of the bridge fell more than 30 m into the Mississippi River, as shown in **Figure 7**. There were more than 100 cars and trucks on the bridge at the time, including a school bus carrying over 50 students.

The failure of this 8-lane, 581-m long interstate bridge came as a surprise. Drivers do not expect a bridge to drop out from underneath them. The design and engineering processes are supposed to ensure that bridge failures do not happen.

Controlled Experiments

After the bridge collapsed, investigators had to use scientific inquiry to determine why the bridge failed. The investigators designed controlled experiments to help them discover exactly what happened. A controlled experiment is a scientific investigation that answers questions, tests hypotheses, and collects data to determine how one factor affects another.

Active Reading   **1. Identify** Highlight the term *controlled experiment* and its description.

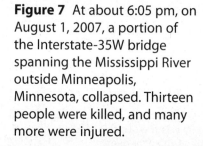

Figure 7 At about 6:05 pm, on August 1, 2007, a portion of the Interstate-35W bridge spanning the Mississippi River outside Minneapolis, Minnesota, collapsed. Thirteen people were killed, and many more were injured.

Identifying Variables and Constants

When conducting an experiment, you must identify factors that can affect the experiment's outcome. A **variable** *is any factor that can have more than one value.* In controlled experiments, there are two kinds of variables. The **independent variable** *is the factor that you want to test. It is changed by the investigator to observe how it affects a dependent variable.* The **dependent variable** *is the factor you observe or measure during an experiment.* **Constants** *are the factors in an experiment that do not change.*

Experimental Groups

A controlled experiment usually has at least two groups. The **experimental group** *is used to study how a change in the independent variable changes the dependent variable.* The **control group** *contains the same factors as the experimental group, but the independent variable is not changed.* Without a control, it is impossible to know whether your experimental observations result from the variable you are testing or some other factor.

This case study will explore how the investigators used scientific inquiry to determine why the bridge collapsed. Notebooks in the margin identify what a scientist might write in a science journal.

Simple Beam Bridges

Before you read about the bridge-collapse investigation, think about the structure of bridges. The simplest type of bridge is a beam bridge, as shown in **Figure 8**. This type of bridge has one horizontal beam across two supports. A disadvantage of beam bridges is that they tend to sag in the middle if they are too long or if the load is too heavy.

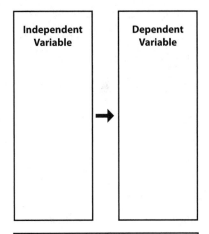

You can change the independent variable to observe how it affects the dependent variable. Without constants, two independent variables could change at the same time, and you would not know which variable affected the dependent variable.

Active Reading 2. **Differentiate** Illustrate the relationship between independent and dependent variables. Then, list two possible constants for a scientific investigation.

Independent Variable		Dependent Variable
	→	

Constants
1.
2.

Figure 8 Simple beam bridges are used to effectively span short distances, such as small creeks.

Figure 9 Truss bridges can span long distances and are strengthened by a series of interconnecting triangles called trusses. ▶

Active Reading
3. Identify As you read the following paragraph, fill in the tags on the photo above to identify the following parts of a truss bridge: support beam, gusset plate, deck of the bridge, end support, and truss.

Truss Bridges

A truss bridge, shown in **Figure 9**, is supported at its ends, by an assembly of interconnected, triangular trusses to strengthen it. The I-35W bridge, shown in **Figure 10**, was a tress bridge designed with straight beams and trusses connected to vertical supports. The beams in the bridge's deck and the supports came together at structures known as gusset plates, shown below on the right. These steel plates joined the triangular and vertical truss elements to the overhead roadway beams that ran along the deck of the bridge. This area, where the truss structure connects to the roadway portion of the bridge at a gusset plate is called a node.

Figure 10 Trusses were a major structural element of the I-35W bridge. The gusset plates at each node in the bridge, shown on the right, are critical pieces that hold the bridge together. ▼

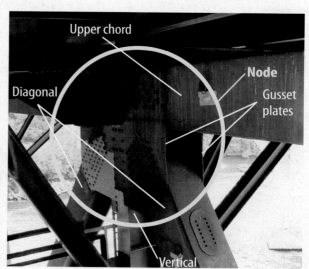

Upper chord

Diagonal

Node
Gusset plates

Vertical

Bridge Failure Observations

After the bridge collapsed, shown in **Figure 11**, the local sheriff's department handled the initial recovery of the bridge structure. Finding, freeing, and identifying victims was a high priority, and unintentional damage to the fallen bridge occurred. However, investigators eventually recovered the entire structure.

Investigators labeled each part with the location where it was found and the date when it was removed. They then moved the pieces to a nearby park where they placed the pieces in their relative original positions. Examining the reassembled structure, investigators found physical evidence to help determine where the breaks in each section occurred.

The investigators found more clues in a motion-activated security camera recording of the bridge collapse. The video showed about 10 seconds of the collapse, which revealed the sequence of events that destroyed the bridge. Investigators used this video to help pinpoint where the failure began.

Scientists often observe and gather information about an object or an event before proposing a hypothesis. This information is recorded or filed for the investigation.

Observations:
• Recovered parts of the collapsed bridge
• A video showing the sequence of events as the bridge fails and falls into the river

Active Reading

4. Sequence Analyze the paragraphs above. Then, number the steps from 1 to 6 that investigators took in the process of recovery and identification of the destroyed bridge structure.

Asking Questions

What factors caused the bridge to fail? Was the original bridge design faulty? Were bridge maintenance and repair poor or lacking? Was there too much weight on the bridge? Or was it a combination of these factors that caused the bridge to fail? Each of these questions was studied to determine why the bridge collapsed.

Asking questions and seeking answers to those questions is a way that scientists formulate hypotheses.

Qualitative data:
A thicker layer of concrete was added to the bridge to protect rods.

Quantitative data:
• The concrete increased the load on the bridge by 13.4 percent.

• The modifications in 1998 increased the load on the bridge by 6.1 percent.

• At the time of the collapse in 2007, the load on the bridge increased by another 20 percent.

Gathering Information and Data

Investigators reviewed the modifications made to the bridge since it opened in 1967. In 1977, engineers noticed that salt used to deice the bridge during winter weather was causing the reinforcement rods in the roadway to weaken. To protect the rods, engineers applied a thicker layer of concrete to the surface of the bridge roadway. Analysis after the collapse revealed that this extra concrete increased the dead load on the bridge by about 13.4 percent. A load can be a force applied to the structure from the structure itself (dead load) or from temporary loads such as traffic, wind gusts, or earthquakes (live load). Investigators recorded this qualitative and quantitative data. **Qualitative data** *uses words to describe what is observed.* **Quantitative data** *uses numbers to describe what is observed.*

In 1998, modifications were made to the bridge when it was noted that the bridge did not meet current safety standards. Analysis showed that these changes further increased the dead load on the bridge by about 6.1 percent.

An Early Hypothesis

At the time of the collapse, the bridge was undergoing additional renovations. Four piles of sand, four piles of gravel, a water tanker filled with over 11,000 L of water, a cement tanker, a concrete mixer, and other equipment, supplies, and workers were assembled on the bridge. This caused the load on the bridge to increase by 20 percent. In addition, normal vehicle traffic was on the bridge. Did these factors overload the bridge, causing the center section to collapse as shown in **Figure 12?** Only a thorough analysis could answer this question.

Active Reading 5. **Analyze** Highlight the factors that contributed to the load on the bridge at the time of its collapse.

Figure 12 The center section of the bridge broke away and fell into the river.

Hypothesis:
The bridge failed because it was overloaded.

Computer Modeling

The analysis of the bridge was conducted using computer-modeling software, as shown in **Figure 13**. Investigators entered data from the Minnesota bridge into a computer to efficiently perform numerous mathematical calculations. After thorough modeling and analysis, it was determined that the bridge was not overloaded.

Revising the Hypothesis

Analysis of routine assessments conducted in 1999 and 2003 provided additional clues as to why the bridge might have failed. At the time, investigators had taken numerous pictures of the bridge structure. The photos revealed bowing of the gusset plates at node U10. Gusset plates are designed to be stronger than the structural parts they connect. The bowing of the plates possibly indicated a problem with the gusset plate design. Previous inspectors and engineers missed this warning sign.

The accident investigators found that some recovered gusset plates were fractured early in the collapse, while others were not damaged. If the bridge had been properly constructed, none of the plates should have failed.

After evaluating the evidence, investigators formulated the hypothesis that the gusset plates failed, which lead to the bridge collapse. Now investigators had to test this hypothesis.

Hypothesis:
1. The bridge failed because it was overloaded.
2. The bridge collapsed because the gusset plates failed.
Prediction:
If a gusset plate is not properly designed, then a heavy load on a bridge will cause the gusset plate to fail.

| **Active Reading** | **6. Summarize** Based on what investigators discovered about the gusset plates, restate the new hypothesis. |

Figure 14 The steel plates, or gusset plates, at the U10 node were too thin for the loads the bridge carried.

Testing the Hypothesis

To calculate the load on the bridge when it collapsed, they estimated the combined weight of the bridge and the traffic on the bridge. The investigators divided the load on the bridge when it collapsed by the known load limits of the bridge to find the demand-to-capacity ratio. The demand-to-capacity ratio provides a measure of a structure's safety.

Analyzing Results

As investigators calculated the demand-to-capacity ratios for each of the main gusset plates, they found that the ratios were particularly high for the U10 node. The U10 plate, shown in **Figure 14,** failed earliest in the bridge collapse. **Table 4** shows the demand-to-capacity ratios for some gusset plates at three nodes. A value greater than 1 means the structure is unsafe. Notice how high the ratios are for the U10 gusset plate compared to the other plates.

Further data showed that the U10 plates were about half the thickness they should have been to support the load they were designed to handle.

Active Reading **7. Infer** Why are evaluation and testing an important part of the design process?

_____

_____

Table 4 **Node-Gusset Plate Analysis**							
Gusset Plate	Thickness (cm)	**Demand-to-Capacity Ratios for the Upper-Node Gusset Plates**					
		Horizontal loads			**Vertical loads**		
U8	3.5	0.05	0.03	0.07	0.31	0.46	0.20
U10	1.3	1.81	1.54	1.83	1.70	1.46	1.69
U12	2.5	0.11	0.11	0.10	0.71	0.37	1.15

Drawing Conclusions

Over the years, modifications to the I-35W bridge added more load to the bridge. On the day of the accident, traffic and the concentration of construction vehicles and materials added still more load. Investigators concluded that if the U10 gusset plates were properly designed, they would have supported the added load. When the investigators examined the original records for the bridge, they were unable to find any detailed gusset plate specifications. They could not determine whether undersized plates were used because of a mistaken calculation or some other error in the design process. The only thing that they could conclude with certainty was that undersized gusset plates could not reliably hold up the bridge.

The Federal Highway Administration and the National Transportation Safety Board published the results of their investigations. These publications provide valuable information that can be used in future designs. These reports are good examples of why it is important to publish results and to share information.

> **Analyzing Results:** The U10 gusset plates should have been twice as thick as they were to support the bridge.

> **Conclusions:** The bridge failed because the appropriate gusset plates were not installed; therefore, the bridge could not carry the increased load.

Active Reading 8. **Conclude** (Circle) the results and conclusions as to why the bridge collapsed.

Lesson Review 3

Inquiry LAB STATION

SC.8.N.1.2, SC.8.N.1.3, SC.8.N.1.4, SC.8.N.1.6, SC.8.N.3.1, SC.8.N.4.1

Try It!

Inquiry Lab *Build and Test a Bridge* at connectED.mcgraw-hill.com

Use Vocabulary

1. **Distinguish** between qualitative data and quantitative data. SC.8.N.1.6

Understand Key Concepts 🔑

2. **Give an example** of a situation in your life in which you depend on testing and evaluation in a product design to keep you safe.

Interpret Graphics

3. **Complete** Sequence scientific inquiry steps used in one part of the case study. LA.8.2.2.3

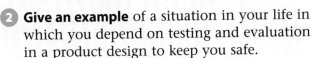

Critical Thinking

4. **Analyze** how the scientific inquiry process differs when engineers design a product, such as a bridge, and when they investigate design failure. SC.8.N.1.5

5. **Evaluate** why the gusset plates were a critical piece in the bridge design. SC.8.N.1.5

Think About It! Scientific inquiry is a multifaceted activity that includes the formulation of investigable questions, the construction of investigations, the collection and evaluation of data, and the communication of results.

 ## Key Concepts Summary

LESSON 1 Scientific Inquiry

- Some steps used during scientific inquiry are making **observations** and **inferences**, developing a **hypothesis**, analyzing results, and drawing conclusions. These steps, among others, can be performed in any order.

- There are many results of scientific inquiry, and a few possible outcomes are the development of new materials and new technology, the discovery of new objects and events, and answers to basic questions.

- **Critical thinking** is comparing what you already know about something to new information and deciding whether or not you agree with the new information.

LESSON 2 Measurement and Scientific Tools

- Scientists developed one universal system of units, the **International System of Units (SI)**, to improve communication among scientists.

- **Scientific notation** is a useful tool for writing large and small numbers in a shorter form.

- Tools such as graduated cylinders and triple-beam balances make scientific investigation easier, more accurate, and repeatable.

LESSON 3 Case Study—The Minneapolis Bridge Failure

- Evaluation and testing are important in the design process for the safety of the consumer and to keep costs of building or manufacturing the product at a reasonable level.

- Scientific inquiry was used throughout the process of determining why the bridge collapsed, including hypothesizing potential reasons for the bridge failure and testing those hypotheses.

Vocabulary

science p. NOS 4
observation p. NOS 6
inference p. NOS 6
hypothesis p. NOS 6
prediction p. NOS 6
scientific theory p. NOS 8
scientific law p. NOS 8
technology p. NOS 9
critical thinking p. NOS 10

description p. NOS 12
explanation p. NOS 12
International System of Units (SI) p. NOS 12
scientific notation p. NOS 15
percent error p. NOS 15

variable p. NOS 21
independent variable p. NOS 21
dependent variable p. NOS 21
constants p. NOS 21
qualitative data p. NOS 21
quantitative data p. NOS 21
experimental group p. NOS 24
control group p. NOS 24

Use Vocabulary

1 The _____ contains the same factors as the experimental group, but the independent variable is not changed.

2 The expression of error as a percentage of the accepted value is _____.

3 The process of studying nature at all levels and the collection of information that is accumulated is _____.

4 The _____ are the factors in the experiment that stay the same.

Fill in the correct answer choice.

🔑 Understand Key Concepts

5 Which is NOT an SI base unit? SC.8.N.2.2

- Ⓐ kilogram
- Ⓒ meter
- Ⓑ liter
- Ⓓ second

6 While analyzing results from an investigation, a scientist calculates a very small number that he or she wants to make easier to use. Which does the scientist use to record the number? SC.8.N.2.2

- Ⓐ explanation
- Ⓒ scientific notation
- Ⓑ inference
- Ⓓ scientific theory

7 Which is NOT true of a scientific law? SC.8.N.3.1

- Ⓐ It can be modified or rejected.
- Ⓑ It states that an event will occur.
- Ⓒ It explains why an event will occur.
- Ⓓ It is based on repeated observations.

Critical Thinking

8 **Write** a brief description of the activity shown in the photo. SC.8.N.1.1

Writing in Science

9 **Apply** On a separate piece of paper, write a paragraph that gives examples of how critical thinking, skepticism, and identifying facts and opinions can help you in your everyday life. Be sure to include a topic sentence and concluding sentence in your paragraph. SC.8.N.2.2

Big Idea Review

10 What is scientific inquiry? Explain why it is a constantly changing process. SC.8.N.2.2

_____

_____

_____

Math Skills   MA.6.A.3.6

11 The accepted scientific value for the density of sucrose is 1.59 g/cm³. The data from three trials to measure the density of sucrose is shown in the table below. Calculate the percent error for each trial.

Trial	Density	Percent Error
Trial 1	1.55 g/cm³	
Trial 2	1.60 g/cm³	
Trial 3	1.58 g/cm³	

Multiple Choice *Bubble the correct answer.*

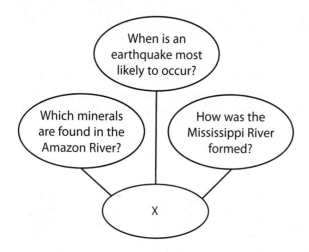

1. The graphic organizer above shows some questions scientists might ask. Which branch of science should go in X? **SC.8.N.1.5**

    (A) chemical science

    (B) Earth science

    (C) life science

    (D) physical science

2. Scientists design experiments in order to test **SC.8.N.2.2**

    (F) conclusions.

    (G) hypotheses.

    (H) observations.

    (I) predictions.

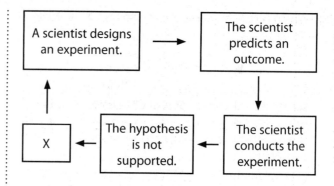

3. The flow chart above shows some of the steps involved in a scientific inquiry. Which would most likely occur at X? **SC.8.N.1.3**

    (A) The scientist analyzes the results.

    (B) The scientist makes a new prediction.

    (C) The scientist modifies or revises the hypothesis.

    (D) The scientist speaks at a science conference.

4. Which statement cannot be tested scientifically? **SC.8.N.2.1**

    (F) Apples are red because they reflect red light waves.

    (G) Apples have more sugar than oranges have.

    (H) Oranges do not grow as well in Canada as they do in Mexico.

    (I) Oranges have a better flavor than apples have.

Name _____   Date _____

Multiple Choice *Bubble the correct answer.*

Use the table below to answers questions 1 and 2.

Melting Point of Aspirin (Accepted value: 135°C)			
	Student R	**Student S**	**Student T**	**Student U**
Trial 1	133.7°C	134.6°C	134.8°C	133.9°C
Trial 2	133.4°C	135.5°C	134.5°C	135.0°C
Trial 3	133.9°C	136.1°C	135.3°C	134.5°C
Average	**133.7°C**	**135.4°C**	**134.9°C**	**134.5°C**

1. According to the information in the table, which student's measurements are the most accurate? **MA.6.S.6.2**

   Ⓐ Student R

   Ⓑ Student S

   Ⓒ Student T

   Ⓓ Student U

2. According to the information in the table, which student's measurements are the most precise? **SC.8.N.1.6**

   Ⓕ Student R

   Ⓖ Student S

   Ⓗ Student T

   Ⓘ Student U

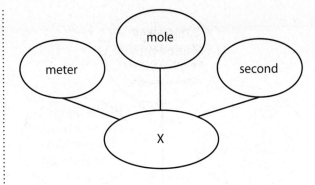

3. The graphic organizer above shows three kinds of measurements. Which description goes in X? **SC.8.N.1.5**

   Ⓐ SI base units

   Ⓑ SI prefixes

   Ⓒ metric units of length

   Ⓓ metric units of mass

4. A triple-beam balance is used to measure **SC.8.N.2.2**

   Ⓕ density.

   Ⓖ mass.

   Ⓗ temperature.

   Ⓘ volume.

Multiple Choice *Bubble the correct answer.*

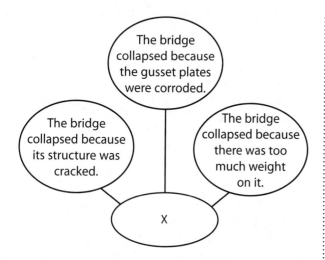

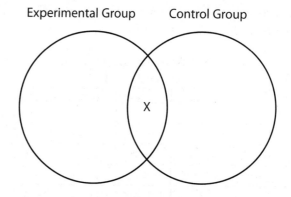

Experimental Group   Control Group

1. In the graphic organizer above, which goes in X? **SC.8.N.1.5**

Ⓐ observations about why the bridge collapsed

Ⓑ possible hypotheses that can be tested on a collapsed bridge

Ⓒ possible reasons why the bridge might have collapsed

Ⓓ predictions about why a bridge might collapse

2. It would be too costly to build actual bridges for testing, so bridge designers use mathematical models instead. These models are **SC.8.N.3.1**

Ⓕ conclusions.

Ⓖ hypotheses.

Ⓗ observations.

Ⓘ predictions.

3. In the Venn diagram above, which does NOT go in X? **SC.8.N.1.1**

Ⓐ constants

Ⓑ measurements

Ⓒ dependent variable

Ⓓ independent variable

4. Which statement is true? **SC.8.N.1.4**

Ⓕ A hypothesis becomes a law if several experiments support it.

Ⓖ A hypothesis is true if one experiment supports it.

Ⓗ A hypothesis is untrue if even one experiment does not support it.

Ⓘ A hypothesis might be true even if several experiments do not support it.

Notes

EARTH in SPACE AND TIME

| 2000 B.C. | | 1600 | | 1700 | 1800 |

1600 B.C.
Babylonian texts show records of people observing Venus without the aid of technology. Its appearance is recorded for 21 years.

265 B.C.
Greek astronomer Timocharis makes the first recorded observation of Mercury.

1610
Galileo Galilei observes the four largest moons of Jupiter through his telescope.

1613
Galileo records observations of the planet Neptune but mistakes it for a star.

1655
Astronomer Christiaan Huygens observes Saturn and discovers its rings, which were previously thought to be large moons on each side.

1781
William Herschel discovers the planet Uranus.

1900

1930
Clyde Tombaugh
discovers Pluto, making
him the first American
to discover a planet.

1971
Mariner 9 visits Mars
and becomes the first
human-made object
to orbit a planet
other than Earth.

2000

2006
After research and
consideration, the
International
Astronomical Union
votes to remove
Pluto from the list
of planets in the
solar system.

? Inquiry
Visit ConnectED for
this unit's
STEM activity.

Models

In 2004, over 200,000 people died as a tsunami swept across the Indian Ocean, shown in **Figure 1.** Researchers have developed different models to study tsunami waves and their effects. A **model** is a representation of an object, a process, an event, or a system that is similar to a physical object or idea. Models are used to study something that is too big or too small, happens too quickly or too slowly, or is too dangerous or too expensive to study directly.

Models of tsunamis help predict how future tsunamis might impact land. Information from these models can help save ecosystems, buildings, and lives.

Active Reading   **1. Determine** <u>Underline</u> the term *model* and its definition. Highlight why models are used.

Types of Models
Mathematical Models and Computer Simulations

A mathematical model represents an event, a process, or a system using equations. A mathematical model can be one equation, for example: speed = distance/time, or several hundred equations.

A computer simulation combines many mathematical models. Simulations allow the user to easily change variables. They often show a change over time or a sequence of events. Computer programs that include animations and graphics use mathematical models.

Researchers from Texas A&M University constructed a tsunami simulation using many mathematical models of Seaside, Oregon, as shown in **Figure 2.** Simulations that model the force of waves hitting buildings are displayed on a computer screen. Researchers change variables, such as the size, the force, or the shape of tsunami waves, to determine how Seaside might be damaged by a tsunami.

Figure 1 A massive wave approaches the shore in the 2004 Indian Ocean tsunami.

Figure 2 This series of images is from an animated simulation model of a tsunami approaching Seaside, Oregon.

Active Reading   **2. Distinguish** Highlight the terms and definitions for mathematical model, physical model, and conceptual model.

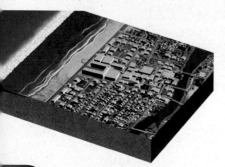  

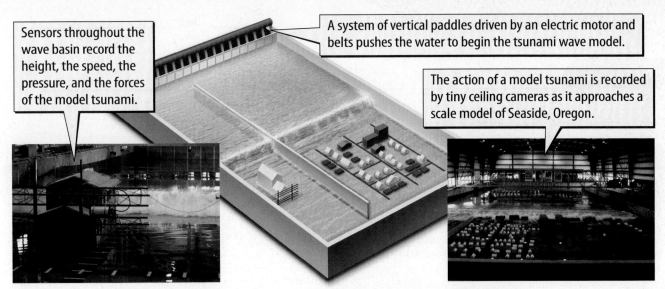

Sensors throughout the wave basin record the height, the speed, the pressure, and the forces of the model tsunami.

A system of vertical paddles driven by an electric motor and belts pushes the water to begin the tsunami wave model.

The action of a model tsunami is recorded by tiny ceiling cameras as it approaches a scale model of Seaside, Oregon.

Figure 3 Researchers study physical models of tsunamis to predict a tsunami's effects.

Physical Models

A physical model is a model that you can see and touch. It shows how parts relate to one another, how something is built, or how complex objects work. Scientists at Oregon State University built physical scale models of Seaside, Oregon, as shown in **Figure 3.** They placed the model at the end of a long wave tank. Sensors in the wave tank and on the model buildings measure and record velocities, forces, and turbulence created by a model tsunami wave. Scientists use these measurements to predict the effects of a tsunami on a coastal town, and to make recommendations for saving lives and preventing damage.

Conceptual Models

Images that represent a process or relationships among ideas are conceptual models. The conceptual model below shows that the United States has a three-part plan for minimizing the effects of tsunamis. Hazard assessment involves identification of areas that are in high risk of tsunamis. Response involves education and public safety. Warning includes a system of sensors that detect the approach of a tsunami.

Inquiry

LAB STATION **Try It!**

SC.8.N.4.1, SC.8.N.4.2

MiniLab *How can you model a tsunami?* at connectED.mcgraw-hill.com

Apply It!

After you complete the lab, answer these questions.

1. **Predict** What might studying the effects of a tsunami on the Seaside model tell scientists?

2. **Synthesize** Based on the knowledge gained by scientists, what safety precautions might be designed for homes, businesses, schools, and events?

Name _____ Date _____

Phases of the Moon

Many people have different ideas about what causes us to see different parts of the Moon (moon phases). Which idea below best matches your thinking?

A: Earth casts a shadow on the Moon that allows us to see only the lit part.

B: The Moon moves into the Sun's shadow, blocking out part of the Moon's light.

C: The part we see depends on where the Moon is in relation to Earth and the Sun.

D: The Sun's movement around Earth causes different parts of the Moon to be reflected.

E: The Moon's rotation causes different parts of the Moon to be reflected back to Earth.

F: None of these; I think there is something else that causes us to see different moon phases.

(Explain) your thinking. Describe your ideas about why we see different phases of the Moon.

FLORIDA
Chapter 1

The Sun-Earth-Moon SYSTEM

FLORIDA BIG IDEAS

1 The Practice of Science

5 Earth in Space and Time

| Think About It! | **What natural phenomena do the motions of Earth and the Moon produce?** |

Look at this time-lapse photograph. The "bites" out of the Sun occurred during a solar eclipse. The Sun's appearance changed in a regular, predictable way as the Moon's shadow passed over a part of Earth.

1 How do you think the Moon's movement can change the Sun's appearance?

_____

_____

_____

2 What predictable changes do you think Earth's movement causes?

_____

_____

_____

3 What other natural phenomena do you think the motions of Earth and the Moon cause?

_____

_____

_____

| Get Ready to Read | **What do you think about the Sun, Earth, and the Moon?** |

Do you agree or disagree with each of these statements? As you read this chapter, see if you change your mind about any of the statements.

AGREE   DISAGREE

1 Earth's movement around the Sun causes sunrises and sunsets. ☐ ☐

2 Earth has seasons because its distance from the Sun changes throughout the year. ☐ ☐

3 The Moon was once a planet that orbited the Sun between Earth and Mars. ☐ ☐

4 Earth's shadow causes the changing appearance of the Moon. ☐ ☐

5 A solar eclipse happens when Earth moves between the Moon and the Sun. ☐ ☐

6 The gravitational pull of the Moon and the Sun on Earth's oceans causes tides. ☐ ☐

 **There's More Online!** Video • Audio • Review • ⓘLab Station • WebQuest • Assessment • Concepts in Motion • Multilingual eGlossary

Earth's MOTION

Vocabulary

orbit p. 12

revolution p. 12

rotation p. 13

rotation axis p. 13

solstice p. 17

equinox p. 17

 Florida NGSSS

LA.8.2.2.3 The student will organize information to show understanding or relationships among facts, ideas, and events (e.g., representing key points within text through charting, mapping, paraphrasing, summarizing, or comparing/contrasting);

MA.6.A.3.6 Construct and analyze tables, graphs, and equations to describe linear functions and other simple relations using both common language and algebraic notation.

SC.8.E.5.1 Recognize that there are enormous distances between objects in space and apply our knowledge of light and space travel to understand this distance.

SC.8.E.5.4 Explore the Law of Universal Gravitation by explaining the role that gravity plays in the formation of planets, stars, and solar systems and in determining their motions.

SC.8.E.5.9 Explain the impact of objects in space on each other including:
1. the Sun on the Earth including seasons and gravitational attraction
2. the Moon on the Earth, including phases, tides, and eclipses, and the relative position of each body.

SC.8.N.1.1 Define a problem from the eighth grade curriculum using appropriate reference materials to support scientific understanding, plan and carry out scientific investigations of various types, such as systematic observations or experiments, identify variables, collect and organize data, interpret data in charts, tables, and graphics, analyze information, make predictions, and defend conclusions.

 Launch Lab

15 minutes

Does Earth's shape affect temperatures on Earth's surface?

Temperatures near Earth's poles are colder than temperatures near the equator. What causes these temperature differences?

Procedure

1. Read and complete a lab safety form.
2. Inflate a **spherical balloon** and tie the balloon closed.
3. Using a **marker,** draw a line around the balloon to represent Earth's equator.
4. Using a **ruler**, place a lit **flashlight** about 8 cm from the balloon so the flashlight beam strikes the equator straight on.
5. Using the marker, trace around the light projected onto the balloon.
6. Have someone raise the flashlight vertically 5–8 cm without changing the direction that the flashlight is pointing. Do not change the position of the balloon. Trace around the light projected onto the balloon again.

Think About This

1. Compare and contrast the shapes you drew on the balloon.

2. At which location on the balloon is the light more spread out? Explain your answer.

3. **Key Concept** Use your model to explain why Earth is warmer near the equator and colder near the poles.

1. From the *International Space Station*, Earth might look like it is just floating, but it is actually traveling around the Sun at more than 100,000 km/h. What natural phenomena do you think Earth's motion might cause?

_____

_____

_____

_____

_____

_____

Earth and the Sun

If you look outside at the ground, trees, and buildings, it does not seem like Earth is moving. Yet Earth is always in motion, spinning in space and traveling around the Sun. As Earth spins, day changes to night and back to day again. The seasons change as Earth travels around the Sun. Summer changes to winter because Earth's motion changes how energy from the Sun spreads out over Earth's surface.

The Sun

The nearest star to Earth is the Sun, which is shown in **Figure 1.** The Sun is approximately 150 million km from Earth. Compared to Earth, the Sun is enormous. The Sun's diameter is more than 100 times greater than Earth's diameter. The Sun's mass is more than 300,000 times greater than Earth's mass.

Deep inside the Sun, nuclei of atoms combine, releasing huge amounts of energy. This process is called nuclear fusion. The Sun releases so much energy from nuclear fusion that the temperature at its core is more than 15,000,000°C. Even at the Sun's surface, the temperature is about 5,500°C. A small part of the Sun's energy reaches Earth as light and thermal energy.

Figure 1 The Sun is a giant ball of hot gases that emits light and energy.

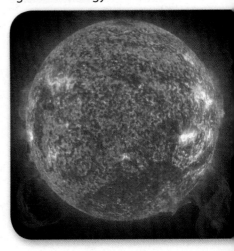

Active Reading   **2. Point Out** (Circle) the temperatures of the Sun's core and its surface.

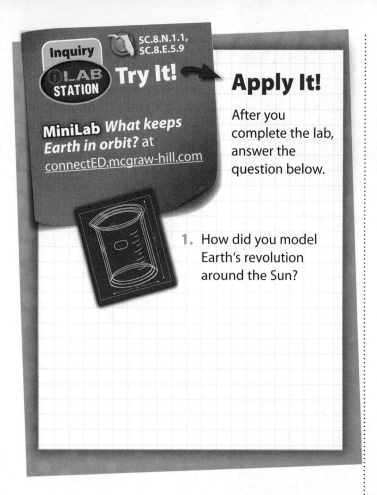

Apply It!

After you complete the lab, answer the question below.

1. How did you model Earth's revolution around the Sun?

Figure 2 Earth moves in a nearly circular orbit. The pull of the Sun's gravity on Earth causes Earth to revolve around the Sun.

Active Reading 3. **Compare** Fill in the blanks in the figure below to show where Earth is farthest from the Sun and where it is closest to the Sun.

Earth's Orbit

As shown in **Figure 2,** Earth moves around the Sun in a nearly circular path. *The path an object follows as it moves around another object is an* **orbit**. *The motion of one object around another object is called* **revolution**. Earth makes one complete revolution around the Sun every 365.24 days.

The Sun's Gravitational Pull

Why does Earth orbit the Sun? The answer is that the Sun's gravity pulls on Earth. The pull of gravity between two objects depends on the masses of the objects and the distance between them. The more mass either object has, or the closer together they are, the stronger the gravitational pull.

The Sun's effect on Earth's motion is illustrated in **Figure 2.** Earth's motion around the Sun is like the motion of an object twirled on a string. The string pulls on the object and makes it move in a circle. If the string breaks, the object flies off in a straight line. In the same way, the pull of the Sun's gravity keeps Earth revolving around the Sun in a nearly circular orbit. If the gravity between Earth and the Sun were to somehow stop, Earth would fly off into space in a straight line.

4. **NGSSS Check Identify** <u>Underline</u> the cause of Earth's orbit around the Sun. SC.8.E.5.4

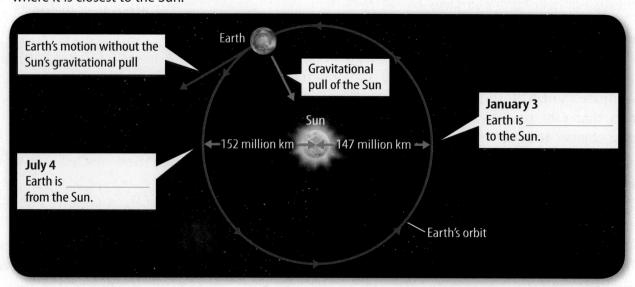

Earth's motion without the Sun's gravitational pull

Earth

Gravitational pull of the Sun

Sun

152 million km   147 million km

January 3
Earth is _____
to the Sun.

July 4
Earth is _____
from the Sun.

Earth's orbit

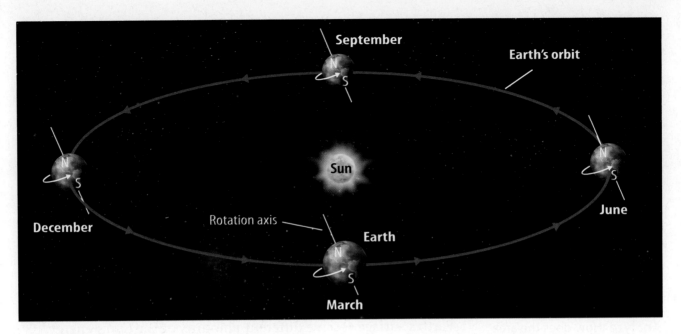

Figure 3 🔑 This diagram shows Earth's orbit, which is nearly circular, from an angle. Earth spins on its rotation axis as it revolves around the Sun. Earth's rotation axis always points in the same direction.

Earth's Rotation

As Earth revolves around the Sun, it spins. *A spinning motion is called* **rotation**. Some spinning objects rotate on a rod or axle. Earth rotates on an imaginary line through its center. *The line on which an object rotates is the* **rotation axis**.

Suppose you could look down on Earth's North Pole and watch Earth rotate. You would see that Earth rotates on its rotation axis in a counterclockwise direction, from west to east. One complete rotation of Earth takes about 24 hours. This rotation helps produce Earth's cycle of day and night. It is daytime on the half of Earth facing toward the Sun and nighttime on the half of Earth facing away from the Sun.

Active Reading **6. Identify** (Circle) the direction of Earth's rotation.

The Sun's Apparent Motion Each day the Sun appears to move from east to west across the sky. It seems as if the Sun is moving around Earth. However, it is Earth's rotation that causes the Sun's apparent motion.

5. Visual Check Analyze Between which months does the north end of Earth's rotation axis point away from the Sun? _____

_____

Earth rotates from west to east. As a result, the Sun appears to move from east to west across the sky. The stars and the Moon also seem to move from east to west across the sky due to Earth's west to east rotation.

To better understand this, imagine riding on a merry-go-round. As you and the ride move, people on the ground appear to be moving in the opposite direction. In the same way, as Earth rotates from west to east, the Sun appears to move from east to west.

Active Reading **7. Apply** Why does the Sun appear to move across the sky from east to west?

_____

The Tilt of Earth's Rotation Axis As shown in **Figure 3**, Earth's rotation axis is tilted. The tilt of Earth's rotation axis is always in the same direction by the same amount. This means that during half of Earth's orbit, the north end of the rotation axis is toward the Sun. During the other half of Earth's orbit, the north end of the rotation axis is away from the Sun.

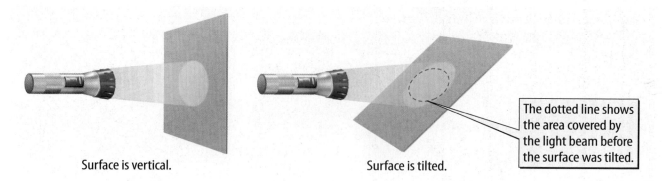

Surface is vertical.

Surface is tilted.

The dotted line shows the area covered by the light beam before the surface was tilted.

Figure 4 The light energy on a surface becomes more spread out as the surface becomes more tilted relative to the light beam.

 8. Visual Check

Analyze How does the area covered by the light differ on the vertical surface and the tilted surface?

_____

_____

_____

9. NGSSS Check

Explain Use the figure to explain why Earth is warmer at the equator and colder at the poles.
SC.8.E.5.9

Temperature and Latitude

As Earth orbits the Sun, only one half of Earth faces the Sun at a time. A beam of sunlight carries energy. The more sunlight that reaches a part of Earth's surface, the warmer that part becomes. Because Earth's surface is curved, different parts of Earth's surface receive different amounts of the Sun's energy.

Energy Received by a Tilted Surface

Suppose you shine a beam of light on a flat card, as shown in **Figure 4.** As you tilt the card relative to the direction of the light beam, light becomes more spread out on the card's surface. As a result, the energy that the light beam carries also spreads out more over the card's surface. An area on the surface within the light beam receives less energy when the surface is more tilted relative to the light beam.

The Tilt of Earth's Curved Surface

Instead of being flat like a card, Earth's surface is curved. Relative to the direction of a beam of sunlight, Earth's surface becomes more tilted as you move away from the equator. As shown in **Figure 5,** the energy in a beam of sunlight tends to become more spread out the farther you travel from the equator. This means that regions near the poles receive less energy than regions near the equator. This makes Earth colder at the poles and warmer at the equator.

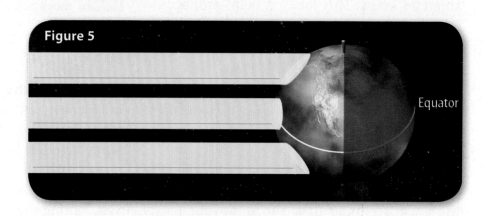

Figure 5

Equator

North end of rotation axis is away from the Sun.

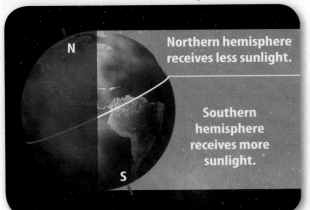

Northern hemisphere receives less sunlight.

Southern hemisphere receives more sunlight.

North end of rotation axis is toward the Sun.

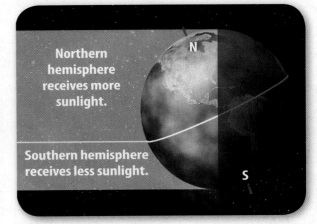

Northern hemisphere receives more sunlight.

Southern hemisphere receives less sunlight.

Figure 6 🔑 The northern hemisphere receives more sunlight in June, and the southern hemisphere receives more sunlight in December.

Active Reading

10. Apply How does the tilt of Earth's rotation axis affect Earth's weather?

Seasons

You might think that summer happens when Earth is closest to the Sun and winter happens when Earth is farthest from the Sun. However, seasonal changes do not depend on Earth's distance from the Sun. In fact, Earth is closest to the Sun in January! Instead, it is the tilt of Earth's rotation axis, combined with Earth's motion around the Sun, that causes the seasons to change.

Spring and Summer in the Northern Hemisphere

During one half of Earth's orbit, the north end of the rotation axis is toward the Sun. Then, the northern hemisphere receives more energy from the Sun than the southern hemisphere, as shown in **Figure 6.** Temperatures increase in the northern hemisphere and decrease in the southern hemisphere. Daylight hours last longer in the northern hemisphere, and nights last longer in the southern hemisphere. This is when spring and summer happen in the northern hemisphere, and fall and winter happen in the southern hemisphere.

Fall and Winter in the Northern Hemisphere

During the other half of Earth's orbit, the north end of the rotation axis is away from the Sun. Then, the northern hemisphere receives less solar energy than the southern hemisphere, as shown in **Figure 6.** Temperatures decrease in the northern hemisphere and increase in the southern hemisphere. This is when fall and winter happen in the northern hemisphere and spring and summer happen in the southern hemisphere.

Math Skills   MA.6.A.3.6

Convert Units

When Earth is 147,000,000 km from the Sun, how far is Earth from the Sun in miles? To calculate the distance in miles, multiply the distance in km by the conversion factor.

$$147{,}000{,}000 \text{ km} \times \frac{0.62 \text{ miles}}{1 \text{ km}}$$
$$= 91{,}100{,}000 \text{ miles}$$

11. Practice When Earth is 152,000,000 km from the Sun, how far is Earth from the Sun in miles?

Figure 7 🔑 **12. Compare** When is the angle between Earth's rotation axis and the Sun the largest? When is it the smallest?

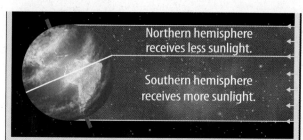

December Solstice
The December solstice is on December 21 or 22.
On this day
- the north end of Earth's rotation axis is away from the Sun;
- days in the northern hemisphere are shortest and nights are longest; winter begins;
- days in the southern hemisphere are longest and nights are shortest; summer begins.

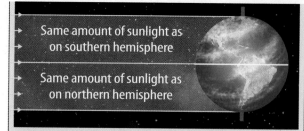

September Equinox
The September equinox is on September 22 or 23.
On this day
- the north end of Earth's rotation axis leans along Earth's orbit;
- there are about 12 hours of daylight and 12 hours of darkness everywhere on Earth;
- autumn begins in the northern hemisphere;
- spring begins in the southern hemisphere.

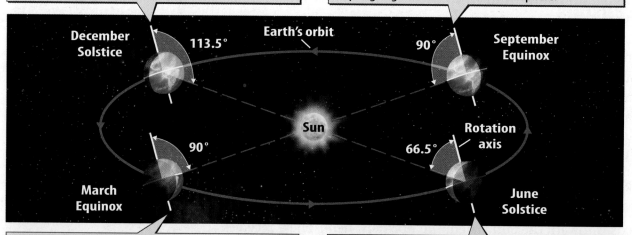

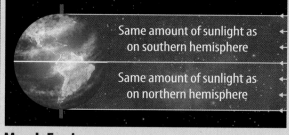

March Equinox
The March equinox is on March 20 or 21.
On this day
- the north end of Earth's rotation axis leans along Earth's orbit;
- there are about 12 hours of daylight and 12 hours of darkness everywhere on Earth;
- spring begins in the northern hemisphere;
- autumn begins in the southern hemisphere.

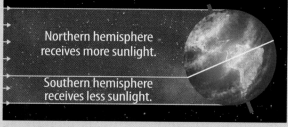

June Solstice
The June solstice is on June 20 or 21.
On this day
- the north end of Earth's rotation axis is toward the Sun;
- days in the northern hemisphere are longest and nights are shortest; summer begins;
- days in the southern hemisphere are shortest and nights are longest; winter begins.

Solstices, Equinoxes, and the Seasonal Cycle

Figure 7 shows that as Earth travels around the Sun, its rotation axis always points in the same direction in space. However, the amount that Earth's rotation axis is toward or away from the Sun changes. This causes the seasons to change in a yearly cycle.

There are four days each year when the direction of Earth's rotation axis is special relative to the Sun. *A* **solstice** *is a day when Earth's rotation axis is the most toward or away from the Sun. An* **equinox** *is a day when Earth's rotation axis is leaning along Earth's orbit, neither toward nor away from the Sun.*

March Equinox to June Solstice When the north end of the rotation axis gradually points more and more toward the Sun, the northern hemisphere gradually receives more solar energy. This is spring in the northern hemisphere.

June Solstice to September Equinox The north end of the rotation axis continues to point toward the Sun but does so less and less. The northern hemisphere starts to receive less solar energy. This is summer in the northern hemisphere.

September Equinox to December Solstice The north end of the rotation axis now points more and more away from the Sun. The northern hemisphere receives less and less solar energy. This is fall in the northern hemisphere.

December Solstice to March Equinox The north end of the rotation axis continues to point away from the Sun but does so less and less. The northern hemisphere starts to receive more solar energy. This is winter in the northern hemisphere.

Changes in the Sun's Apparent Path Across the Sky

Figure 8 shows how the Sun's apparent path through the sky changes from season to season in the northern hemisphere. The Sun's apparent path through the sky in the northern hemisphere is lowest on the December solstice and highest on the June solstice.

Active Reading

FOLDABLES® LA.8.2.2.3

Make a bound book with four full pages. Label the pages with the names of the solstices and equinoxes. Use each page to organize information about each season.

WORD ORIGIN

equinox
from Latin *equinoxium*, means "equality of night and day"

Figure 8 As the seasons change, the path of the Sun across the sky changes.

✓ 13. **Visual Check**
Relate In the figure below, fill in the blanks with the season that starts on each solstice or equinox.

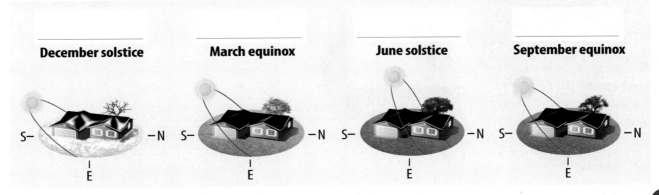

| December solstice | March equinox | June solstice | September equinox |

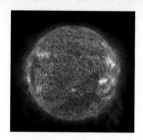

The gravitational pull of the Sun causes Earth to revolve around the Sun in a near-circular orbit.

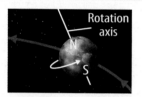

Earth's rotation axis is tilted and always points in the same direction in space.

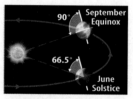

Equinoxes and solstices are days when the direction of Earth's rotation axis relative to the Sun is special.

Inquiry SC.8.N.1.1 SC.8.E.5.9

⚲LAB STATION Try It!

Skill Lab *How does Earth's tilted rotation axis affect the seasons?* at connectED.mcgraw-hill.com

Use Vocabulary

1 **Distinguish** between Earth's rotation and Earth's revolution.

_____

_____

2 When a(n) _____ occurs, the northern hemisphere and the southern hemisphere receive the same amount of sunlight.

Understand Key Concepts 🔑

3 What is caused by the tilt of Earth's rotational axis?

(A) Earth's orbit   (C) Earth's revolution

(B) Earth's seasons   (D) Earth's rotation

4 **Contrast** the amount of sunlight received by an area near the equator and a same-sized area near the South Pole. SC.8.E.5.9

_____

_____

5 **Contrast** the Sun's gravitational pull on Earth when Earth is closest to the Sun and when Earth is farthest from the Sun. SC.8.E.5.4

_____

_____

Interpret Graphics

6 **Summarize** Fill in the season table below for the northern hemisphere. LA.8.2.2.3

Season	Starts on Solstice or Equinox?	How Rotation Axis Leans
Summer		
Fall		
Winter		
Spring		

Critical Thinking

7 **Defend** The December solstice is often called the winter solstice. Do you think this is an appropriate label? Defend your answer.

_____

_____

Math Skills   MA.6.A.3.6

8 The Sun's diameter is about 1,390,000 km. What is the Sun's diameter in miles?

Moon Over Florida LA.8.2.2.3, SC.8.E.5.9, SC.8.N.1.1

How does the Moon appear from Florida? After you complete this chapter, observe
the Moon for a full lunar cycle and record your observations. What time did the
Moon rise each night? What phases of the Moon did you see? Include the names
and sketches of the lunar phases.

Night	Moon Rise Time	Lunar Phase	Night	Moon Rise Time	Lunar Phase
1			15		
2			16		
3			17		
4			18		
5			19		
6			20		
7			21		
8			22		
9			23		
10			24		
11			25		
12			26		
13			27		
14			28		

Lunar Phase Sketches

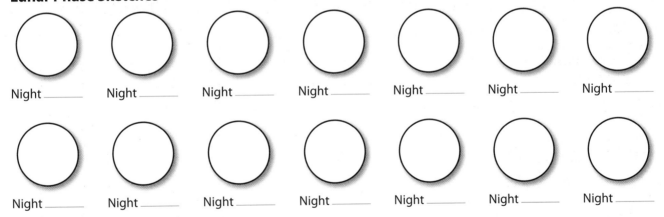

Night _____   Night _____   Night _____   Night _____   Night _____   Night _____   Night _____

Night _____   Night _____   Night _____   Night _____   Night _____   Night _____   Night _____

 Extension

If you live near one of Florida's beaches, also record the times of high tide and low
tide. How does the phase of the Moon affect the tides?

Earth's MOON

 How does the Moon move around Earth?

 Why does the Moon's appearance change?

Vocabulary

maria p. 22

phase p. 24

waxing phase p. 24

waning phase p. 24

 Florida NGSSS

LA.8.2.2.3 The student will organize information to show understanding or relationships among facts, ideas, and events (e.g., representing key points within text through charting, mapping, paraphrasing, summarizing, or comparing/contrasting);

SC.8.E.5.4 Explore the Law of Universal Gravitation by explaining the role that gravity plays in the formation of planets, stars, and solar systems and in determining their motions.

SC.8.E.5.9 Explain the impact of objects in space on each other including:
1. the Sun on the Earth including seasons and gravitational attraction
2. the Moon on the Earth, including phases, tides, and eclipses, and the relative position of each body.

SC.8.N.1.1 Define a problem from the eighth grade curriculum using appropriate reference materials to support scientific understanding, plan and carry out scientific investigations of various types, such as systematic observations or experiments, identify variables, collect and organize data, interpret data in charts, tables, and graphics, analyze information, make predictions, and defend conclusions.

SC.8.E.5.9

Inquiry Launch Lab

15 minutes

Why does the Moon appear to change shape?

The Sun is always shining on Earth and the Moon. However, the Moon's shape seems to change from night to night and day to day. What could cause the Moon's appearance to change?

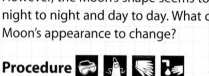

Procedure

1. Read and complete a lab safety form.
2. Place a **ball** on a level surface.
3. Position a **flashlight** so that the light beam shines fully on one side of the ball. Stand behind the flashlight.
4. Make a drawing of the ball's appearance.
5. Stand behind the ball, facing the flashlight, and repeat step 4.
6. Stand to the left of the ball and repeat step 4.

Data and Observations

Think About This

1. What caused the ball's appearance to change?

2. **Key Concept** What do you think produces the Moon's changing appearance in the sky?

Inquiry Two Planets?

1. The smaller body is Earth's Moon, not a planet. Just as Earth moves around the Sun, the Moon moves around Earth. What changes do you think are caused by the Moon's motion around Earth?

_____

_____

_____

_____

_____

Seeing the Moon

Imagine what people thousands of years ago thought when they looked up at the Moon. They might have wondered why the Moon shines and why it seems to change shape. They probably would have been surprised to learn that the Moon does not emit light at all. Unlike the Sun, the Moon is a solid object that does not emit its own light. You only see the Moon because light from the Sun reflects off the Moon and into your eyes. Some facts about the Moon, such as its mass, size, and distance from Earth, are shown in **Table 1.**

Active Reading

2. Point Out <u>Underline</u> the reason you are able to see the Moon.

Active Reading

FOLDABLES LA.8.2.2.3

Use two sheets of paper to make a bound book. Use it to organize information about the lunar cycle. Each page of your book should represent one week of the lunar cycle.

Table 1	**Moon Data**			
Mass	**Diameter**	**Average distance from Earth**	**Time for one rotation**	**Time for one revolution**
1.2% of Earth's mass	27% of Earth's diameter	384,000 km	27.3 days	27.3 days

Figure 9 The Moon probably formed 4.5 billion years ago when a large object collided with Earth. Material ejected from the collision eventually clumped together and formed the Moon.

Active Reading **3. Sequence** Put the images in the correct sequence describing the formation of the Moon by writing *1, 2* or *3* in the circles.

The particles gradually clump together and form the Moon.

An object the size of Mars crashes into the semi-molten Earth about 4.5 billion years ago.

The impact ejects vaporized rock into space. As the rock cools, it forms a ring of particles around Earth.

The Moon's Formation

The most widely accepted idea for the Moon's formation is the giant impact hypothesis, shown in **Figure 9.** According to this hypothesis, shortly after Earth formed about 4.6 billion years ago, an object about the size of the planet Mars collided with Earth. The impact ejected vaporized rock that formed a ring around Earth. Eventually, the material in the ring cooled and clumped together, forming the Moon.

The Moon's Surface

The surface of the Moon was shaped early in its history. Examples of common features on the Moon's surface are shown in **Figure 10.**

Craters The Moon's craters were formed when objects from space crashed into the Moon. Light-colored streaks called rays extend outward from some craters.

Most of the impacts that formed the Moon's craters occurred more than 3.5 billion years ago, long before dinosaurs lived on Earth. Earth was also heavily bombarded by objects from space during this time. However, on Earth, wind, water, and plate tectonics erased the craters. The Moon has no atmosphere, water, or plate tectonics, so craters formed billions of years ago on the Moon have hardly changed.

Maria *The large, dark, flat areas on the Moon are called* **maria** (MAR ee uh). The maria formed after most impacts on the Moon's surface had stopped. Maria formed when lava flowed up through the Moon's crust and solidified. The lava covered many of the Moon's craters and other features. When this lava solidified, it was dark and flat.

Active Reading **4. Recall** Underline How did maria form on the Moon's surface?

Highlands The light-colored highlands are too high for the lava that formed the maria to reach. The highlands are older than the maria and are covered with craters.

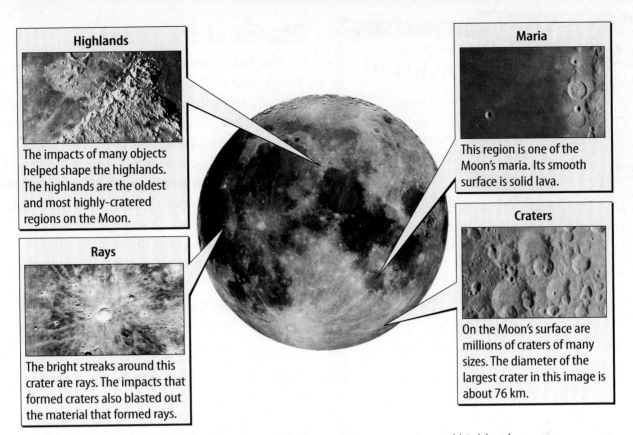

Highlands

The impacts of many objects helped shape the highlands. The highlands are the oldest and most highly-cratered regions on the Moon.

Rays

The bright streaks around this crater are rays. The impacts that formed craters also blasted out the material that formed rays.

Maria

This region is one of the Moon's maria. Its smooth surface is solid lava.

Craters

On the Moon's surface are millions of craters of many sizes. The diameter of the largest crater in this image is about 76 km.

Figure 10 The Moon's surface features include craters, rays, maria, and highlands.

The Moon's Motion

While Earth is revolving around the Sun, the Moon is revolving around Earth. The gravitational pull of Earth on the Moon causes the Moon to move in an orbit around Earth. The Moon makes one revolution around Earth every 27.3 days.

 5. NGSSS Check Explain <u>Underline</u> What produces the Moon's revolution around Earth?
SC.8.E.5.4, SC.8.E.5.9

The Moon also rotates as it revolves around Earth. One complete rotation of the Moon also takes 27.3 days. This means the Moon makes one rotation in the same amount of time that it makes one revolution around Earth. **Figure 11** shows that, because the Moon makes one rotation for each revolution of Earth, the same side of the Moon always faces Earth. This side of the Moon is called the near side. The side of the Moon that cannot be seen from Earth is called the far side of the Moon.

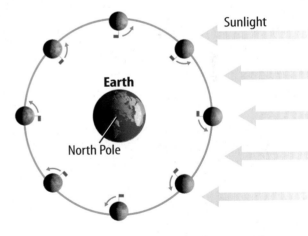

Figure 11 Many people think the Moon has a dark side and a light side. Half of the Moon is always in sunlight, but the lit portion changes as the Moon rotates and revolves around Earth.

6. Visual Check Apply Why does the same side of the Moon always face Earth?

_____

_____

_____

MiniLab *How can the Moon be rotating if the same side of the Moon always faces Earth?* at connectED.mcgraw-hill.com

Apply It!

After you complete the lab, answer the question below.

1. Write an analogy to explain how the same side of the Moon is always facing Earth.

Phases of the Moon

The Sun is always shining on half of the Moon, just as the Sun is always shining on half of Earth. However, as the Moon moves around Earth, usually only part of the Moon's near side is lit. *The lit part of the Moon or a planet that can be seen from Earth is called a* **phase**. As shown in **Figure 12,** the motion of the Moon around Earth causes the phase of the Moon to change. The sequence of phases is the lunar cycle. One lunar cycle takes 29.5 days or slightly more than four weeks to complete.

 7. **NGSSS Check Identify** Underline what produces the phases of the Moon. SC.8.E.5.9

Waxing Phases

During the **waxing phases,** *more of the Moon's near side is lit each night.*

Week 1—First Quarter As the lunar cycle begins, a sliver of light can be seen on the Moon's western edge. Gradually the lit part becomes larger. By the end of the first week, the Moon is at its first quarter phase. In this phase, the Moon's entire western half is lit.

Week 2—Full Moon During the second week, more and more of the near side becomes lit. When the Moon's near side is completely lit, it is at the full moon phase.

Waning Phases

During the **waning phases,** *less of the Moon's near side is lit each night.* As seen from Earth, the lit part is now on the Moon's eastern side.

Week 3—Third Quarter During this week, the lit part of the Moon becomes smaller until only the eastern half of the Moon is lit. This is the third quarter phase.

Week 4—New Moon During this week, less and less of the near side is lit. When the Moon's near side is completely dark, it is at the new moon phase.

The Lunar Cycle 🔑

Figure 12 As the Moon revolves around Earth, the part of the Moon's near side that is lit changes. The figure below shows how the Moon would look at different places in its orbit.

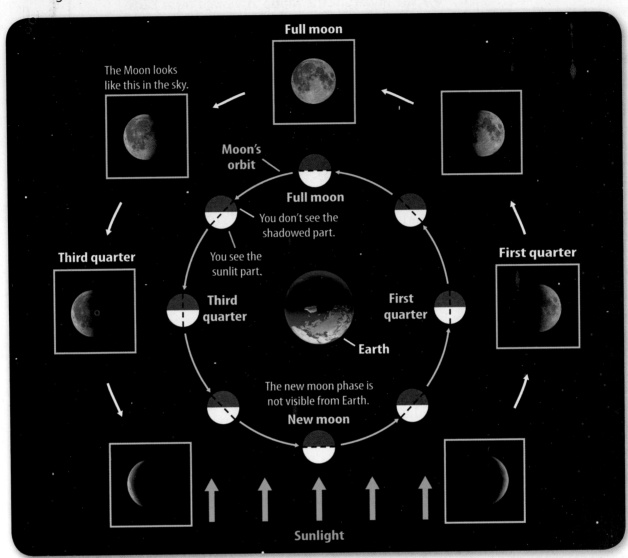

The Moon at Midnight

The Moon's motion around Earth causes the Moon to rise, on average, about 50 minutes later each day. The figure below shows how the Moon looks at midnight during three phases of the lunar cycle.

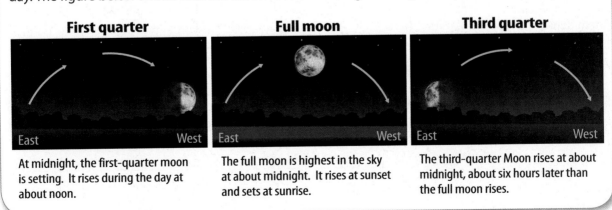

First quarter

At midnight, the first-quarter moon is setting. It rises during the day at about noon.

Full moon

The full moon is highest in the sky at about midnight. It rises at sunset and sets at sunrise.

Third quarter

The third-quarter Moon rises at about midnight, about six hours later than the full moon rises.

According to the giant impact hypothesis, a large object collided with Earth about 4.5 billion years ago to form the Moon.

Features like maria, craters, and highlands formed on the Moon's surface early in its history.

The Moon's phases change in a regular pattern during the Moon's lunar cycle.

Use Vocabulary

1. The lit part of the Moon as viewed from Earth is a(n) _____.

2. When the lit side of the Moon appears to be getting larger, the lunar phase is described as _____.

3. When the lit side of the Moon appears to be getting smaller, the lunar phase is described as _____.

Understand Key Concepts 🔑

4. Which phase occurs when the Moon is between the Sun and Earth?

   (A) first quarter   (C) new moon

   (B) full moon   (D) third quarter

5. **Reason** Why does the Moon have phases?

   _____

   _____

   _____

Interpret Graphics

6. **Organize Information** Fill in the table below with descriptions of the lunar surface. LA.8.2.2.3

Crater	Ray
Maria	Highland

Critical Thinking

7. **Reflect** Imagine the Moon rotates twice in the same amount of time the Moon orbits Earth once. Would you be able to see the Moon's far side from Earth? SC.8.E.5.9

   _____

   _____

Return to the Moon

Exploring Earth's Moon is a step toward exploring other planets and building outposts in space.

"The United States undertook a series of human spaceflight missions from 1961–1975 called the Apollo program. The goal of the program was to land humans on the Moon and bring them safely back to Earth. Six of the missions reached this goal. The Apollo program was a huge success, but it was just the beginning.

NASA began another space program that had a goal to return astronauts to the Moon to live and work. However, before that could happen, scientists needed to know more about conditions on the Moon and what materials are available there.

Collecting data was the first step. In 2009, NASA launched the *Lunar Reconnaissance Orbiter (LRO)* spacecraft. The LRO spent a year orbiting the Moon's two poles. It collected detailed data that scientists can use to make maps of the Moon's features and resources, such as deep craters that formed on the Moon when comets and asteroids slammed into it billions of years ago. Some scientists predicted that these deep craters contain frozen water.

One of the instruments launched with the *LRO* was the *Lunar Crater Observation and Sensing Satellite (LCROSS)*. *LCROSS* observations confirmed the scientists' predictions that water exists on the Moon. A rocket launched from *LCROSS* impacted the Cabeus crater near the Moon's south pole. The material that was ejected after the rocket's impact included water.

NASA's goal of returning astronauts to the Moon was delayed, and their missions now focus on exploring Mars instead. But the discoveries made on the Moon will help scientists develop future missions that could take humans farther into the solar system.

Apollo SPACE PROGRAM

The Apollo Space Program included 17 missions. Here are some milestones:

January 27 1967
Apollo 1 Fire killed all three astronauts on board during a launch simulation for the first piloted flight to the Moon.

December 21–27 1968
Apollo 8 First manned spacecraft orbits the Moon.

July 16–24 1969
Apollo 11 First humans, Neil Armstrong and Buzz Aldrin, walk on the Moon.

July 1971
Apollo 15 Astronauts drive the first rover on the Moon.

December 7–19 1972
Apollo 17 The first phase of human exploration of the Moon ended with this last lunar landing mission.

It's Your Turn

BRAINSTORM As a group, brainstorm the different occupations that would be needed to successfully operate a base on the Moon or another planet. Discuss the tasks that a person would perform in each occupation.

Eclipses and TIDES

Vocabulary

umbra p. 29

penumbra p. 29

solar eclipse p. 30

lunar eclipse p. 32

tide p. 33

 Florida NGSSS

LA.8.2.2.3 The student will organize information to show understanding or relationships among facts, ideas, and events (e.g., representing key points within text through charting, mapping, paraphrasing, summarizing, or comparing/contrasting);

SC.8.E.5.9 Explain the impact of objects in space on each other including:

1. the Sun on the Earth including seasons and gravitational attraction

2. the Moon on the Earth, including phases, tides, and eclipses, and the relative position of each body.

SC.8.N.1.1 Define a problem from the eighth grade curriculum using appropriate reference materials to support scientific understanding, plan and carry out scientific investigations of various types, such as systematic observations or experiments, identify variables, collect and organize data, interpret data in charts, tables, and graphics, analyze information, make predictions, and defend conclusions.

 Launch Lab

15 minutes

How do shadows change?

You can see a shadow when an object blocks a light source. What happens to an object's shadow when the object moves?

Procedure

1. Read and complete a lab safety form.

2. Select an **object** provided by your teacher.

3. Shine a **flashlight** on the object, projecting its shadow on the wall.

4. While holding the flashlight in the same position, move the object closer to the wall—away from the light. Then, move the object toward the light. Record your observations.

Data and Observations

Think About This

1. Compare and contrast the shadows created in each situation. Did the shadows have dark parts and light parts? Did these parts change?

2. **Key Concept** Imagine you look at the flashlight from behind your object, looking from the darkest and lightest parts of the object's shadow. How much of the flashlight could you see from each location?

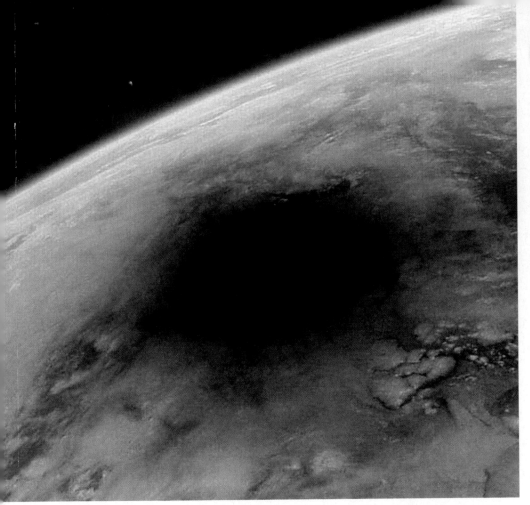

Inquiry **What is this dark spot?**

1. Cosmonauts took this photo from aboard the *Mir* orbiting space station. What do you think caused the shadow on Earth?

_____

_____

_____

_____

Shadows—The Umbra and the Penumbra

A shadow results when one object blocks the light that another object emits or reflects. When a tree blocks light from the Sun, it casts a shadow. If you want to stand in the shadow of a tree, the tree must be in a line between you and the Sun.

If you go outside on a sunny day and look carefully at a shadow on the ground, you might notice that the edges of the shadow are not as dark as the rest of the shadow. Light from the Sun and other wide sources casts shadows with two distinct parts, as shown in **Figure 13.** *The* **umbra** *is the central, darker part of a shadow where light is totally blocked. The* **penumbra** *is the lighter part of a shadow where light is partially blocked.* If you stood within an object's penumbra, you would be able to see only part of the light source. If you stood within an object's umbra, you would not see the light source at all.

WORD ORIGIN

penumbra

from Latin *paene*, means "almost"; and *umbra*, means "shade, shadow"

✓ **2. Visual Check**

Interpret Label the umbra and the penumbra of the shadow below.

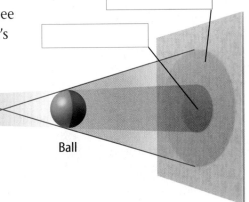

Figure 13 The shadow that a wide light source produces has two parts—the umbra and the penumbra. The light source cannot be seen from within the umbra. The light source can be partially seen from within the penumbra. ▶

Light source

Ball

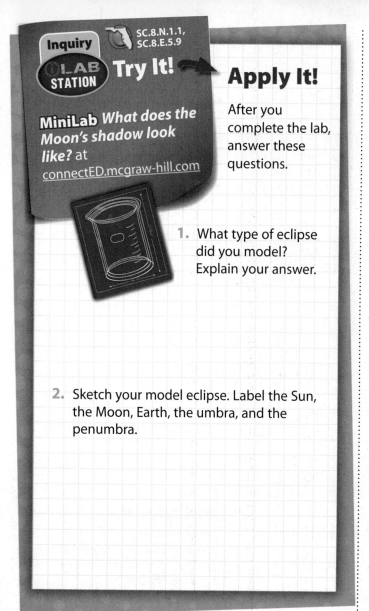

Inquiry SC.8.N.1.1, SC.8.E.5.9

iLAB STATION **Try It!**

MiniLab *What does the Moon's shadow look like?* at connectED.mcgraw-hill.com

Apply It!

After you complete the lab, answer these questions.

1. What type of eclipse did you model? Explain your answer.

2. Sketch your model eclipse. Label the Sun, the Moon, Earth, the umbra, and the penumbra.

Solar Eclipses

As the Sun shines on the Moon, the Moon casts a shadow that extends out into space. Sometimes the Moon passes between Earth and the Sun. This can only happen during the new moon phase. When Earth, the Moon, and the Sun are lined up, the Moon casts a shadow on Earth's surface, as shown in **Figure 14.** You can see the Moon's shadow in the photo at the beginning of this lesson. *When the Moon's shadow appears on Earth's surface, a* **solar eclipse** *is occurring.*

3. **NGSSS Check** **Explain** Why does a solar eclipse occur only during a new moon? SC.8.E.5.9

_____

_____

As Earth rotates, the Moon's shadow moves along Earth's surface, as shown in **Figure 14.** The type of eclipse you see depends on whether you are in the path of the umbra or the penumbra. If you are outside the umbra and penumbra, you cannot see a solar eclipse at all.

Total Solar Eclipses

You can only see a total solar eclipse from within the Moon's umbra. During a total solar eclipse, the Moon appears to cover the Sun completely, as shown in **Figure 15** on the next page. Then, the sky becomes dark enough that you can see stars. A total solar eclipse lasts no longer than about 7 minutes.

Figure 14 🔑 A solar eclipse occurs only when the Moon moves directly between Earth and the Sun. The Moon's shadow moves across Earth's surface.

4. **Visual Check** **Label** Where would a total solar eclipse be seen, and where would a partial solar eclipse be seen?

Earth

Moon

Penumbra

Umbra

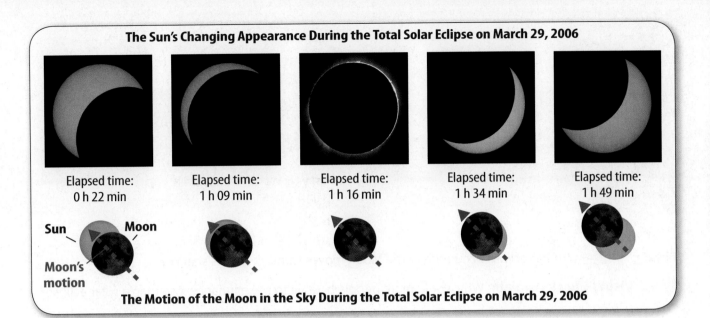

The Sun's Changing Appearance During the Total Solar Eclipse on March 29, 2006

Elapsed time: 0 h 22 min

Elapsed time: 1 h 09 min

Elapsed time: 1 h 16 min

Elapsed time: 1 h 34 min

Elapsed time: 1 h 49 min

Sun

Moon

Moon's motion

The Motion of the Moon in the Sky During the Total Solar Eclipse on March 29, 2006

Partial Solar Eclipses

You can only see a total solar eclipse from within the Moon's umbra, but you can see a partial solar eclipse from within the Moon's much larger penumbra. The stages of a partial solar eclipse are similar to the stages of a total solar eclipse, except that the Moon never completely covers the Sun.

Why don't solar eclipses occur every month?

Solar eclipses only can occur during a new moon, when Earth and the Sun are on opposite sides of the Moon. However, solar eclipses do not occur during every new-moon phase. **Figure 16** shows why. The Moon's orbit is tilted slightly compared to Earth's orbit. As a result, during most new moons, Earth is either above or below the Moon's shadow. However, every so often the Moon is in a line between the Sun and Earth. Then the Moon's shadow passes over Earth and a solar eclipse occurs.

Figure 15 This sequence of photographs shows how the Sun's appearance changed during a total solar eclipse in 2006.

Active Reading

5. Recall <u>Underline</u> why a solar eclipse does not occur every month.

Figure 16 A solar eclipse occurs only when the Moon crosses Earth's orbit and is in a direct line between Earth and the Sun.

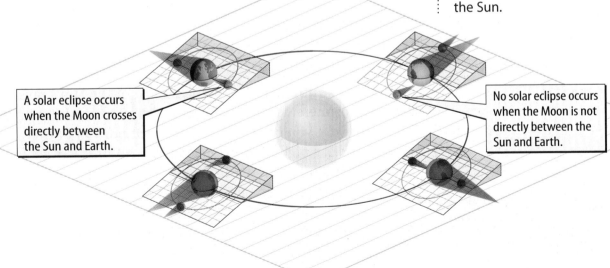

A solar eclipse occurs when the Moon crosses directly between the Sun and Earth.

No solar eclipse occurs when the Moon is not directly between the Sun and Earth.

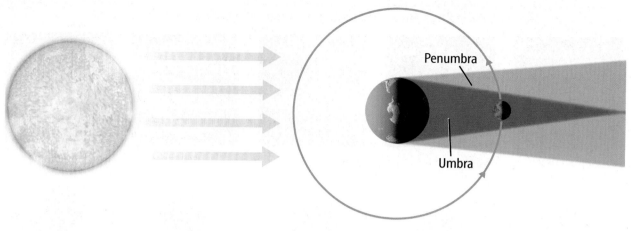

Figure 17 🔑 A lunar eclipse occurs when the Moon moves through Earth's shadow.

Penumbra

Umbra

✔ **6. Visual Check Explain** Why would more people be able to see a lunar eclipse than a solar eclipse?

_____

Active Reading

FOLDABLES® LA.8.2.2.3

Make a two-tab book from a sheet of notebook paper. Label the tabs Solar Eclipse and Lunar Eclipse. Use it to organize your notes on eclipses.

Solar Eclipse   Lunar Eclipse

Active Reading **8. Describe** What color does the Moon appear to be during a total lunar eclipse?

_____

Lunar Eclipses

Just like the Moon, Earth casts a shadow into space. As the Moon revolves around Earth, it sometimes moves into Earth's shadow, as shown in **Figure 17.** *A **lunar eclipse** occurs when the Moon moves into Earth's shadow.* Then Earth is in a line between the Sun and the Moon. This means that a lunar eclipse can occur only during the full moon phase.

Like the Moon's shadow, Earth's shadow has an umbra and a penumbra. Different types of lunar eclipses occur depending on which part of Earth's shadow the Moon moves through. Unlike solar eclipses, you can see any lunar eclipse from any location on the side of Earth facing the Moon.

When the entire Moon moves through Earth's umbra, a total lunar eclipse occurs. **Figure 18** on the next page shows how the Moon's appearance changes during a total lunar eclipse. The Moon's appearance changes as it gradually moves into Earth's penumbra, then into Earth's umbra, back into Earth's penumbra, and then out of Earth's shadow entirely.

You can still see the Moon even when it is completely within Earth's umbra. Although Earth blocks most of the Sun's rays, Earth's atmosphere deflects some sunlight into Earth's umbra. This is also why you can often see the unlit portion of the Moon on a clear night. Astronomers often call this Earthshine. This reflected light has a reddish color and gives the Moon a reddish tint during a total lunar eclipse.

 7. NGSSS Check State When can a lunar eclipse occur? SC.8.E.5.9

_____

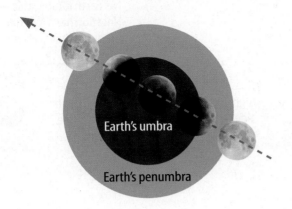

Figure 18 If the entire Moon passes through Earth's umbra, the Moon gradually darkens until a dark shadow covers it completely.

Earth's umbra

Earth's penumbra

✓ 9. **Visual Check** **Contrast** How would a total lunar eclipse look different from a total solar eclipse?

_____

_____

Partial Lunar Eclipses

When only part of the Moon passes through Earth's umbra, a partial lunar eclipse occurs. The stages of a partial lunar eclipse are similar to those of a total lunar eclipse, shown in **Figure 18,** except the Moon is never completely covered by Earth's umbra. The part of the Moon in Earth's penumbra appears only slightly darker, while the part of the Moon in Earth's umbra appears much darker.

Why don't lunar eclipses occur every month?

Lunar eclipses can occur only during a full moon phase, when the Moon and the Sun are on opposite sides of Earth. However, lunar eclipses do not occur during every full moon because of the tilt of the Moon's orbit with respect to Earth's orbit. During most full moons, the Moon is slightly above or slightly below Earth's penumbra.

Tides

The positions of the Moon and the Sun also affect Earth's oceans. If you have spent time near Florida's coast, you might have seen how the ocean's height, or sea level, rises and falls twice each day. _A_ **tide** _is the daily rise and fall of sea level._ Examples of tides are shown in **Figure 19.** It is primarily the Moon's gravity that causes Earth's oceans to rise and fall twice each day.

Figure 19 In the Bay of Fundy, high tides can be more than 10 m higher than low tides.

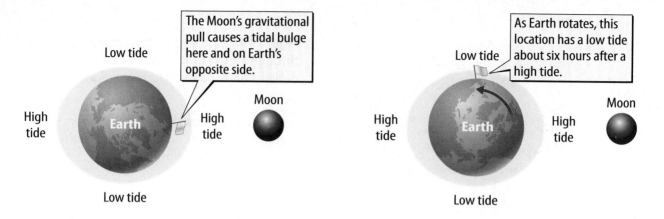

The Moon's gravitational pull causes a tidal bulge here and on Earth's opposite side.

Low tide

High tide · Earth · High tide

Moon

Low tide

As Earth rotates, this location has a low tide about six hours after a high tide.

Low tide

High tide · Earth · High tide

Moon

Low tide

Figure 20 In this view looking down on Earth's North Pole, the flag moves into a tidal bulge as Earth rotates. A coastal area has a high tide about once every 12 hours.

The Moon's Effect on Earth's Tides

The difference in the strength of the Moon's gravity on opposite sides of Earth causes Earth's tides. The Moon's gravity is slightly stronger on the side of Earth closer to the Moon and slightly weaker on the side of Earth opposite the Moon. These differences cause tidal bulges in the oceans on opposite sides of Earth, as shown in **Figure 20**. High tides occur at the tidal bulges, and low tides occur between them.

The Sun's Effect on Earth's Tides

Because the Sun is so far away from Earth, its effect on tides is about half that of the Moon. **Figure 21** shows how the positions of the Sun and the Moon affect Earth's tides.

Spring Tides During the full moon and new moon phases, spring tides occur. This is when the Sun's and the Moon's gravitational effects combine and produce higher high tides and lower low tides.

Neap Tides A week after a spring tide, a neap tide occurs. Then the Sun, Earth, and the Moon form a right angle. When this happens, the Sun's effect on tides reduces the Moon's effect. High tides are lower and low tides are higher at neap tides.

10. NGSSS Check Compare Why is the Sun's effect on tides less than the Moon's effect? SC.8.E.5.9

_____

_____

Figure 21 A spring tide occurs when the Sun, Earth, and the Moon are in a line. A neap tide occurs when the Sun and the Moon form a right angle with Earth.

11. Visual Check Relate Fill in the blanks with the correct phases of the Moon.

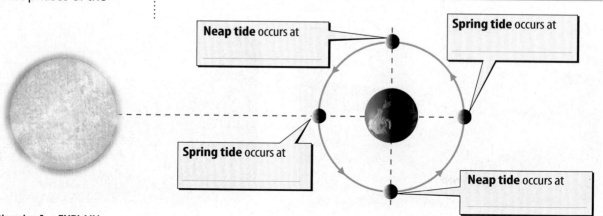

Neap tide occurs at _____

Spring tide occurs at _____

Spring tide occurs at _____

Neap tide occurs at _____

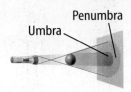

Shadows from a wide light source have two distinct parts.

The Moon's shadow produces solar eclipses. Earth's shadow produces lunar eclipses.

The positions of the Moon and the Sun in relation to Earth cause gravitational differences that produce tides.

Inquiry SC.8.N.1.1 SC.8.N.5.9

①LAB STATION Try It!

Inquiry Lab *Phases of the Moon* at connectED.mcgraw-hill.com

Use Vocabulary

1. **Distinguish** between an umbra and a penumbra.

_____

_____

2. **Use the term** *tide* in a sentence.

_____

Understand Key Concepts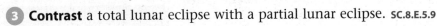

3. **Contrast** a total lunar eclipse with a partial lunar eclipse. SC.8.E.5.9

_____

_____

4. Which could occur during a total solar eclipse? SC.8.E.5.9
   - Ⓐ first quarter moon
   - Ⓒ neap tide
   - Ⓑ full moon
   - Ⓓ spring tide

Interpret Graphics

5. **Conclude** What type of eclipse does the figure above illustrate? Explain your answer. SC.8.E.5.9

_____

_____

_____

6. **Categorize Information** Fill in the graphic organizer below to identify two bodies that affect Earth's tides. SC.8.E.5.9

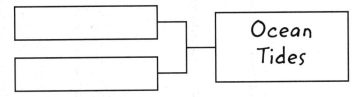

Ocean Tides

Critical Thinking

7. **Summarize** How do the Sun and the Moon affect Earth's tides?

_____

_____

Think About It! Gravity causes objects in space to impact each other. Earth's motion around the Sun causes seasons. The Moon's motion around Earth causes phases of the Moon. Earth and the Moon's motions together cause eclipses and ocean tides.

🔑 Key Concepts Summary

Vocabulary

LESSON 1 Earth's Motion

- The gravitational pull of the Sun on Earth causes Earth to revolve around the Sun in a nearly circular **orbit**.

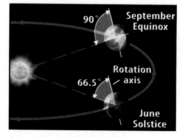

- Areas on Earth's curved surface become more tilted with respect to the direction of sunlight the farther you travel from the equator. This causes sunlight to spread out closer to the poles, making Earth colder at the poles and warmer at the equator.

- As Earth revolves around the Sun, the tilt of Earth's **rotation axis** produces changes in how sunlight spreads out over Earth's surface. These changes in the concentration of sunlight cause the seasons.

orbit p. 12
revolution p. 12
rotation p. 13
rotation axis p. 13
solstice p. 17
equinox p. 17

LESSON 2 Earth's Moon

- The gravitational pull of Earth on the Moon makes the Moon revolve around Earth. The Moon rotates once as it makes one complete orbit around Earth.

- The lit part of the Moon that you can see from Earth—the Moon's **phase**—changes during the lunar cycle as the Moon revolves around Earth.

maria p. 22
phase p. 24
waxing phase p. 24
waning phase p. 24

LESSON 3 Eclipses and Tides

- When the Moon's shadow appears on Earth's surface, a **solar eclipse** occurs.

- When the Moon moves into Earth's shadow, a **lunar eclipse** occurs.

- The gravitational pull of the Moon and the Sun on Earth produces **tides**, the rise and fall of sea level that occurs twice each day.

umbra p. 29
penumbra p. 29
solar eclipse p. 30
lunar eclipse p. 32
tide p. 33

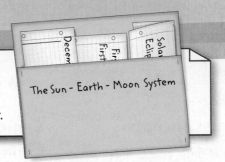

Active Reading

FOLDABLES® **Chapter Project**

Assemble your Lesson Foldables as shown to make a Chapter Project. Use the project to review what you have learned in this chapter.

The Sun - Earth - Moon System

Use Vocabulary

Distinguish between the terms in each of the following pairs.

1 revolution, orbit

_____

_____

2 rotation, rotation axis

_____

_____

3 solstice, equinox

_____

4 waxing phases, waning phases

_____

_____

5 umbra, penumbra

_____

_____

6 solar eclipse, lunar eclipse

_____

_____

7 tide, phase

_____

_____

Link Vocabulary and Key Concepts

Use vocabulary terms from the previous page to complete the concept map.

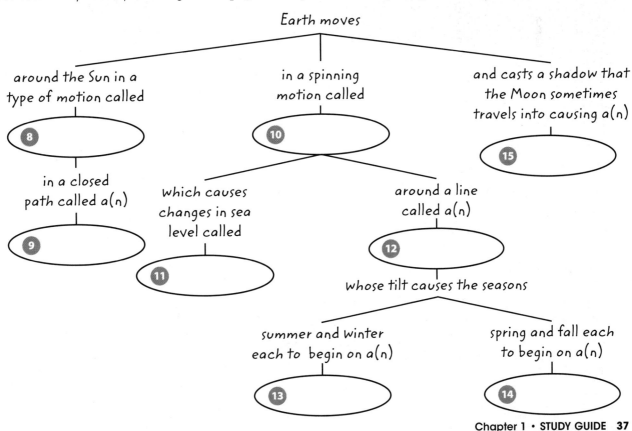

Earth moves

around the Sun in a type of motion called

8

in a spinning motion called

10

and casts a shadow that the Moon sometimes travels into causing a(n)

15

in a closed path called a(n)

9

which causes changes in sea level called

11

around a line called a(n)

12

whose tilt causes the seasons

summer and winter each to begin on a(n)

13

spring and fall each to begin on a(n)

14

Fill in the correct answer choice.

🔑 Understand Key Concepts

1. Which property of the Sun most affects the strength of gravitational attraction between the Sun and Earth? SC.8.E.5.4

    Ⓐ mass
    Ⓑ radius
    Ⓒ shape
    Ⓓ temperature

2. Which would be different if Earth rotated from east to west but at the same rate? SC.8.E.5.9

    Ⓐ the amount of energy striking Earth
    Ⓑ the days on which solstices occur
    Ⓒ the direction of the Sun's apparent motion across the sky
    Ⓓ the number of hours in a day

3. In the image below, which season is the northern hemisphere experiencing? SC.8.E.5.9

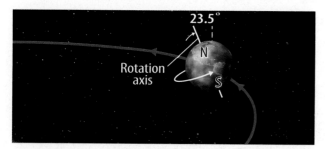

    Ⓐ fall
    Ⓑ spring
    Ⓒ summer
    Ⓓ winter

4. Which best explains why Earth is colder at the poles than at the equator? SC.8.5.9

    Ⓐ Earth is farther from the Sun at the poles than at the equator.
    Ⓑ Earth's orbit is not a perfect circle.
    Ⓒ Earth's rotation axis is tilted.
    Ⓓ Earth's surface is more tilted at the poles than at the equator.

5. How are the revolutions of the Moon and Earth alike? SC.8.E.5.4

    Ⓐ Both are produced by gravity.
    Ⓑ Both are revolutions around the Sun.
    Ⓒ Both orbits are the same size.
    Ⓓ Both take the same amount of time.

Critical Thinking

6. **Relate** the ways Earth moves and how each affects Earth. SC.8.E.5.9

Earth's Motions	Effects of Earth's Motions

7. **Summarize** Why is the same side of the Moon always visible from Earth? SC.8.E.5.9

    _____

    _____

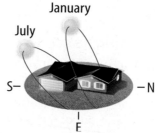

8. **Interpret Graphics** The figure above shows the Sun's position in the sky at noon in January and July. Is the house located in the northern hemisphere or the southern hemisphere? Explain. SC.8.E.5.9

    _____

    _____

    _____

9. **Illustrate** On a separate piece of paper, make a diagram of the Moon's phases. Include labels and explanations with your drawing. SC.8.E.5.9

10. Differentiate between a total solar eclipse and a partial solar eclipse. SC.8.E.5.9

    _____

    _____

    _____

11 Generalize the reason that solar and lunar eclipses do not occur every month. SC.8.E.5.9

_____
_____
_____

12 Differentiate Explain the two types of tides and the cause of each type of tide. SC.8.E.5.9

_____
_____

Writing in Science

13 Survey a group of at least ten people to determine how many know the cause of Earth's seasons. On a separate piece of paper, write a summary of your results including a main idea, supporting details, and a concluding sentence. SC.8.E.5.9

Big Idea Review

14 At the South Pole, the Sun does not appear in the sky for six months out of the year. When does this happen? What is happening at the North Pole during these months? Explain why Earth's poles receive so little solar energy. SC.8.E.5.9

_____
_____
_____
_____

15 What phenomena do the motions of Earth and the Moon produce? SC.8.E.5.9

_____
_____
_____
_____

Math Skills   MA.6.A.3.6

Convert Units

16 When the Moon is 384,000 km from Earth, how far is the Moon from Earth in miles?

17 If you travel 345 mi from Jacksonville to Miami, how many kilometers do you travel?

18 The nearest star other than the Sun is about 40 trillion km away. About how many miles away is the nearest star other than the Sun?

Fill in the correct answer choice.

Multiple Choice

1 What keeps Earth in orbit around the Sun? **SC.8.E.5.4**

(A) tilt of its axis

(B) gravitational attraction to the Sun

(C) gravitational attraction to the Moon

(D) gravitational attraction to the Sun and the Moon

Use the diagram below to answer question 2.

Time 1

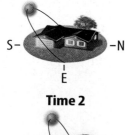

Time 2

2 What happens between times 1 and 2 in the diagram above? **SC.8.E.5.9**

(F) Days grow shorter and shorter.

(G) The season changes from fall to winter.

(H) The region begins to point away from the Sun.

(I) The region gradually receives more solar energy.

3 How many times larger is the Sun's diameter than Earth's diameter? **SC.8.E.5.9**

(A) about 10 times larger

(B) about 100 times larger

(C) about 1,000 times larger

(D) about 10,000 times larger

4 Which diagram illustrates the Moon's third quarter phase? **SC.8.E.5.9**

(F)

(G)

(H)

(I)

5 Which accurately describes Earth's position and orientation during summer in the northern hemisphere? **SC.8.E.5.9**

(A) Earth is at its closest point to the Sun.

(B) Earth's hemispheres receive equal amounts of solar energy.

(C) The north end of Earth's rotational axis leans toward the Sun.

(D) The Sun emits a greater amount of light and heat energy.

6 Why does the Sun's energy warm Earth more at the equator than at the poles? **SC.8.E.5.9**

(F) The equator has tropical rain forests.

(G) Sunlight is less spread out near the equator.

(H) Sunlight is more spread out near the equator.

(I) The equator lacks ice, which reflects the Sun's energy.

7 During one lunar cycle, the Moon SC.8.E.5.9

(A) completes its east-to-west path across the sky exactly once.

(B) completes its entire sequence of phases.

(C) progresses only from the new-moon phase to the full-moon phase.

(D) revolves around Earth twice.

Use the diagram below to answer question 8.

Moon

8 What does the flag in the diagram above represent? SC.8.E.5.9

(F) high tide

(G) low tide

(H) neap tide

(I) spring tide

9 During which lunar phase might a solar eclipse occur? SC.8.E.5.9

(A) first quarter moon

(B) full moon

(C) new moon

(D) third quarter moon

10 Which does the entire Moon pass through during a partial lunar eclipse? SC.8.E.5.9

(F) Earth's penumbra

(G) Earth's umbra

(H) the Moon's penumbra

(I) the Moon's umbra

Use the diagram below to answer questions 11 and 12.

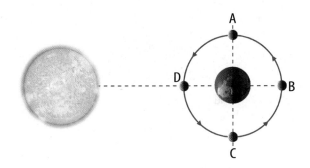

11 Where are the neap tides indicated in the above diagram? SC.8.E.5.9

(A) A and B

(B) B and C

(C) A and C

(D) C and D

12 Where are the spring tides indicated in the above diagram? SC.8.E.5.9

(F) B and C

(G) C and D

(H) B and D

(I) A and C

NEED EXTRA HELP?

If You Missed Question...	1	2	3	4	5	6	7	8	9	10	11	12
Go to Lesson...	1	1	1	2	1	1	2	3	3	3	3	3

Multiple Choice *Bubble the correct answer.*

1. Which image shows the December solstice?
SC.8.E.5.9

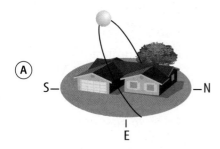

(A) S— —N
                E

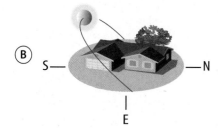

(B) S— —N
                E

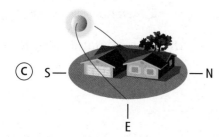

(C) S— —N
                E

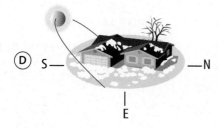

(D) S— —N
                E

2. Which process produces large amounts of energy within the Sun? **SC.8.E.5.5**

(F) nuclear fission

(G) nuclear fusion

(H) Earth's motion around the Sun

(I) the Sun's gravitational pull

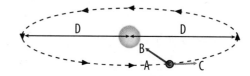

3. Which arrow in the image above indicates the path Earth would take if the Sun disappeared? **SC.8.E.5.4**

(A) A

(B) B

(C) C

(D) D

4. What causes the seasons? **SC.8.E.5.9**

(F) the changes in the tilt of Earth's rotation axis

(G) the force of the Sun's gravitational pull on Earth

(H) the great distance Earth is from the Sun

(I) the tilt of Earth's axis as it moves around the Sun

Multiple Choice *Bubble the correct answer.*

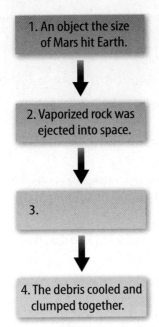

1. An object the size of Mars hit Earth.

↓

2. Vaporized rock was ejected into space.

↓

3.

↓

4. The debris cooled and clumped together.

1. The above four steps show how the Moon was formed. Which statement below is step 3? **SC.8.E.5.4**

- (A) Debris fell back to Earth, which caused craters.

- (B) Maria were formed by the impacts of large objects.

- (C) The Sun burned up most of the rocky debris.

- (D) Vaporized rock formed a ring around Earth.

2. The Moon is visible because it **SC.8.E.5.9**

- (F) reflects light from Mars.

- (G) reflects light from Earth.

- (H) reflects light from the stars.

- (I) reflects light from the Sun.

Use the image below to answer questions 3 and 4.

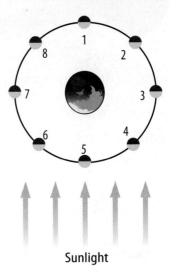

Sunlight

3. Which numbers indicate the phases during which the Moon is waning? **SC.8.E.5.9**

- (A) 1, 5

- (B) 2, 3, 4

- (C) 5, 6, 7, 8

- (D) 2, 3, 4, 5, 6, 7

4. Which number indicates how the Moon is seen from Earth during the second week of the lunar cycle? **SC.8.E.5.9**

- (F) 1

- (G) 3

- (H) 5

- (I) 7

Multiple Choice *Bubble the correct answer.*

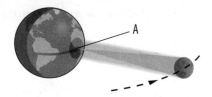

1. According to the image above, what type of eclipse is occurring at the point labeled A on the globe? **SC.8.E.5.9**

   (A) partial lunar eclipse

   (B) partial solar eclipse

   (C) total lunar eclipse

   (D) total solar eclipse

2. Why don't solar eclipses happen every month? **SC.8.E.5.9**

   (F) Earth's orbit around the Sun is at an angle, which keeps Earth out of the Moon's shadow.

   (G) Earth's tilt on its rotation axis keeps it from falling into the Moon's shadow during most months.

   (H) The Moon's orbit is irregular, and most of the time the Moon is too far away to cast a shadow on Earth.

   (I) The Moon's orbit is tilted compared to Earth's orbit, so Earth is not in the Moon's shadow most months.

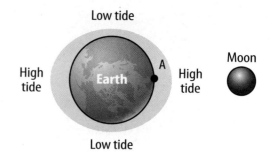

3. According to the image above, how will the tide at Point A change over the next 12 hours? **SC.8.E.5.9**

   (A) The tide will change from low tide to high tide.

   (B) The tide will change from high tide to low tide.

   (C) The tide will change from high tide to low tide and back to high tide.

   (D) The tide will change from low tide to high tide and back to low tide.

4. A partial solar eclipse is seen at locations in **SC.8.E.5.9**

   (F) Earth's penumbra.

   (G) Earth's umbra.

   (H) the Moon's penumbra.

   (I) the Moon's umbra.

Notes

Name _____   Date _____

Objects in
Our Solar
System

A system consists of parts that make up a whole. What are the different parts that make up our solar system? Put an *X* next to each of the objects you think are part of our solar system.

_____ planets   _____ the Sun   _____ nearby stars other than the Sun

_____ distant stars   _____ constellations   _____ asteroids

_____ comets   _____ moons   _____ human-made satellites

_____ galaxies   _____ black holes   _____ universe

Explain your thinking. Describe what determines whether an object is part of our solar system.

The Solar

FLORIDA BIG IDEAS

1 The Practice of Science

3 The Role of Theories, Laws, Hypotheses, and Models

5 Earth in Space and Time

Think About It!

What kinds of objects are in the solar system?

This photo, taken by the Cassini spacecraft, shows part of Saturn's rings and two of its moons. Saturn is a planet that orbits the Sun. The moons, tiny Epimetheus and much larger Titan, orbit Saturn. Besides planets and moons, many other objects are in the solar system.

1 How would you describe a planet such as Saturn?

_____

_____

_____

2 How do you think astronomers classify the objects they discover?

_____

_____

_____

3 What types of objects do you think make up the solar system?

_____

_____

_____

Get Ready to Read

What do you think about the solar system?

Before you read, decide if you agree or disagree with each of these statements. As you read this chapter, see if you change your mind about any of the statements.

	AGREE	DISAGREE
1 Astronomers measure distances between space objects using astronomical units.	☐	☐
2 Gravitational force keeps planets in orbit around the Sun.	☐	☐
3 Earth is the only inner planet that has a moon.	☐	☐
4 Venus is the hottest planet in the solar system.	☐	☐
5 The outer planets also are called the gas giants.	☐	☐
6 The atmospheres of Saturn and Jupiter are mainly water vapor.	☐	☐
7 Asteroids and comets are mainly rock and ice.	☐	☐
8 A meteoroid is a meteor that strikes Earth.	☐	☐

 **Connect**ED **There's More Online!**
Video • Audio • Review • ⓘLab Station • WebQuest • Assessment • Concepts in Motion • Multilingual eGlossary

The Structure of the SOLAR SYSTEM

Vocabulary

asteroid p. 53

comet p. 53

astronomical unit p. 54

period of revolution p. 54

period of rotation p. 54

 Florida NGSSS

LA.8.2.2.3 The student will organize information to show understanding or relationships among facts, ideas, and events (e.g., representing key points within text through charting, mapping, paraphrasing, summarizing, or comparing/contrasting);

SC.8.E.5.1 Recognize that there are enormous distances between objects in space and apply our knowledge of light and space travel to understand this distance.

SC.8.E.5.3 Distinguish the hierarchical relationships between planets and other astronomical bodies relative to solar system, galaxy, and universe, including distance, size, and composition.

SC.8.E.5.4 Explore the Law of Universal Gravitation by explaining the role that gravity plays in the formation of planets, stars, and solar systems and in determining their motions.

SC.8.E.5.7 Compare and contrast the properties of objects in the Solar System including the Sun, planets, and moons to those of Earth, such as gravitational force, distance from the Sun, speed, movement, temperature, and atmospheric conditions.

SC.8.E.5.8 Compare various historical models of the Solar System, including geocentric and heliocentric.

SC.8.N.1.1 Define a problem from the eighth grade curriculum using appropriate reference materials to support scientific understanding, plan and carry out scientific investigations of various types, such as systematic observations or experiments, identify variables, collect and organize data, interpret data in charts, tables, and graphics, analyze information, make predictions, and defend conclusions.

 Launch Lab

10 minutes

How do you know which distance unit to use?

You can use different units to measure distance. For example, millimeters might be used to measure the length of a bolt, and kilometers might be used to measure the distance between cities. In this lab, you will investigate why some units are easier to use than others for certain measurements.

Procedure

1. Read and complete a lab safety form.
2. Use a **centimeter ruler** to measure the length of a **pencil** and the thickness of this **book.** Record the measurements below.
3. Use the centimeter ruler to measure the width of your classroom. Then measure the width of the room using a **meterstick.** Record the measurements below.

Data and Observations

Think About This

1. Why are meters easier to use than centimeters for measuring the classroom?

2. **Key Concept** Why do you think astronomers might need a unit larger than a kilometer to measure distances in the solar system?

Inquiry Are these stars?

1. Did you know that shooting stars are not actually stars? The bright streaks are small, rocky particles burning up as they enter Earth's atmosphere. Why is the term *shooting star* misleading? These particles are part of the solar system and are often associated with comets. What types of objects do you think make up the solar system?

_____

_____

_____

_____

_____

_____

What is the solar system?

Have you ever made a wish on a star? If so, you might have wished on a planet instead of a star. Sometimes, as shown in **Figure 1,** the first starlike object you see at night is not a star at all. It's Venus, the planet closest to Earth.

It's hard to tell the difference between planets and stars in the night sky because they all appear as tiny lights. Thousands of years ago, observers noticed that a few of these tiny lights moved, but others did not. The ancient Greeks called these objects planets, which means "wanderers." Astronomers now know that the planets do not wander about the sky; the planets move around the Sun. The Sun and the group of objects that move around it make up the solar system.

Active Reading

2. Recall What objects do the planets in the solar system move around?

_____

A few of the tiny lights that you can see in the night sky are part of our solar system. Almost all the other specks of light are stars. They are much farther away than any objects in our solar system. Astronomers have discovered that some of those stars have planets moving around them.

Figure 1 When looking at the night sky, you will likely see stars and planets. In the photo below, the planet Venus is the bright object seen above the Moon.

Objects in the Solar System

The center of the solar system is the Sun, a **star**. Inside the Sun, a process called nuclear fusion produces an enormous amount of energy. It emits some of this energy as light. The Sun is the largest object in the solar system. Its diameter is about 1.4 million km—ten times the diameter of the largest planet, Jupiter. Its mass makes up about 99 percent of the entire solar system's mass. The Sun's great mass applies gravitational forces to objects in the solar system. Gravitational forces cause the planets and other objects to move around, or **orbit**, the Sun.

The Law of Universal Gravitation

In the late 1600s, an English scientist and mathematician, Sir Isaac Newton, developed the law of universal gravitation. This law states that all objects are attracted to each other by a gravitational force. The strength of the force depends on the mass of each object and the distance between their centers.

 3. NGSSS Check Explain What is the law of universal gravitation?. SC.8.E.5.4

_____

_____

Gravity and orbits Newton realized that gravity's attractive force could explain why planets orbit the Sun, as shown in **Figure 2.** He realized that the planets did not travel in a straight line because of gravitational attraction. Without gravity the planets would continue to travel in a straight line and would not orbit the Sun. The same is true for Earth and the Moon, the other planets and their moons, stars, and all other orbiting objects in the universe.

Figure 2 The orbits of the inner and outer planets are shown to scale. The Sun and the planets are not to scale.

Active Reading **4. Identify** What keeps the planets in their orbits?

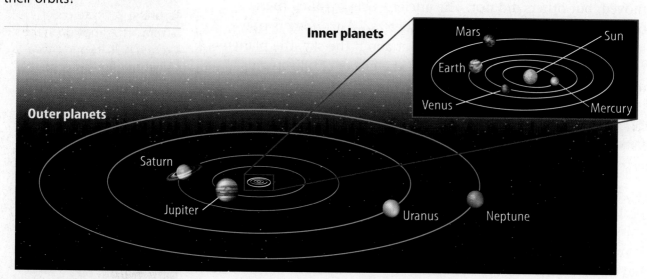

Inner planets

Mars · Sun · Earth · Venus · Mercury

Outer planets

Saturn · Jupiter · Uranus · Neptune

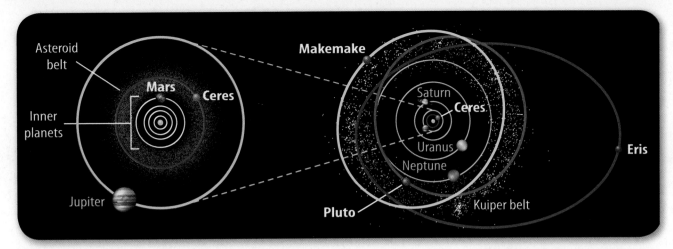

Objects That Orbit the Sun

Planets Astronomers classify eight objects that orbit the Sun as planets, as shown in **Figure 2**. An object is a planet only if it orbits the Sun and has a nearly spherical shape. Also, the mass of a planet must be much larger than the total mass of all other objects whose orbits are close by.

Inner Planets and Outer Planets As shown in **Figure 2,** the four planets closest to the Sun are the inner planets—Mercury, Venus, Earth, and Mars. These planets are made mainly of rocky materials. The four planets farthest from the Sun are the outer planets. The outer planets, Jupiter, Saturn, Uranus (YOOR uh nus), and Neptune, are made mainly of ice and gases. The outer planets are much larger than Earth and are sometimes called gas giants.

 6. **NGSSS Check** **Explain** Highlight how the inner planets differ from the outer planets. **SC.8.E.5.7**

Dwarf Planets Some objects in the solar system are classified as dwarf planets. A dwarf planet is a spherical object that orbits the Sun with many objects orbiting near it. But, unlike a planet, a dwarf planet does not have more mass than objects in nearby orbits. **Figure 3** shows the locations of four dwarf planets. Dwarf planets are made of rock and ice and are smaller than Earth.

Asteroids *Millions of small, rocky objects called* **asteroids** *orbit the Sun in the asteroid belt between the orbits of Mars and Jupiter.* The asteroid belt is shown in **Figure 3**. Asteroids range in size from less than a meter to several hundred kilometers in length. Unlike planets and dwarf planets, asteroids, such as the one shown in **Figure 4,** usually are not spherical.

Comets *A* **comet** *is made of gas, dust, and ice and moves around the Sun in an oval-shaped orbit.* Comets come from the outer parts of the solar system. There might be 1 trillion comets orbiting the Sun.

Figure 3 Ceres, a dwarf planet, orbits the Sun as planets do. The orbit of Ceres is in the asteroid belt between Mars and Jupiter.

5. **Visual Check**
Locate Which dwarf planet is farthest from the Sun?

WORD ORIGIN

asteroid
from Greek *asteroeides*, means "resembling a star"

Figure 4 The asteroid Gaspra orbits the Sun in the asteroid belt. Its odd shape is about 19 km long and 11 km wide.

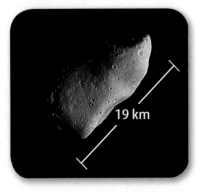

19 km

Active Reading

FOLDABLES® LA.8.2.2.3

Make a tri-fold book from a sheet of paper and label it as shown. Use it to summarize information about the types of objects that make up the solar system.

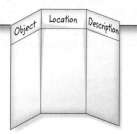

Object | Location | Description

Table 1 Because the distances of the planets from the Sun are so large, it is easier to express these distances using astronomical units rather than kilometers.

Active Reading **8. Summarize** What causes planets to orbit the Sun?

The Astronomical Unit

On Earth, distances are often measured in meters (m) or kilometers (km). Objects in the solar system, however, are so far apart that astronomers use a larger distance unit. *An* **astronomical unit** *(AU) is the average distance from Earth to the Sun—about 150 million km.* **Table 1** lists each planet's average distance from the Sun in km and AU.

7. NGSSS Check **Define** What is an astronomical unit and why is is used? SC.8.E.5.1, SC.8.E.5.7

Table 1 Average Distance of the Planets from the Sun		
Planet	**Average Distance (km)**	**Average Distance (AU)**
Mercury	57,910,000	0.39
Venus	108,210,000	0.72
Earth	149,600,000	1.00
Mars	227,920,000	1.52
Jupiter	778,570,000	5.20
Saturn	1,433,530,000	9.58
Uranus	2,872,460,000	19.20
Neptune	4,495,060,000	30.05

The Motion of the Planets

Have you ever swung a ball on the end of a string in a circle over your head? In some ways, the motion of a planet around the Sun is like the motion of that ball. As shown in **Figure 5** on the next page, the Sun's gravitational force pulls each planet toward the Sun. This force is similar to the pull of the string that keeps the ball moving in a circle. The Sun's gravitational force pulls on each planet and keeps it moving along a curved path around the Sun.

Revolution and Rotation

Objects in the solar system move in two ways. They orbit, or revolve, around the Sun. *The time it takes an object to travel once around the Sun is its* **period of revolution**. Earth's period of revolution is one year. The objects also spin, or rotate, as they orbit the Sun. *The time it takes an object to complete one rotation is its* **period of rotation**. Earth has a period of rotation of one day.

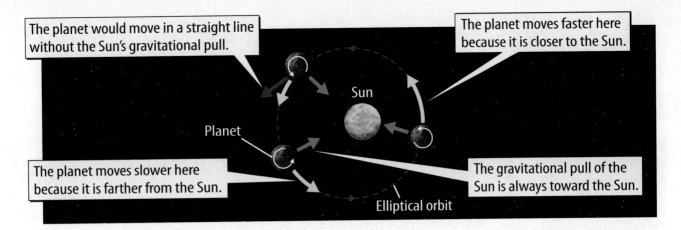

The planet would move in a straight line without the Sun's gravitational pull.

The planet moves faster here because it is closer to the Sun.

Sun

Planet

The planet moves slower here because it is farther from the Sun.

The gravitational pull of the Sun is always toward the Sun.

Elliptical orbit

Planetary Orbits and Speeds

Earth was once thought to be the center of our solar system. In this geocentric model, the Sun, the Moon and planets revolved in circular orbits around a stationary Earth. In the early 1500s, Nicholas Copernicus proposed that Earth and other planets revolve in circular orbits around a stationary Sun, a heliocentric model.

In 1600s, Johannes Kepler discovered that planets' orbits are ellipses, not circles. An ellipse contains two fixed points, called foci (singular, *focus*). Foci are equal distance from the ellipse's center and determine its shape. As shown in **Figure 5,** the Sun is at one focus. As a planet revolves, the distance between it and the Sun changes. Kepler also discovered that a planet's speed increases as it gets nearer to the Sun.

 10. **NGSSS Check Cite** Underline the shape of the planets' orbits. SC.8.E.5.7

Figure 5 🔑 Planets and other objects in the solar system revolve around the Sun because of its gravitational pull on them.

Active Reading 9. **Name** What is this model of the solar system called?

_____

_____

SC.8.N.1.1
SC.8.E.5.7

Inquiry

LAB STATION Try It!

MiniLab *How can you model an elliptical orbit?* at connectED.mcgraw-hill.com

Apply It! After you complete the lab, answer these questions.

1. What is an ellipse? What is another example of an ellipse?

2. Which ellipse will have a shorter period of revolution?

Visual Summary

The solar system contains the Sun, the inner planets, the outer planets, the dwarf planets, asteroids, and comets.

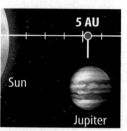

An astronomical unit (AU) is a unit of distance equal to about 150 million km.

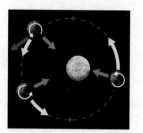

The speeds of the planets change as they move around the Sun in elliptical orbits.

Use Vocabulary

1 **Compare and contrast** a *period of revolution* and a *period of rotation*.
SC.8.E.5.7

2 **Define** *dwarf planet* in your own words.

3 **Distinguish** between an *asteroid* and a *comet*. SC.8.E.5.3

Understand Key Concepts 🔑

4 **Summarize** how and why planets orbit the Sun and how and why a planet's speed changes in orbit. SC.8.E.5.4

5 Which statement is true? SC.8.E.5.8
   (A) Earth's revolution is circular.
   (B) Earth's revolution is elliptical.
   (C) Earth's distance from the Sun is constant.
   (D) Earth's average distance from the Sun is 2 astronomical units.

Interpret Graphics

6 **Take Notes** List information about five objects or group of objects in the solar system mentioned in the lesson. SC.8.E.5.7

Object	Description

Critical Thinking

7 **Evaluate** How would the speed of a planet be different if its orbit were a circle instead of an ellipse? SC.8.E.5.7

Meteors are pieces of a comet or an asteroid that heat up as they fall through Earth's atmosphere. Meteors that strike Earth are called meteorites.

History from Space

Meteorites give a peek back in time.

About 4.6 billion years ago, Earth and the other planets did not exist. In fact, there was no solar system. Instead, a large disk of gas and dust, known as the solar nebula, swirled around a forming Sun, as shown in the top picture to the right. How did the planets and other objects in the solar system form?

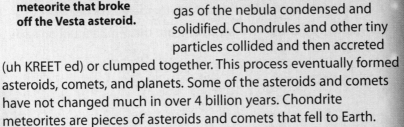

Denton Ebel is looking for the answer. He is a geologist at the American Museum of Natural History in New York City. Ebel explores the hypothesis that over millions of years, tiny particles in the solar nebula clumped together and formed the asteroids, comets, and planets that make up our solar system.

Denton Ebel holds a meteorite that broke off the Vesta asteroid.

The solar nebula contained tiny particles called chondrules (KON drewls). They formed when the hot gas of the nebula condensed and solidified. Chondrules and other tiny particles collided and then accreted (uh KREET ed) or clumped together. This process eventually formed asteroids, comets, and planets. Some of the asteroids and comets have not changed much in over 4 billion years. Chondrite meteorites are pieces of asteroids and comets that fell to Earth. The chondrules within the meteorites are the oldest solid material in our solar system.

For Ebel, chondrite meteorites contain information about the formation of the solar system. Did the materials in the meteorite form throughout the solar system and then accrete? Or did asteroids and comets form and accrete near the Sun, drift outward to where they are today, and then grow larger by accreting ice and dust? Ebel's research is helping to solve the mystery of how our solar system formed.

Accretion Hypothesis

According to the accretion hypothesis, the solar system formed in stages.

First there was a solar nebula. The Sun formed when gravity caused the nebula to collapse.

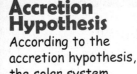

The rocky inner planets formed from accreted particles.

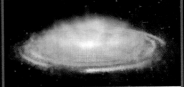

The gaseous outer planets formed as gas, ice, and dust condensed and accreted.

It's Your Turn

TIME LINE Work in groups. Learn more about the history of Earth from its formation until life begins to appear. Create a time line showing major events. Present your time line to the class

The Inner PLANETS

Vocabulary

terrestrial planet p. 59

greenhouse effect p. 61

Florida NGSSS

LA.8.2.2.3 The student will organize information to show understanding or relationships among facts, ideas, and events (e.g., representing key points within text through charting, mapping, paraphrasing, summarizing, or comparing/contrasting);

MA.6.A.3.6 Construct and analyze tables, graphs, and equations to describe linear functions and other simple relations using both common language and algebraic notation.

SC.8.E.5.1 Recognize that there are enormous distances between objects in space and apply our knowledge of light and space travel to understand this distance.

SC.8.E.5.3 Distinguish the hierarchical relationships between planets and other astronomical bodies relative to solar system, galaxy, and universe, including distance, size, and composition.

SC.8.E.5.7 Compare and contrast the properties of objects in the Solar System including the Sun, planets, and moons to those of Earth, such as gravitational force, distance from the Sun, speed, movement, temperature, and atmospheric conditions.

SC.8.N.1.1 Define a problem from the eighth grade curriculum using appropriate reference materials to support scientific understanding, plan and carry out scientific investigations of various types, such as systematic observations or experiments, identify variables, collect and organize data, interpret data in charts, tables, and graphics, analyze information, make predictions, and defend conclusions.

 Launch Lab

20 minutes

What affects the temperature on the inner planets?

Mercury and Venus are closer to the Sun than Earth. What determines the temperature on these planets? Let's find out.

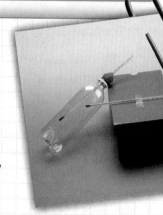

Procedure 🥽 🧤

1. Read and complete a lab safety form.

2. Insert a **thermometer** into a **clear 2-L plastic bottle.** Wrap **modeling clay** around the lid to hold the thermometer in the center of the bottle. Form an airtight seal with the clay.

3. Rest the bottle against the side of a **shoe box** in direct sunlight. Lay a second **thermometer** on top of the box next to the bottle so that the bulbs are at about the same height. The thermometer bulb should not touch the box. Secure the thermometer in place using **tape.**

4. Read the thermometers and record the temperatures below.

5. Wait 15 minutes and then read and record the temperature on each thermometer.

Data and Observations

Think About This

1. How did the temperature of the two thermometers compare?

2. **Key Concept** What do you think caused the difference in temperature?

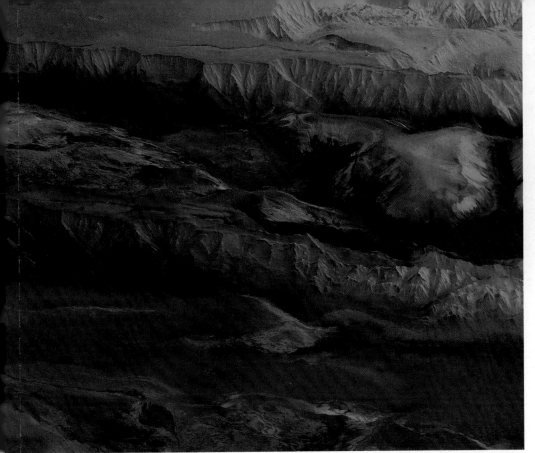

inquiry **Where is this?**

1. This spectacular landscape is the surface of Mars, one of the inner planets. Other inner planets have similar rocky surfaces. Surprisingly, there are planets in the solar system that have no solid surface on which to stand. What can scientists learn by analyzing the appearance of a planet's surface?

_____

_____

_____

_____

_____

Planets Made of Rock

Imagine that you are walking outside. How would you describe the ground? You might say it is dusty or grassy. If you live near a lake or an ocean, you might say sandy or wet. But beneath the ground or lake or ocean is a layer of solid rock.

The inner planets—Mercury, Venus, Earth, and Mars—are the planets closest to the Sun, as shown in **Figure 6.** *Earth and the other inner planets are also called the* **terrestrial planets**. Like Earth, the other inner planets also are made of rock and metallic materials and have a solid outer layer. Inner planets, however, have different sizes, atmospheres, and surfaces.

WORD ORIGIN

terrestrial
from Latin *terrestris*, means "earthly"

Figure 6 The inner planets are roughly similar in size. Earth is about two and half times larger than Mercury. All inner planets have a solid outer layer.

Inner Planets 🔑

Mercury   Venus   Earth   Mars

Active Reading **2. Identify** (Circle) the smallest inner planet.

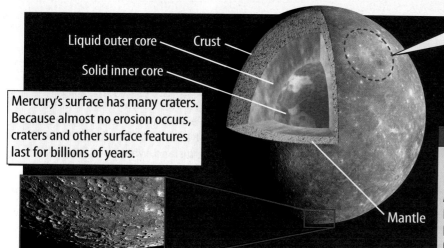

Liquid outer core — Crust

Solid inner core —

The Caloris Basin is about 1,550 km across. It is one of the largest impact craters in the solar system. It was formed billions of years ago by the impact of an object about 100 km in diameter.

Mercury's surface has many craters. Because almost no erosion occurs, craters and other surface features last for billions of years.

Mantle

Mercury Data

Mass: 5.5% of Earth's mass
Diameter: 38.3% of Earth's diameter
Average distance from Sun: 0.39 AU
Period of rotation: 59 days
Period of revolution: 88 days
Number of moons: 0

Figure 7 🔑 The *Messenger* space probe flew by Mercury in 2008 and photographed the planet's cratered surface.

Active Reading 3. **Contrast** What is one difference between Mercury's inner and outer cores?

_____

_____

_____

Active Reading

FOLDABLES® LA.8.2.2.3

Make a four-door book. Label each door with the name of an inner planet. Use the book to organize your notes on the inner planets.

Mercury

The smallest planet and the planet closest to the Sun is Mercury, shown in **Figure 7.** Mercury has no atmosphere. A planet has an atmosphere when its gravity is strong enough to hold gases close to its surface. The strength of a planet's gravity depends on the planet's mass. Because Mercury's mass is so small, its gravity is not strong enough to hold onto an atmosphere. Without an atmosphere there is no wind that moves energy from place to place across the planet's surface. This results in temperatures as high as 450°C on the side of Mercury facing the Sun and as cold as −170°C on the side facing away from the Sun.

Mercury's Surface

Impact craters cover the surface of Mercury. There are also smooth plains of solidified lava from long-ago eruptions. Long, high cliffs occur also. These might have formed when the planet cooled quickly, causing the surface to wrinkle and crack. Without an atmosphere, almost no erosion occurs on the surface. As a result, features that formed billions of years ago have changed very little.

Mercury's Structure

The structures of the inner planets are similar. Like all inner planets, Mercury has a core made of iron and nickel. Surrounding the core is a layer called the mantle. It is mainly made of silicon and oxygen. The crust is a thin, rocky layer above the mantle. Mercury's large core might have been formed by a collision with a large object during Mercury's formation.

4. **NGSSS Check Compare** How are the inner planets similar? SC.8.E.5.7

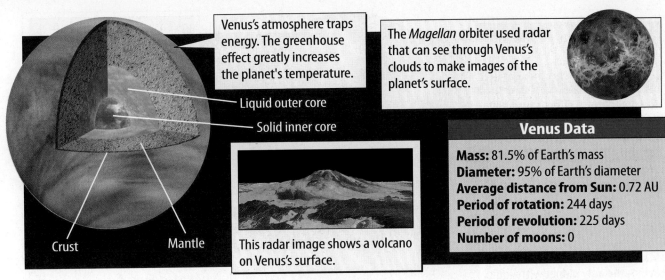

Venus's atmosphere traps energy. The greenhouse effect greatly increases the planet's temperature.

The *Magellan* orbiter used radar that can see through Venus's clouds to make images of the planet's surface.

Liquid outer core

Solid inner core

Crust   Mantle

This radar image shows a volcano on Venus's surface.

Venus Data

Mass: 81.5% of Earth's mass
Diameter: 95% of Earth's diameter
Average distance from Sun: 0.72 AU
Period of rotation: 244 days
Period of revolution: 225 days
Number of moons: 0

Venus

The second planet from the Sun is Venus, as shown in **Figure 8.** It is about the same size as Earth. Venus spins so slowly that its period of rotation is longer than its period of revolution. This means that a day on Venus is longer than a year. Unlike most planets, Venus rotates from east to west. Several space probes have flown by or landed on Venus.

Venus's Atmosphere

The atmosphere of Venus is about 97 percent carbon dioxide. It is so dense that the atmospheric pressure on Venus is about 90 times greater than on Earth. Even though Venus has almost no water in its atmosphere or on its surface, a thick layer of clouds covers the planet. Unlike the clouds of water vapor on Earth, the clouds on Venus are made of acid.

The Greenhouse Effect on Venus

With an average temperature of about 460°C, Venus is the hottest planet in the solar system. The high temperatures are caused by the greenhouse effect. *The **greenhouse effect** occurs when a planet's atmosphere traps solar energy and causes the surface temperature to increase.* Carbon dioxide in Venus's atmosphere traps some of the solar energy that is absorbed and then emitted by the planet. This heats up the planet. Without the greenhouse effect, Venus would be almost 450°C cooler.

Venus's Structure and Surface

Venus's internal structure, as shown in **Figure 8,** is similar to Earth's. Radar images show that more than 80 percent of Venus's surface is covered by solidified lava. Much of this lava might have been produced by volcanic eruptions that occurred about half a billion years ago.

Figure 8 Because a thick layer of clouds covers Venus, its surface has not been seen. Between 1990 and 1994, the *Magellan* space probe mapped the surface using radar.

Active Reading **5. Relate** How does Venus's day compare to Venus's year?

6. NGSSS Check
Explain Why is Venus hotter than Mercury? SC.8.E.5.7

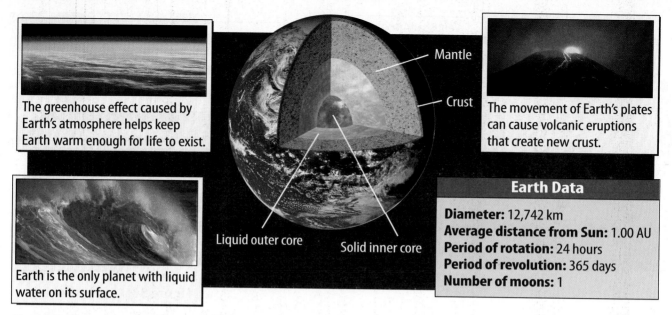

The greenhouse effect caused by Earth's atmosphere helps keep Earth warm enough for life to exist.

Earth is the only planet with liquid water on its surface.

Mantle

Crust

Liquid outer core   Solid inner core

The movement of Earth's plates can cause volcanic eruptions that create new crust.

Earth Data

Diameter: 12,742 km
Average distance from Sun: 1.00 AU
Period of rotation: 24 hours
Period of revolution: 365 days
Number of moons: 1

Figure 9 🔑 Earth has more water in its atmosphere and on its surface than the other inner planets. Earth's surface is younger than the surfaces of the other inner planets because new crust is constantly forming.

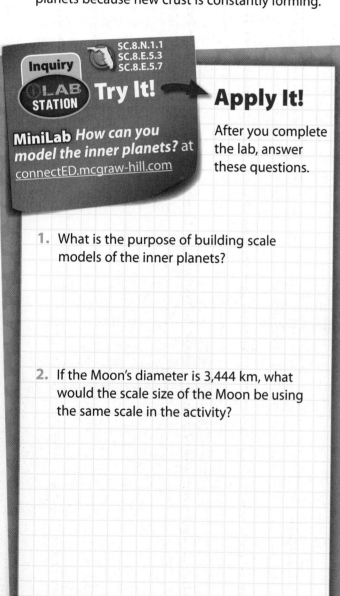

SC.8.N.1.1
SC.8.E.5.3
SC.8.E.5.7

Inquiry

🔵 **LAB STATION** **Try It!**

MiniLab *How can you model the inner planets?* at connectED.mcgraw-hill.com

Apply It!

After you complete the lab, answer these questions.

1. What is the purpose of building scale models of the inner planets?

2. If the Moon's diameter is 3,444 km, what would the scale size of the Moon be using the same scale in the activity?

Earth

Earth, shown in **Figure 9,** is the third planet from the Sun. Unlike Mercury and Venus, Earth has a moon.

Earth's Atmosphere

A mixture of gases and a small amount of water vapor make up most of Earth's atmosphere. They produce a greenhouse effect that increases Earth's average surface temperature. This effect and Earth's distance from the Sun warm Earth enough for large bodies of liquid water to exist. Earth's atmosphere also absorbs much of the Sun's radiation and protects the surface below. Earth's protective atmosphere, the presence of liquid water, and the planet's moderate temperature range support a variety of life.

Earth's Structure

As shown in **Figure 9,** Earth has a solid inner core surrounded by a liquid outer core. The mantle surrounds the liquid outer core. Above the mantle is Earth's crust. It is broken into large pieces, called plates, that constantly slide past, away from, or into each other. The crust is made mostly of oxygen and silicon and is constantly created and destroyed.

Active Reading 7. **State** Highlight why there is life on Earth.

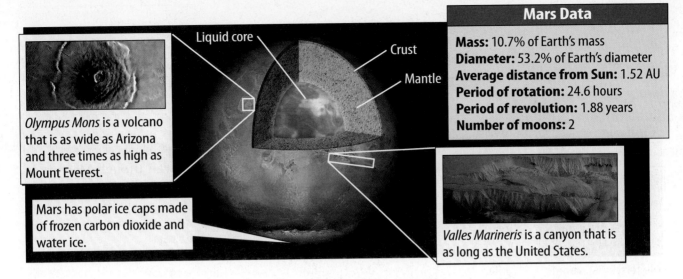

Liquid core

Crust

Mantle

Olympus Mons is a volcano that is as wide as Arizona and three times as high as Mount Everest.

Mars has polar ice caps made of frozen carbon dioxide and water ice.

Valles Marineris is a canyon that is as long as the United States.

Mars Data

Mass: 10.7% of Earth's mass
Diameter: 53.2% of Earth's diameter
Average distance from Sun: 1.52 AU
Period of rotation: 24.6 hours
Period of revolution: 1.88 years
Number of moons: 2

Mars

The fourth planet from the Sun is Mars, shown in **Figure 10.** Mars is about half the size of Earth. It has two very small and irregularly shaped moons. These moons might be asteroids that were captured by Mars's gravity.

Many space probes have visited Mars. Most of them have searched for signs of water that might indicate the presence of living organisms. Images of Mars show features that might have been made by water, such as the gullies in **Figure 11.** So far no evidence of liquid water or life has been found.

Active Reading 8. **Recall** Underline what scientists search for that may indicate life on another planet.

Mars's Atmosphere

The atmosphere of Mars is about 95 percent carbon dioxide. It is thin and much less dense than Earth's atmosphere. Temperatures range from about −125°C at the poles to about 20°C at the equator during a martian summer. Winds on Mars sometimes produce great dust storms that last for months.

Mars's Surface

The reddish color of Mars is because its soil contains iron oxide, a compound in rust. Some of Mars's major surface features are shown in **Figure 10.** The enormous canyon Valles Marineris is about 4,000 km long. The martian volcano Olympus Mons is the largest known mountain in the solar system. Mars also has polar ice caps made of frozen carbon dioxide and ice.

The southern hemisphere of Mars is covered with craters. The northern hemisphere is smoother and appears to be covered by lava flows. Some scientists have proposed that the lava flows were caused by the impact of an object about 2,000 km in diameter.

Figure 10 🔑 Mars is a small, rocky planet with deep canyons and tall mountains.

9. **NGSSS Check Summarize** Describe the atmosphere of each inner planet. SC.8.E.5.7

_____

_____

_____

_____

_____

Figure 11 Gullies such as these might have been formed by the flow of liquid water.

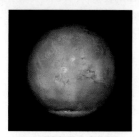

The terrestrial planets include Mercury, Venus, Earth, and Mars.

The inner planets are made of rocks but have different characteristics. Earth is the only planet with liquid water.

The greenhouse effect increases the surface temperature of Venus.

Inquiry SC.8.N.1.1 SC.8.E.5.7

LAB STATION Try It!

Skill Lab *What can we learn about planets by graphing their characteristics?* at connectED.mcgraw-hill.com

Use Vocabulary

1 **Define** *greenhouse effect* in your own words. SC.8.E.5.7

Understand Key Concepts 🔑

2 **Explain** why Venus is hotter than Mercury, even though Mercury is closer to the Sun. SC.8.E.5.7

3 **Infer** Why could rovers be used to explore Mars, but not Venus?

4 Which of the inner planets has the greatest mass? SC.8.E.5.7
  (A) Mercury   (C) Earth
  (B) Venus   (D) Mars

5 **Relate** Describe the relationship between an inner planet's distance from the Sun and its period of revolution. SC.8.E.5.7

6 **Compare and Contrast** Fill in the table below to compare and contrast properties of Venus and Earth. SC.8.E.5.7

Planet	Similarities	Differences
Venus		
Earth		

Critical Thinking

7 **Imagine** How might the temperatures on Mercury be different if it had the same mass as Earth? Explain.

8 **Judge** Do you think the inner planets should be explored or should the money be spent on other things? Justify your opinion.

Sort facts about Earth and Venus. Place the number of each fact in the Venn diagram.

1. has a greenhouse effect

2. has extremely high temperatures

3. has water in its atmosphere

4. year is longer than day

5. atmosphere mostly carbon dioxide

6. rotates counter-clockwise

7. rotates clockwise

8. a terrestrial planet

9. an inner planet

10. has a moon

11. has water on its surface

12. can support life

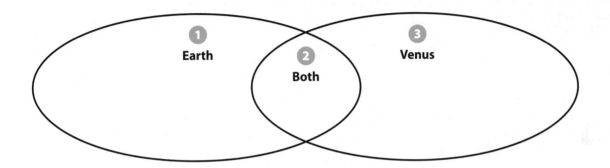

Summarize information about the inner planets. Place a check mark in each box that applies to each planet.

	Mercury	Venus	Earth	Mars
Atmosphere				
Inner and outer core				
Liquid outer core				
Liquid core, only				
Solid inner core				
Atmosphere 90% CO_2				
Cratered surface				
Liquid water on surface				
Ice on surface				
A moon or moons				
Mantle and crust				
Signs of volcanic action				

The Outer
PLANETS

ESSENTIAL QUESTIONS

 How are the outer planets similar?

 What are the outer planets made of?

Vocabulary

Galilean moons p. 69

 Launch Lab

15 minutes

How do we see distant objects in the solar system?

Some of the outer planets were discovered hundreds of years ago. Why weren't all planets discovered?

Object	Distance from Sun (cm)
Sun	0
Jupiter	39
Saturn	71
Uranus	143
Neptune	295

Procedure

1. Read and complete a lab safety form.
2. Use a **meterstick, masking tape,** and the **data table** to mark and label the position of each object on the tape on the floor along a straight line.
3. Shine a **flashlight** from "the Sun" horizontally along the tape.
4. Have a partner hold a page of this **book** in the flashlight beam at each planet location. Record your observations below.

Data and Observations

Think About This

1. What happens to the image of the page as you move away from the flashlight?

2. **Key Concept** Why do you think it is more difficult to observe the outer planets than the inner planets?

 Florida NGSSS

LA.8.2.2.3 The student will organize information to show understanding or relationships among facts, ideas, and events (e.g., representing key points within text through charting, mapping, paraphrasing, summarizing, or comparing/contrasting);

MA.6.A.3.6 Construct and analyze tables, graphs, and equations to describe linear functions and other simple relations using both common language and algebraic notation.

SC.8.E.5.3 Distinguish the hierarchical relationships between planets and other astronomical bodies relative to solar system, galaxy, and universe, including distance, size, and composition.

SC.8.E.5.4 Explore the Law of Universal Gravitation by explaining the role that gravity plays in the formation of planets, stars, and solar systems and in determining their motions.

SC.8.E.5.7 Compare and contrast the properties of objects in the Solar System including the Sun, planets, and moons to those of Earth, such as gravitational force, distance from the Sun, speed, movement, temperature, and atmospheric conditions.

inquiry **What's below?**

1. Clouds often prevent airplane pilots from seeing the ground below. Similarly, clouds block the view of Jupiter's surface. What do you think is below Jupiter's colorful cloud layer?

The Gas Giants

Have you ever seen water drops on the outside of a glass of ice? They form because water vapor in the air changes to a liquid on the cold glass. Gases also change to liquids at high pressures. These properties of gases affect the outer planets.

The outer planets, shown in **Figure 12,** are called the gas giants because they are primarily made of hydrogen and helium. These elements are usually gases on Earth.

The outer planets have strong gravitational forces because of their huge sizes. These forces apply tremendous pressure to the atmosphere of each planet and change gases to liquids. Thus, the outer planets mainly have liquid interiors. In general, an outer planet has a thick gas and liquid layer covering a small solid core.

 2. NGSSS Check **Compare** How are the outer planets similar? **SC.8.E.5.7**

Figure 12 🔑 The outer planets are primarily made of gases and liquids.

Active Reading **3. Identify** (Circle) the largest outer planet.

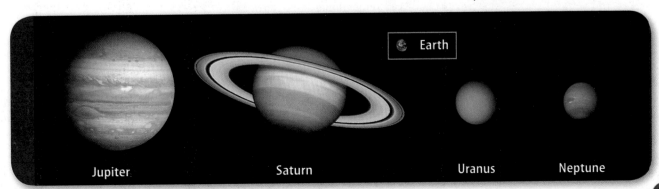

Earth

Jupiter   Saturn   Uranus   Neptune

Active Reading

FOLDABLES® LA.8.2.2.3

Make a four-door book. Label each door with the name of an outer planet. Use the book to organize your notes on the outer planets.

Jupiter

The largest planet in the solar system, Jupiter, is shown in **Figure 13.** Its diameter is more than 11 times larger than the diameter of Earth. Its mass is more than twice the mass of all the other planets combined. One way to understand just how big Jupiter is is to realize that more than 1,000 Earths would fit within this gaseous planet's volume.

Jupiter takes almost 12 Earth years to complete one orbit. Yet, it spins faster than any other planet. Its period of rotation is less than 10 hours. Like all the outer planets, Jupiter has a ring system.

Jupiter's Atmosphere

The atmosphere on Jupiter is about 90 percent hydrogen and 10 percent helium and is about 1,000 km deep. Within the atmosphere are layers of dense, colorful clouds. Because Jupiter rotates so quickly, these clouds stretch into colorful, swirling bands. The Great Red Spot on the planet's surface is a storm of swirling gases.

Jupiter's Structure

Overall, Jupiter is about 80 percent hydrogen and 20 percent helium with small amounts of other materials. The planet is a ball of gas swirling around a thick liquid layer that conceals a solid core. About 1,000 km below the outer edge of the cloud layer, the pressure is so great that the hydrogen gas changes to liquid. This thick layer of liquid hydrogen covers Jupiter's core. Scientists do not know for sure what makes up the core. They suspect that the core is made of rock and iron. The core might be as large as Earth and could have 10 times more mass.

Figure 13 Jupiter is mainly hydrogen and helium. Throughout most of the planet, the pressure is high enough to change the hydrogen gas into a liquid.

Active Reading **5. Explain** Why is Jupiter's period of revolution so much longer than the inner planets' periods of revolution?

4. **NGSSS Check** **List** What makes up each of Jupiter's three distinct layers? SC.8.E.5.3

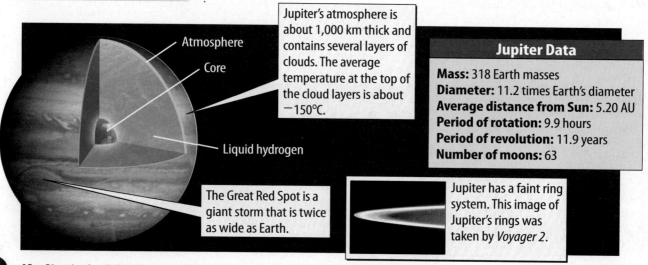

Atmosphere

Core

Jupiter's atmosphere is about 1,000 km thick and contains several layers of clouds. The average temperature at the top of the cloud layers is about −150°C.

Liquid hydrogen

The Great Red Spot is a giant storm that is twice as wide as Earth.

Jupiter Data

Mass: 318 Earth masses
Diameter: 11.2 times Earth's diameter
Average distance from Sun: 5.20 AU
Period of rotation: 9.9 hours
Period of revolution: 11.9 years
Number of moons: 63

Jupiter has a faint ring system. This image of Jupiter's rings was taken by *Voyager 2*.

The Moons of Jupiter

Jupiter has at least 63 moons, more than any other planet. Jupiter's four largest moons were first spotted by Galileo Galilei in 1610. *The four largest moons of Jupiter—Io, Europa, Ganymede, and Callisto—are known as the* **Galilean moons**. The Galilean moons all are made of rock and ice. The moons Ganymede, Callisto, and Io are larger than Earth's Moon. Collisions between Jupiter's moons and meteorites likely resulted in the particles that make up the planet's faint rings.

Saturn

Saturn is the sixth planet from the Sun. Like Jupiter, Saturn rotates rapidly and has horizontal bands of clouds. Saturn is about 90 percent hydrogen and 10 percent helium. It is the least-dense planet. Its density is less than that of water.

Saturn's Structure

Saturn is made mostly of hydrogen and helium with small amounts of other materials. As shown in **Figure 14,** Saturn's structure is similar to Jupiter's structure—an outer gas layer, a thick layer of liquid hydrogen, and a solid core.

The ring system around the planet is the largest and most complex in the solar system. Saturn has seven bands of rings, each containing thousands of narrower ringlets. The main ring system is over 70,000 km wide, but it is likely less than 30 m thick. The ice particles in the rings are possibly from a moon that was shattered in a collision with another icy object.

 7. **NGSSS Check** Describe (Circle) what makes up Saturn and its ring system. **SC.8.E.5.3**

Math Skills MA.6.A.3.6

Ratios

A ratio is a quotient—it is one quantity divided by another. Ratios can be used to compare distances. For example, Jupiter is **5.20** AU from the Sun, and Neptune is **30.05** from the Sun. Divide the larger distance by the smaller distance:

$$\frac{30.05}{5.20} = 5.78$$

Neptune is 5.78 times farther from the Sun than Jupiter.

6. **Practice**
How many times farther from the Sun is Uranus (distance = 19.20 AU) than Saturn (distance = 9.58 AU)?

Figure 14 🔑 Like Jupiter, Saturn is mainly hydrogen and helium. Saturn's rings are one of the most noticeable features of the solar system.

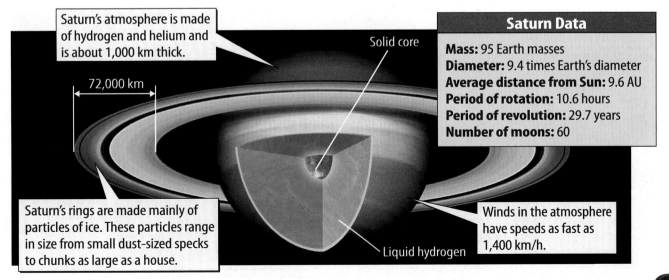

Saturn's atmosphere is made of hydrogen and helium and is about 1,000 km thick.

Solid core

72,000 km

Saturn's rings are made mainly of particles of ice. These particles range in size from small dust-sized specks to chunks as large as a house.

Liquid hydrogen

Winds in the atmosphere have speeds as fast as 1,400 km/h.

Saturn Data

Mass: 95 Earth masses
Diameter: 9.4 times Earth's diameter
Average distance from Sun: 9.6 AU
Period of rotation: 10.6 hours
Period of revolution: 29.7 years
Number of moons: 60

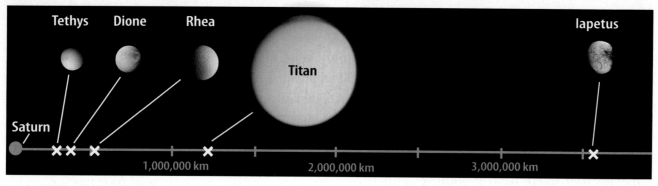

Tethys　　Dione　　Rhea　　　　　　　　　　　　　　　　　　Iapetus

Titan

Saturn

1,000,000 km　　　　2,000,000 km　　　　3,000,000 km

Figure 15 The five largest moons of Saturn are shown above drawn to scale.

 8. Recognize Circle the largest moon of Saturn.

Saturn's Moons

Saturn has at least 60 moons. The five largest moons, Titan, Rhea, Dione, Iapetus, and Tethys, are shown in **Figure 15.** Most of Saturn's moons are chunks of ice less than 10 km in diameter. However, Titan is larger than the planet Mercury. Titan is the only moon in the solar system with a dense atmosphere. In 2005, the *Cassini* orbiter released the *Huygens* (HOY guns) **probe** that landed on Titan's surface.

Uranus

Uranus, shown in **Figure 16,** is the seventh planet from the Sun. It has a system of narrow, dark rings and a diameter about four times that of Earth. *Voyager 2* is the only space probe to explore Uranus. The probe flew by the planet in 1986.

Uranus has a deep atmosphere composed mostly of hydrogen and helium. The atmosphere also contains a small amount of methane. Beneath the atmosphere is a thick, slushy layer of water, ammonia, and other materials. Uranus might also have a solid, rocky core.

10. NGSSS Check Identify What are the substances that make up the atmosphere and the thick slushy layer on Uranus? **SC.8.E.5.7**

WORD ORIGIN

probe
from Medieval Latin *proba*, means "examination"

Figure 16 Uranus is mainly gas and liquid with a small, solid core. Methane gas in the atmosphere gives Uranus a bluish color.

Active Reading 9. Explain What is different about Uranus's rotation axis?

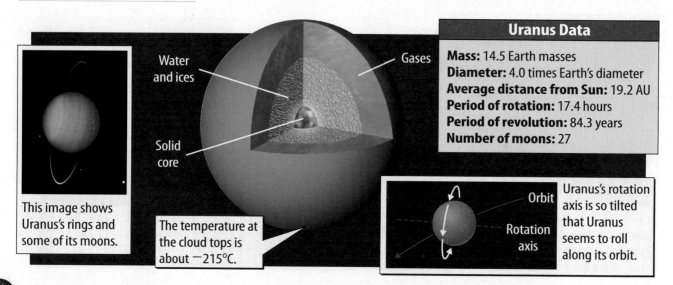

Water and ices

Gases

Solid core

This image shows Uranus's rings and some of its moons.

The temperature at the cloud tops is about −215°C.

Orbit

Rotation axis

Uranus's rotation axis is so tilted that Uranus seems to roll along its orbit.

Uranus Data
Mass: 14.5 Earth masses
Diameter: 4.0 times Earth's diameter
Average distance from Sun: 19.2 AU
Period of rotation: 17.4 hours
Period of revolution: 84.3 years
Number of moons: 27

Uranus's Axis and Moons

Figure 16 shows that Uranus has a tilted axis of rotation. In fact, it is so tilted that the planet moves around the Sun like a rolling ball. This sideways tilt might have been caused by a collision with an Earth-sized object.

Uranus has at least 27 moons. The two largest moons, Titania and Oberon, are considerably smaller than Earth's moon. Titania has an icy cracked surface that once might have been covered by an ocean.

Neptune

Neptune, shown in **Figure 17,** was discovered in 1846. Like Uranus, Neptune's atmosphere is mostly hydrogen and helium, with a trace of methane. Its interior also is similar to the interior of Uranus. Neptune's interior is partially frozen water and ammonia with a rock and iron core.

Neptune has at least 13 moons and a faint, dark ring system. Its largest moon, Triton, is made of rock with an icy outer layer. It has a surface of frozen nitrogen and geysers that erupt nitrogen gas.

 11. **NGSSS Check** **Contrast** How does the atmosphere and interior of Nepture compare with that of Uranus? SC.8.E.5.7

_____

_____

_____

Inquiry LAB STATION **Try It!** SC.8.N.1.1, LA.8.2.2.3

MiniLab *How do Saturn's moons affect its rings?* at connectED.mcgraw-hill.com

Apply It!

After you complete the lab, answer these questions.

1. What planets have ring systems that would be affected by their moons?

2. What is special about Saturn's largest moon?

Figure 17 🔑 The atmosphere of Neptune is similar to that of Uranus—mainly hydrogen and helium with a trace of methane. The dark circular areas on Neptune are swirling storms. Winds on Neptune sometimes exceed 1,000 km/h.

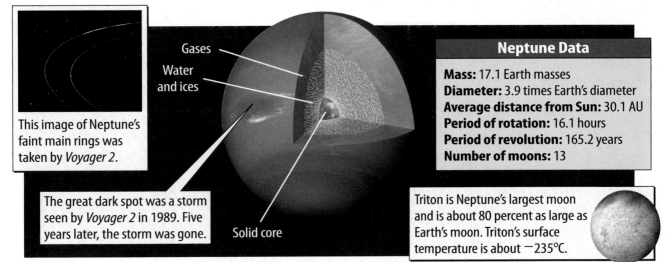

This image of Neptune's faint main rings was taken by *Voyager 2*.

Gases

Water and ices

The great dark spot was a storm seen by *Voyager 2* in 1989. Five years later, the storm was gone.

Solid core

Neptune Data

Mass: 17.1 Earth masses
Diameter: 3.9 times Earth's diameter
Average distance from Sun: 30.1 AU
Period of rotation: 16.1 hours
Period of revolution: 165.2 years
Number of moons: 13

Triton is Neptune's largest moon and is about 80 percent as large as Earth's moon. Triton's surface temperature is about −235°C.

All of the outer planets are primarily made of materials that are gases on Earth. Colorful clouds of gas cover Saturn and Jupiter.

Jupiter is the largest outer planet. Its four largest moons are known as the Galilean moons.

Uranus has an unusual tilt, possibly due to a collision with a large object.

Use Vocabulary

1 **Identify** What are the four Galilean moons of Jupiter? SC.8.E.5.3

_____

_____

Understand Key Concepts 🔑

2 **Contrast** How are the rings of Saturn different from the rings of Jupiter? SC.8.E.5.3

_____

_____

3 Which planet's rings probably formed from a collision between an icy moon and another icy object?

(A) Jupiter   (C) Saturn

(B) Neptune   (D) Uranus

Interpret Graphics

4 **Organize** List the outer planets.

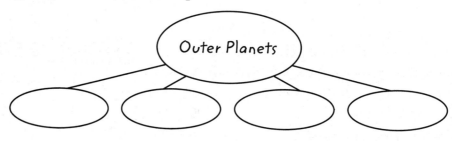

Critical Thinking

5 **Predict** what would happen to Jupiter's atmosphere if its gravitational force suddenly decreased. Explain.

_____

_____

_____

Math Skills   MA.6.A.3.6

6 **Calculate** Mars is about 1.52 AU from the Sun, and Saturn is about 9.58 AU from the Sun. How many times farther from the Sun is Saturn than Mars?

Pluto

What in the world is it?

Since Pluto's discovery in 1930, students have learned that the solar system has nine planets. But in 2006, the number of planets was changed to eight. What happened?

Neil deGrasse Tyson is an astrophysicist at the American Museum of Natural History in New York City. He and his fellow Museum scientists were among the first to question Pluto's classification as a planet. One reason was that Pluto is smaller than six moons in our solar system, including Earth's moon. Another reason was that Pluto's orbit is more oval-shaped, or elliptical, than the orbits of other planets. Also, Pluto has the most tilted orbit of all planets—17 degrees out of the plane of the solar system. Finally, unlike other planets, Pluto is mostly ice.

Tyson also questioned the definition of a planet—an object that orbits the Sun. Then shouldn't comets be planets? In addition, he noted that when Ceres, an object orbiting the Sun between Jupiter and Mars, was discovered in 1801, it was classified as a planet. But, as astronomers discovered more objects like Ceres, it was reclassified as an asteroid. Then, during the 1990s, many space objects similar to Pluto were discovered. They orbit the Sun beyond Neptune's orbit in a region called the Kuiper belt.

These new discoveries led Tyson and others to conclude that Pluto should be reclassified. In 2006, the International Astronomical Union agreed. Pluto was reclassified as a dwarf planet—an object that is spherical in shape and orbits the Sun in a zone with other objects. Pluto lost its rank as smallest planet, but became "king of the Kuiper belt."

Pluto TIME LINE

1930
Astronomer Clyde Tombaugh discovers a ninth planet, Pluto.

1992
The first object is discovered in the Kuiper belt.

July 2005
Eris—a Pluto-sized object—is discovered in the Kuiper belt.

January 2006
NASA launches *New Horizons* spacecraft, expected to reach Pluto in 2015.

August 2006
Pluto is reclassified as a dwarf planet.

Neil deGrasse Tyson is director of the Hayden Planetarium at the American Museum of Natural History.

This illustration shows what Pluto might look like if you were standing on one of its moons.

It's Your Turn

RESEARCH With a group, identify the different types of objects in our solar system. Consider size, composition, location, and whether the objects have moons. Propose at least two different ways to group the objects.

Dwarf Planets and OTHER OBJECTS

ESSENTIAL QUESTIONS

 What is a dwarf planet?

 What are the characteristics of comets and asteroids?

 How does an impact crater form?

Vocabulary

meteoroid p. 78

meteor p. 78

meteorite p. 78

impact crater p. 78

 Florida NGSSS

LA.8.2.2.3 The student will organize information to show understanding or relationships among facts, ideas, and events (e.g., representing key points within text through charting, mapping, paraphrasing, summarizing, or comparing/contrasting);

SC.8.E.5.1 Recognize that there are enormous distances between objects in space and apply our knowledge of light and space travel to understand this distance.

SC.8.E.5.3 Distinguish the hierarchical relationships between planets and other astronomical bodies relative to solar system, galaxy, and universe, including distance, size, and composition.

SC.8.E.5.4 Explore the Law of Universal Gravitation by explaining the role that gravity plays in the formation of planets, stars, and solar systems and in determining their motions.

SC.8.E.5.7 Compare and contrast the properties of objects in the Solar System including the Sun, planets, and moons to those of Earth, such as gravitational force, distance from the Sun, speed, movement, temperature, and atmospheric conditions.

SC.8.N.1.1 Define a problem from the eighth grade curriculum using appropriate reference materials to support scientific understanding, plan and carry out scientific investigations of various types, such as systematic observations or experiments, identify variables, collect and organize data, interpret data in charts, tables, and graphics, analyze information, make predictions, and defend conclusions.

SC.8.N.3.1 Select models useful in relating the results of their own investigations.

 Launch Lab

15 minutes

How might asteroids and moons form?

In this activity, you will explore one way moons and asteroids might have formed.

Procedure

1. Read and complete a lab safety form.

2. Form a small ball from **modeling clay** and roll it in **sand**.

3. Press a thin layer of modeling clay around a **marble**.

4. Tie equal lengths of **string** to each ball. Hold the strings so the balls are above a **sheet of paper**.

5. Have someone pull back the marble so that its string is parallel to the tabletop and then release it. Record the results below.

Data and Observations

Think About This

1. If the collision you modeled occurred in space, what would happen to the sand?

2. Key Concept Infer one way scientists propose moons and asteroids formed.

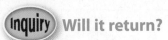 **Will it return?**

1. You would probably remember a sight like this. This image of comet c/2006 P1 was taken in 2007. The comet is no longer visible from Earth. Why do you think comets appear then reappear hundreds to millions of years later?

Dwarf Planets

Ceres was discovered in 1801 and was called a planet until similar objects were discovered near it. Then it was called an asteroid. For decades after Pluto's discovery in 1930, it was called a planet. Then, similar objects were discovered, and Pluto lost its planet classification. What type of object is Pluto?

In 2006, the International Astronomical Union (IAU) adopted a new category—dwarf planets. The IAU defines a dwarf planet as an object that orbits a star. When a dwarf planet formed, there was enough mass and gravity that it made a sphere. A dwarf planet has objects similar in mass orbiting near it or crossing its orbital path. Astronomers classify Pluto, Ceres, Eris, Makemake, and Haumea (how May ah) as dwarf planets. **Figure 18** shows four dwarf planets.

 2. NGSSS Check Describe What are the characteristics of a dwarf planet? SC.8.E.5.3

Figure 18 Four dwarf planets and the Moon are shown to scale.

3. Visual Check
Distinguish How do dwarf planets compare in size to the Moon?

Dwarf Planets

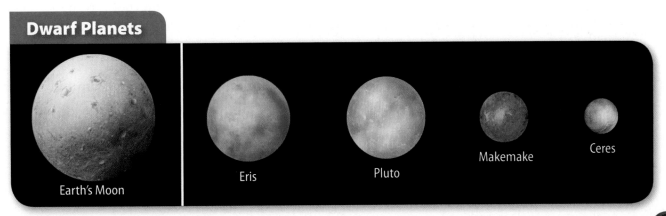

Earth's Moon

Eris

Pluto

Makemake

Ceres

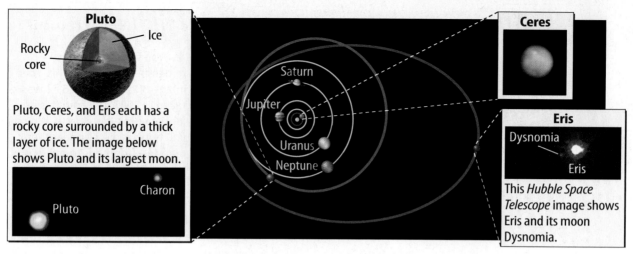

Pluto

Rocky core — Ice

Pluto, Ceres, and Eris each has a rocky core surrounded by a thick layer of ice. The image below shows Pluto and its largest moon.

Charon

Pluto

Ceres

Eris

Dysnomia — Eris

This *Hubble Space Telescope* image shows Eris and its moon Dysnomia.

Figure 19 Because most dwarf planets are so far from Earth, astronomers do not have detailed images of them.

4. Visual Check
Determine Which dwarf planet orbits closest to Earth?

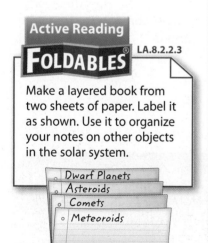
Ceres

Ceres, shown in **Figure 19**, orbits the Sun in the asteroid belt. With a diameter of about 950 km, Ceres is about one-fourth the size of the Moon. It is the smallest dwarf planet. Ceres might have a rocky core surrounded by a layer of frozen water and a thin, dusty crust.

Pluto

Pluto is about two-thirds the size of the Moon. Pluto is so far from the Sun that its period of revolution is about 248 years. Like Ceres, Pluto has a rocky core surrounded by ice. With an average surface temperature of about -230°C, Pluto is so cold that it is covered with frozen nitrogen.

Pluto has three known moons. The largest moon, Charon, has a diameter that is about half the diameter of Pluto. Pluto also has two smaller moons, Hydra and Nix.

Eris

The largest dwarf planet, Eris, was discovered in 2003. Its orbit lasts about 557 years. Currently, Eris is three times farther from the Sun than Pluto is. The structure of Eris is probably similar to Pluto. Dysnomia (dis NOH mee uh) is the only known moon of Eris.

Makemake and Haumea

In 2008, the IAU designated two new objects as dwarf planets: Makemake and Haumea. Though smaller than Pluto, Makemake is one of the largest objects in a region of the solar system called the Kuiper (KI puhr) belt. The Kuiper belt extends from about the orbit of Neptune to about 50 AU from the Sun. Haumea is also in the Kuiper belt and is smaller than Pluto.

Active Reading
5. Name Which dwarf planet is the largest? Which dwarf planet is the smallest?

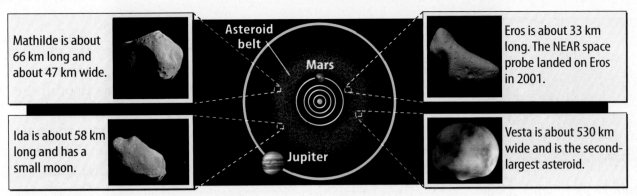

Figure 20 🔑 The asteroids that orbit the Sun in the asteroid belt are many sizes and shapes.

Mathilde is about 66 km long and about 47 km wide.

Ida is about 58 km long and has a small moon.

Asteroid belt

Mars

Jupiter

Eros is about 33 km long. The NEAR space probe landed on Eros in 2001.

Vesta is about 530 km wide and is the second-largest asteroid.

Asteroids

Recall from Lesson 1 that asteroids are pieces of rock and ice. Most asteroids orbit the Sun in the asteroid belt. The asteroid belt is between the orbits of Mars and Jupiter, as shown in **Figure 20**. Hundreds of thousands of asteroids have been discovered. The largest asteroid, Pallas, is over 500 km in diameter.

Asteroids are chunks of rock and ice that never clumped together like the rocks and ice that formed the inner planets. Some astronomers suggest that the strength of Jupiter's gravitational field might have caused the chunks to collide so violently that they broke apart instead of sticking together. This means that asteroids are objects left over from the formation of the solar system.

 6. NGSSS Check Identify Where do the orbits of most asteroids occur? SC.8.E.5.3

Comets

Recall that comets are mixtures of rock, ice, and dust. The particles in a comet are loosely held together by the gravitational attractions among the particles. As shown in **Figure 21**, comets orbit the Sun in long elliptical orbits.

The Structure of Comets

The solid, inner part of a comet is its nucleus, as shown in **Figure 21**. As a comet moves closer to the Sun, it heats and can develop a bright tail. Heating changes the ice in the comet into a gas. Energy from the Sun pushes some of the gas and dust away from the nucleus and makes it glow. This produces the comet's bright tail and glowing nucleus, called a coma.

 7. NGSSS Check Define What are the characteristics of a comet? SC.8.E.5.3

Figure 21 🔑 When energy from the Sun strikes the gas and dust in the comet's nucleus, it can create a two-part tail. The gas tail always points away from the Sun.

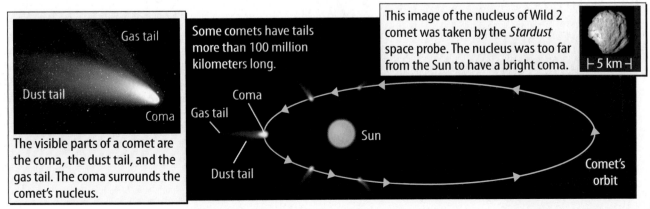

Gas tail

Dust tail

Coma

The visible parts of a comet are the coma, the dust tail, and the gas tail. The coma surrounds the comet's nucleus.

Some comets have tails more than 100 million kilometers long.

Coma

Gas tail

Dust tail

Sun

Comet's orbit

This image of the nucleus of Wild 2 comet was taken by the *Stardust* space probe. The nucleus was too far from the Sun to have a bright coma.

⊢ 5 km ⊣

Figure 22 🗝 When a large meteorite strikes, it can form a giant impact crater like this 1.2-km wide crater in Arizona.

Short-Period and Long-Period Comets

A short-period comet takes less than 200 Earth years to orbit the Sun. Most short-period comets come from the Kuiper belt. A long-period comet takes more than 200 Earth years to orbit the Sun. Long-period comets come from an area at the outer edge of the solar system called the Oort cloud. It surrounds the solar system and extends about 100,000 AU from the Sun. Some long-period comets take millions of years to orbit the Sun.

Meteoroids

Every day, many millions of particles called meteoroids enter Earth's atmosphere. *A* **meteoroid** *is a small, rocky particle that moves through space.* Most meteoroids are only about as big as a grain of sand. As a meteoroid passes through Earth's atmosphere, friction makes the meteoroid and the air around it hot enough to glow. *A* **meteor** *is a streak of light in Earth's atmosphere made by a glowing meteoroid.* Most meteors burn up in the atmosphere. However, some meteors are large enough that they reach Earth's surface before they burn up completely. When this happens, they are called meteorites. *A* **meteorite** *is a meteoroid that strikes a planet or a moon.*

When a large meteoroite strikes a moon or planet, it often forms a bowl-shaped depression such as the one shown in **Figure 22**. *An* **impact crater** *is a round depression formed on the surface of a planet, moon, or other space object by the impact of a meteorite.* There are more than 170 impact craters on Earth.

Active Reading 8. **Summarize** What causes an impact crater to form?

_____

_____

Inquiry 🔬**LAB STATION** **Try It!** SC.8.N.3.1

MiniLab *How do impact craters form?* at connectED.mcgraw-hill.com

Apply It! After you complete the lab, answer these questions.

1. What did the flour and cornmeal represent in this activity?

2. Infer why there are so many more craters present on the Moon than on Earth.

An asteroid, such as Ida, is a chunk of rock and ice that orbits the Sun.

Comets, which are mixture of rock, ice, and dust, orbit the Sun. A comet's tail is caused by its interaction with the Sun.

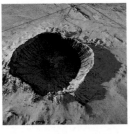

When a large meteorite strikes a planet or moon, it often makes an impact crater.

Inquiry **Try It!**
SC.8.N.1.3,
SC.8.E.5.1,
SC.8.E.5.3

Inquiry Lab *Scaling down the Solar System* at connectED.mcgraw-hill.com

Use Vocabulary

1 **Define** *impact crater* in your own words.

2 **Distinguish** between a meteorite and a meteoroid. SC.8.E.5.3

3 **Use the term** *meteor* in a complete sentence. SC.8.E.5.3

Understand Key Concepts 🔑

4 Which produces an impact crater?
- (A) comet
- (B) meteor
- (C) meteorite
- (D) planet

5 **Reason** Are you more likely to see a meteor or a meteoroid? Explain. SC.8.E.5.3

6 **Differentiate** between objects located in the asteroid belt and objects located in the Kuiper belt. SC.8.E.5.3

Interpret Graphics

7 **Organize** List the major characteristics of a dwarf planet. SC.8.E.5.7

Object	Defining Characteristic
Dwarf Planet	

Critical Thinking

8 **Evaluate** Do you agree with the decision to reclassify Pluto as a dwarf planet? Defend your opinion. SC.8.E.5.7

Chapter 2 Study Guide

Think About It! Gravity and energy influence the formation of objects in the solar system including planets, dwarf planets, comets, asteroids, and other small solar system bodies.

 ## Key Concepts Summary

	Vocabulary

LESSON 1 The Structure of the Solar System

- The inner planets are made mainly of solid materials. The outer planets, which are larger than the inner planets, have thick gas and liquid layers covering a small solid core.

- Astronomers measure vast distances in space in **astronomical units**; an astronomical unit is about 150 million km.

- The speed of each planet changes as it moves along its elliptical orbit around the Sun.

asteroid p. 53

comet p. 53

astronomical unit p. 54

period of revolution p. 54

period of rotation p. 54

LESSON 2 The Inner Planets

- The inner planets—Mercury, Venus, Earth, and Mars—are made of rock and metallic materials.

- The **greenhouse effect** makes Venus the hottest planet.

- Mercury has no atmosphere. The atmospheres of Venus and Mars are almost entirely carbon dioxide. Earth's atmosphere is a mixture of gases and a small amount of water vapor.

terrestrial planet p. 59

greenhouse effect p. 61

LESSON 3 The Outer Planets

- The outer planets—Jupiter, Saturn, Uranus, and Neptune—are primarily made of hydrogen and helium.

- Jupiter and Saturn have thick cloud layers, but are mainly liquid hydrogen. Saturn's rings are largely particles of ice. Uranus and Neptune have thick atmospheres of hydrogen and helium.

Galilean moons p. 69

LESSON 4 Dwarf Planets and Other Objects

- A dwarf planet is an object that orbits a star, has enough mass to pull itself into a spherical shape, and has objects similar in mass orbiting near it.

- An asteroid is a small rocky object that orbits the Sun. Comets are made of rock, ice, and dust and orbit the Sun in highly elliptical paths.

- The impact of a **meteorite** forms an **impact crater**.

meteoroid p. 78

meteor p. 78

meteorite p. 78

impact crater p. 78

Active Reading

FOLDABLES® Chapter Project

Assemble your lesson Foldables as shown to make a Chapter Project. Use the project to review what you have learned in this chapter.

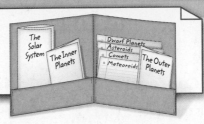

Use Vocabulary

Match each phrase with the correct vocabulary term from the Study Guide.

1 the time it takes an object to complete one rotation on its axis

_____

2 the average distance from Earth to the Sun

_____

3 the time it takes an object to travel once around the Sun

_____

4 an increase in temperature caused by energy trapped by a planet's atmosphere

_____

5 an inner planet

_____

6 the four largest moons of Jupiter

_____

7 a streak of light in Earth's atmosphere made by a glowing meteoroid

_____

Link Vocabulary and Key Concepts

Use vocabulary terms to complete the concept map.

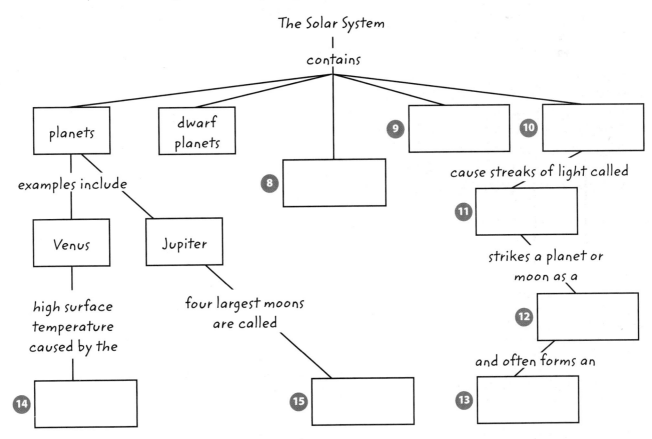

Chapter 2 — Review

Fill in the correct answer choice.

🔑 Understand Key Concepts

1 Which solar system object is the largest? SC.8.E.5.3
- Ⓐ Jupiter
- Ⓑ Neptune
- Ⓒ the Sun
- Ⓓ Saturn

2 Which best describes the asteroid belt? SC.8.E.5.3
- Ⓐ another name for the Oort cloud
- Ⓑ the region where comets originate
- Ⓒ large chunks of gas, dust, and ice
- Ⓓ millions of small, rocky objects

3 Which describes a planet's speed as it orbits the Sun? SC.8.E.5.7
- Ⓐ It constantly decreases.
- Ⓑ It constantly increases.
- Ⓒ It does not change.
- Ⓓ It increases then decreases.

4 The diagram below shows a planet's orbit around the Sun. What does the blue arrow represent? SC.8.E.5.4
- Ⓐ the gravitational pull of the Sun
- Ⓑ the planet's orbital path
- Ⓒ the planet's path if Sun did not exist
- Ⓓ the planet's speed

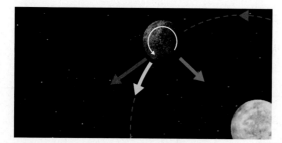

5 Which describes the greenhouse effect? SC.8.E.5.7
- Ⓐ effect of gravity on temperature
- Ⓑ energy emitted by the Sun
- Ⓒ energy trapped by atmosphere
- Ⓓ reflection of light from a planet

6 How are the terrestrial planets similar? SC.8.E.5.7
- Ⓐ similar densities
- Ⓑ similar diameters
- Ⓒ similar periods of rotation
- Ⓓ similar rocky surfaces

Critical Thinking

7 **Relate** changes in speed during a planet's orbit to the shape of the orbit and the gravitational pull of the Sun. SC.8.E.5.4

8 **Compare** In what ways are planets and dwarf planets similar? SC.8.E.5.3

9 **Apply** Like Venus, Earth's atmosphere contains carbon dioxide. What might happen on Earth if the amount of carbon dioxide in the atmosphere increases? Explain. SC.8.E.5.7

10 **Defend** A classmate states that life will someday be found on Mars. Defend the statement and offer a reason why life might exist on Mars. SC.8.E.5.7

11 **Infer** whether a planet with active volcanoes would have more or fewer craters than a planet without active volcanoes. Explain. SC.8.E.5.7

12 **Evaluate** The Huygens probe transmitted data about Titan for 90 min. In your opinion, was this worth the effort of sending the probe? SC.8.E.5.1

Asteroid belt — Mars — Jupiter

13 **Support** Use the diagram of the asteroid belt to support the explanation of how the belt formed. SC.8.E.5.4

_____

_____

_____

14 **Explain** why Jupiter's moon Ganymede is not considered a dwarf planet, even though it is bigger than Mercury. SC.8.E.5.3

_____

_____

_____

_____

Writing in Science

16 **Create** a pamphlet that describes how the International Astronomical Union classifies planets, dwarf planets, and small solar system objects. SC.8.E.5.3

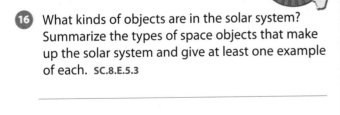

Big Idea Review

16 What kinds of objects are in the solar system? Summarize the types of space objects that make up the solar system and give at least one example of each. SC.8.E.5.3

_____

_____

_____

_____

17 Describe what Saturn and its rings are made of and explain why the other two objects are moons. SC.8.E.5.3

_____

_____

_____

Math Skills   MA.6.A.3.6

Use Ratios

Inner Planet Data			
Planet	Diameter (% of Earth's diameter)	Mass (% of Earth's mass)	Average Distance from Sun (AU)
Mercury	38.3	5.5	0.39
Venus	95	81.5	0.72
Earth	100	100	1.00
Mars	53.2	10.7	1.52

18 Use the table above to calculate how many times farther from the Sun Mars is compared to Mercury.

19 Calculate how much greater Venus's mass is compared to Mercury's mass.

Fill in the correct answer choice.

Multiple Choice

1 Which is a terrestrial planet? SC.8.E.5.3

Ⓐ Ceres

Ⓑ Neptune

Ⓒ Pluto

Ⓓ Venus

2 An astronomical unit (AU) is the average distance SC.8.E.5.1

Ⓕ between Earth and the Moon.

Ⓖ from Earth to the Sun.

Ⓗ to the nearest star in the galaxy.

Ⓘ to the edge of the solar system.

3 Which is NOT a characteristic of ALL planets? SC.8.E.5.3

Ⓐ exceed the total mass of nearby objects

Ⓑ have a nearly spherical shape

Ⓒ have one or more moons

Ⓓ make an elliptical orbit around the Sun

Use the diagram below to answer question 4.

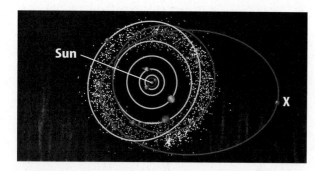

4 Which object in the solar system is marked by an *X* in the diagram? SC.8.E.5.3

Ⓕ asteroid

Ⓖ meteoroid

Ⓗ dwarf planet

Ⓘ outer planet

Use the diagram of Saturn below to answer questions 5 and 6.

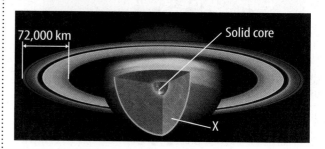

5 The thick inner layer marked *X* in the diagram above is made of which material? SC.8.E.5.7

Ⓐ carbon dioxide

Ⓑ gaseous helium

Ⓒ liquid hydrogen

Ⓓ molten rock

6 In the diagram, Saturn's rings are shown to be 72,000 km in width. Approximately how thick are Saturn's rings? SC.8.E.5.7

Ⓕ 30 m

Ⓖ 1,000 km

Ⓗ 14,000 km

Ⓘ 1 AU

7 Which are NOT found on Mercury's surface? SC.8.E.5.7

Ⓐ high cliffs

Ⓑ impact craters

Ⓒ lava flows

Ⓓ sand dunes

8 What is the primary cause of the extremely high temperatures on the surface of Venus? SC.8.E.5.7

Ⓕ heat rising from the mantle

Ⓖ lack of an atmosphere

Ⓗ proximity to the Sun

Ⓘ the greenhouse effect

Use the diagram below to answer question 9.

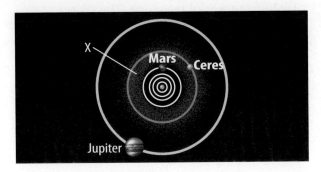

9 In the diagram above, which region of the solar system is marked by an *X*? SC.8.E.5.3

(A) the asteroid belt

(B) the dwarf planets

(C) the Kuiper belt

(D) the Oort cloud

10 What is a meteorite? SC.8.E.5.7

(F) a surface depression formed by collision with a rock from space

(G) a fragment of rock that strikes a planet or a moon

(H) a mixture of ice, dust, and gas with a glowing tail

(I) a small rocky particle that moves through space

11 What gives Mars its reddish color? SC.8.E.5.7

(A) ice caps of frozen carbon dioxide

(B) lava from Olympus Mons

(C) liquid water in gullies

(D) soil rich in iron oxide

12 What explains the motion of planets around the Sun? SC.8.E.5.4

(F) period of rotation

(G) greenhouse effect

(H) period of revolution

(I) Law of Universal Gravitation

Use the image below to answer question 13.

13 What type of object is NOT shown in this illustration of the solar system? SC.8.E.5.7

(A) a star

(B) gas giants

(C) dwarf planets

(D) terrestrial planets

14 Why do astronomers use astronomical units? SC.8.E.5.1

(F) Astronomical units are easy to convert to feet and inches.

(G) Distances between objects in the solar system are enormous.

(H) Periods of revolution and rotation can only be compared in astronomical units.

(I) Temperature differences on some planets in the solar system are too extreme.

NEED EXTRA HELP?

If You Missed Question...	1	2	3	4	5	6	7	8	9	10	11	12	13	14
Go to Lesson...	2	1	1	1	3	3	2	2	1,4	4	2	1	4	1

Multiple Choice *Bubble the correct answer.*

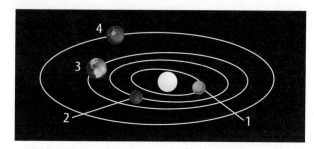

1. The image above shows the inner planets. Which planet is represented by the number 2? **SC.8.E.5.7**

- (A) Earth
- (B) Mars
- (C) Mercury
- (D) Venus

2. Nuclear fusion, a process inside the Sun, produces an enormous amount of **SC.8.N.1.1**

- (F) energy.
- (G) water.
- (H) carbon dioxide.
- (I) carbon monoxide.

3. An astronomer observes a planetary object using a powerful telescope. The object appears to be in orbit around the Sun. It is rocky and has an irregular shape. What is the object? **SC.8.E.5.3**

- (A) asteroid
- (B) comet
- (C) meteor
- (D) planet

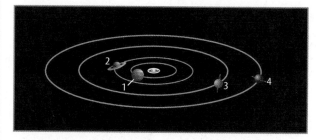

4. The diagram above shows the outer planets. Which planet is represented by the number 3? **SC.8.E.5.7**

- (F) Jupiter
- (G) Neptune
- (H) Saturn
- (I) Uranus

Multiple Choice *Bubble the correct answer.*

Earth Data	
Diameter:	12,742 km
Average distance from Sun:	J
Period of rotation:	K
Period of revolution:	L
Number of moons:	1

1. Which are the correct values missing from the table above? **SC.8.E.5.7**

    Ⓐ J: 0.39 AU; K: 59 days; L: 88 days

    Ⓑ J: 0.72 AU; K: 244 days; L: 225 days

    Ⓒ J: 1.00 AU; K: 24 hours; L: 365 days

    Ⓓ J: 1.52 AU; K: 24.6 hours; L: 1.88 years

2. Corey is calculating how much he would weigh on the different terrestrial planets. He would weigh the least on **SC.8.E.5.7**

    Ⓕ Earth.

    Ⓖ Mars.

    Ⓗ Mercury.

    Ⓘ Venus.

3. How is Venus similar to Earth? **SC.8.E.5.7**

    Ⓐ Both have the same day length.

    Ⓑ Both rotate in the same direction.

    Ⓒ Both have ample water on their surfaces.

    Ⓓ Both have a solid inner core and a liquid outer core.

4. Which of the inner planets has the highest percentage of carbon dioxide in its atmosphere? **SC.8.E.5.7**

    Ⓕ Earth

    Ⓖ Mars

    Ⓗ Mercury

    Ⓘ Venus

Multiple Choice *Bubble the correct answer.*

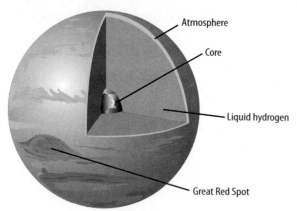

Atmosphere

Core

Liquid hydrogen

Great Red Spot

1. Which planet is shown in the image above?
SC.8.E.5.3

(A) Jupiter

(B) Neptune

(C) Saturn

(D) Uranus

2. Which outer planet has the most complex ring system in the solar system? **SC.8.E.5.3**

(F) Jupiter

(G) Neptune

(H) Saturn

(I) Uranus

3. What are the outer planets mostly made of?
SC.8.E.5.7

(A) hydrogen and helium

(B) rock and hydrogen

(C) helium and metallic materials

(D) rock and metallic materials

4. What is unusual about Uranus, as seen in the image above? **SC.8.E.5.7**

(F) the axis of rotation

(G) the distinctive rings

(H) the gases that it is made of

(I) the nearly spherical shape

Benchmark Mini-Assessment Chapter 2 • Lesson 4

Multiple Choice *Bubble the correct answer.*

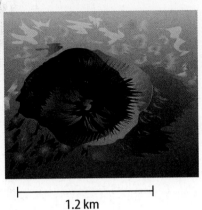

1.2 km

1. What type of object created the depression on Earth's surface that is shown in the image above? **SC.8.E.5.7**

   (A) asteroid

   (B) comet

   (C) meteor

   (D) meteorite

2. A dwarf planet is likely to **SC.8.E.5.3**

   (F) have a bright tail and a glowing nucleus.

   (G) have chunks of rock and ice that never clump together.

   (H) have a liquid hydrogen layer that surrounds a solid core.

   (I) have objects similar in mass orbiting nearby.

3. A space probe traveling from Earth will hit the asteroid belt after passing which planet? **SC.8.E.5.1**

   (A) Mars

   (B) Saturn

   (C) Uranus

   (D) Venus

4. Which image below is a solar system object that may have originated in the Oort cloud? **SC.8.E.5.3**

(F)

(G)

(H)

(I)

Name _____   Date _____

PAGE KEELEY
SCIENCE PROBES

Describing Stars

If you look up into the sky on a clear, dark night you will see many stars. Put an X next to each statement that describes the stars that are in the night sky.

____ **A.** Our Sun is a star.
____ **B.** Stars move across the night sky.
____ **C.** Stars have no mass.
____ **D.** All stars are the same brightness.
____ **E.** The inner core of a star is solid.
____ **F.** It is common for two stars to orbit each other.
____ **G.** Stars are in galaxies.
____ **H.** Stars are on the other side of Earth only during the daytime.

____ **I.** Stars eventually become supernovae and explode.
____ **J.** You see a star as it was when light left it, not as it is now.
____ **K.** Stars live forever.
____ **L.** Stars are very far away.
____ **M.** Stars are larger than planets.
____ **N.** Stars twinkle.
____ **O.** Stars are the same color but different temperatures.

Explain your ideas about stars. Provide reasons to support why you think some of the statements describe stars and why some do not describe stars.

Stars and
GALAXIES

FLORIDA BIG IDEAS

1 **The Practice of Science**
3 **The Role of Theories, Laws, Hypotheses, and Models**
5 **Earth in Space and Time**

The Big Idea

Think About It!

What makes up the universe, and how does gravity affect the universe?

What can't you see? This photograph shows a small part of the universe. You can see many stars and galaxies in this image. But the universe also contains many things you cannot see.

1 How do you think scientists study the universe?

_____
_____
_____

2 What do you think makes up the universe?

_____
_____
_____

3 How do you think gravity affects the universe?

_____
_____
_____

Get Ready to Read

What do you think about stars and galaxies?

Before you read, decide if you agree or disagree with each of these statements. As you read this chapter, see if you change your mind about any of the statements.

	AGREE	DISAGREE
1 The night sky is divided into constellations.	☐	☐
2 A light-year is a measurement of time.	☐	☐
3 Stars shine because there are nuclear reactions in their cores.	☐	☐
4 Sunspots appear dark because they are cooler than nearby areas.	☐	☐
5 The more matter a star contains, the longer it is able to shine.	☐	☐
6 Gravity plays an important role in the formation of stars.	☐	☐
7 Most of the mass in the universe is in stars.	☐	☐
8 The Big Bang theory is an explanation of the beginning of the universe.	☐	☐

 **ConnectED**

There's More Online!
Video • Audio • Review • ⓘLab Station • WebQuest • Assessment • Concepts in Motion • Multilingual eGlossary   93

The View from EARTH

Vocabulary

spectroscope p. 97

astronomical unit p. 98

light-year p. 98

apparent magnitude p. 99

luminosity p. 99

 Florida NGSSS

LA.8.2.2.3 The student will organize information to show understanding or relationships among facts, ideas, and events (e.g., representing key points within text through charting, mapping, paraphrasing, summarizing, or comparing/contrasting);

MA.6.A.3.6 Construct and analyze tables, graphs, and equations to describe linear functions and other simple relations using both common language and algebraic notation.

SC.8.E.5.1 Recognize that there are enormous distances between objects in space and apply our knowledge of light and space travel to understand this distance.

SC.8.E.5.5 Describe and classify specific physical properties of stars: apparent magnitude (brightness), temperature (color), size, and luminosity (absolute brightness).

SC.8.E.5.10 Assess how technology is essential to science for such purposes as access to outer space and other remote locations, sample collection, measurement, data collection and storage, computation, and communication of information.

SC.8.E.5.11 Identify and compare characteristics of the electromagnetic spectrum such as wavelength, frequency, use, and hazards and recognize its application to an understanding of planetary images and satellite photographs.

Also covers: SC.8.N.1.1, SC.8.N.3.1

 SC.8.E.5.11

 Launch Lab

20 minutes

How can you "see" invisible energy?

Procedure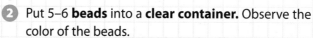

1. Read and complete a lab safety form.
2. Put 5–6 **beads** into a **clear container.** Observe the color of the beads.
3. In a darkened room, shine light from a **flashlight** onto the beads for several seconds. Record your observations below. Repeat this step, exposing the beads to light from an **incandescent lightbulb** and a **fluorescent light**. Record your observations.
4. Stand outside in a shady spot for several seconds. Then expose the beads to direct sunlight. Record your observations.

Data and Observations

Think About This

1. How did the light from the different sources affect the color of the beads?

2. What do you think made the beads change color?

3. **Key Concept** How do you think invisible forms of light help scientists understand stars and other objects in the sky?

 **Where is this?**

1. This view of the night sky is similar to what the night sky looked like to your ancestors. Where might you still be able to see a sky like this at night? Why isn't the view from most cities and towns like this?

Looking at the Night Sky

Have you ever looked up at the sky on a clear, dark night and seen countless stars? If you have, you are lucky. Few people see a sky like that shown above. Lights from towns and cities make the night sky too bright for faint stars to be seen.

If you look at a clear night sky for a long time, the stars seem to move. But what you are really seeing is Earth's movement. Earth spins, or rotates, once every 24 hours. Day turns to night and then back to day as Earth rotates. Because Earth rotates from west to east, objects in the sky rise in the east and set in the west.

Earth spins on its axis, an imaginary line from the North Pole to the South Pole. The star Polaris is almost directly above the North Pole. As Earth spins, stars near Polaris appear to travel in a circle around Polaris, as shown in **Figure 1**. These stars never set when viewed from the northern hemisphere. They are always present in the night sky.

Active Reading

2. Explain What causes the apparent movement of stars in the night sky?

Figure 1 The stars around Polaris appear as streaks of light in this time-lapse photograph.

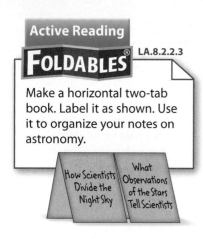

Make a horizontal two-tab book. Label it as shown. Use it to organize your notes on astronomy.

How Scientists Divide the Night Sky | What Observations of the Stars Tell Scientists

Naked-Eye Astronomy

You don't need expensive equipment to view the sky. *Naked-eye astronomy* means gazing at the sky with just your eyes, without binoculars or a telescope. Long before the telescope was invented, people observed stars to tell time and find directions. They learned about planets, seasons, and astronomical events merely by watching the sky. As you practice naked-eye astronomy, remember never to look directly at the Sun—it could damage your eyes.

Constellations

As people in ancient cultures gazed at the night sky, they saw patterns. The patterns resembled people, animals, or objects, such as the hunter and the dragon shown in **Figure 2**. The Greek astronomer Ptolemy (TAH luh mee) identified dozens of star patterns nearly 2,000 years ago. Today, these patterns and others like them are known as ancient constellations.

Present-day astronomers use many ancient constellations to divide the sky into 88 regions. Some of these regions, which are also called constellations, are shown in the sky map in **Figure 2**. Dividing the sky helps scientists communicate to others what area of sky they are studying.

Figure 2 🔑 Most modern constellations contain an ancient constellation.

☑ 3. **Visual Check**
Identify Which constellation would probably be the closest to Polaris? Why?

Active Reading 4. **Describe** How do astronomers divide the night sky?

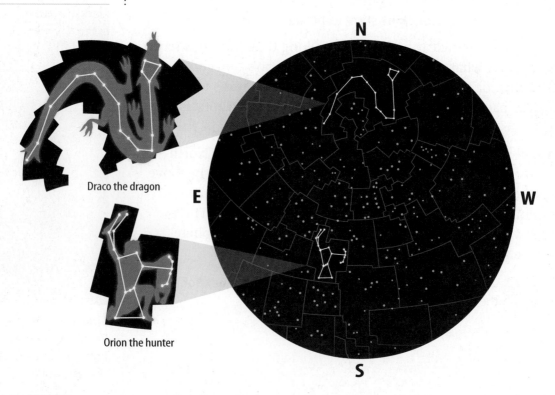

Draco the dragon

Orion the hunter

N

E

W

S

Electromagnetic Spectrum

Figure 3 Different parts of the electromagnetic spectrum have different wavelengths and different energies. You can see only a small part of the energy in these wavelengths.

Active Reading 5. **Locate** (Circle) which wavelength has the highest energy.

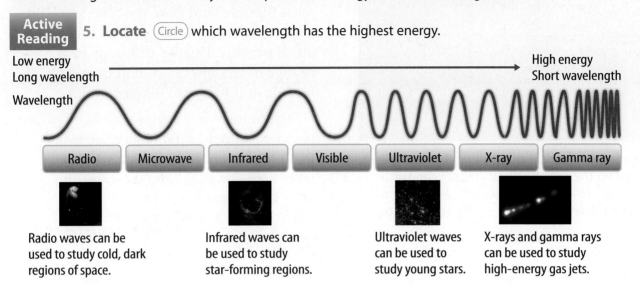

Low energy
Long wavelength

High energy
Short wavelength

Wavelength

| Radio | Microwave | Infrared | Visible | Ultraviolet | X-ray | Gamma ray |

Radio waves can be used to study cold, dark regions of space.

Infrared waves can be used to study star-forming regions.

Ultraviolet waves can be used to study young stars.

X-rays and gamma rays can be used to study high-energy gas jets.

Telescopes

Telescopes can collect much more light than the human eye can detect. Visible light is just one part of the electromagnetic spectrum. As shown in **Figure 3,** the electromagnetic spectrum is a continuous range of wavelengths. Longer wavelengths have low energy. Shorter wavelengths have high energy. Different objects in space emit different ranges of wavelengths. The range of wavelengths that a star emits is the star's spectrum (plural, spectra).

Spectroscopes

Scientists study the spectra of stars using an instrument called a spectroscope. *A* **spectroscope** *spreads light into different wavelengths.* Using spectroscopes, astronomers can study stars' characteristics, including temperatures, compositions, and energies. For example, newly formed stars emit mostly radio and infrared waves, which have low energy. Exploding stars emit mostly high-energy ultraviolet waves and X-rays.

6. **NGSS Check** **Describe** What can astronomers learn from a star's spectrum? SC.8.E.5.5

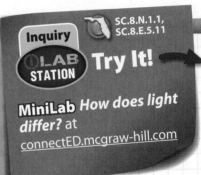

Inquiry SC.8.N.1.1, SC.8.E.5.11

LAB STATION **Try It!**

MiniLab *How does light differ?* at connectED.mcgraw-hill.com

Apply It!

After you complete the lab, answer this question.

1. How would knowing the wavelengths of energy being emitted by stars be helpful?

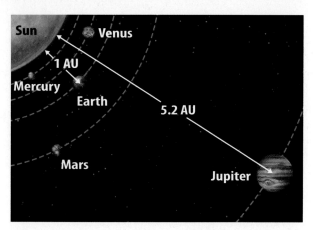

Figure 4 Measurements in the solar system are based on the average distance between Earth and the Sun—1 astronomical unit (AU). The most distant planet, Neptune, is 30 AU from the Sun.

WORD ORIGIN

parallax
from Greek *parallaxis*, means "alteration"

Math Skills   MA.6.A.3.6, SC.8.E.5.1

Use Proportions

Proportions can be used to calculate distances to astronomical objects. Light can travel nearly 10 trillion km in 1 year (y). How many years would it take light to reach Earth from a star that is 100 trillion km away?

1. Set up a proportion.

$$\frac{10 \text{ trillion km}}{1 \text{ y}} = \frac{100 \text{ trillion km}}{x \text{ y}}$$

2. Cross multiply.

10 trillion km $\times (x)$ y = 100 trillion km $\times$ 1 y

3. Solve for x by dividing both sides by 10 trillion km.

$$x = \frac{100 \text{ trillion km}}{10 \text{ trillion km}} = 10 \text{ y}$$

8. Practice

How many years would it take light to reach Earth from a star 60 trillion km away?

Measuring Distances

Hold up your thumb at arm's length. Close one eye, and look at your thumb. Now open that eye, and close the other eye. Did your thumb seem to jump? This is an example of parallax. Parallax is the apparent change in an object's position caused by looking at it from two different points.

Astronomers use angles created by parallax to measure how far objects are from Earth. Instead of the eyes being the two points of view, they use two points in Earth's orbit around the Sun.

Active Reading   7. **Define** `Highlight` the explanation of parallax.

Distances Within the Solar System

Because the universe is too large to be measured easily in meters or kilometers, astronomers use other units of measurement. For distances within the solar system, they use **astronomical units (AU)**. An astronomical unit is the average distance between Earth and the Sun, about 150 million km. Astronomical units are convenient to use in the solar system because distances easily can be compared to the distance between Earth and the Sun, as shown in **Figure 4**.

Distances Beyond the Solar System

Astronomers measure distances to objects beyond the solar system using a larger distance unit—the light-year. Despite its name, a light-year measures distance, not time. *A **light-year** is the distance light travels in 1 year.* Light travels at a rate of about 300,000 km/s. That means 1 light-year is about 10 trillion km! Proxima Centauri, the nearest star to the Sun, is 4.2 light-years away.

Looking Back in Time

Because it takes time for light to travel, you see a star not as it is today, but as it was when light left it. At 4.2 light-years away, Proxima Centauri appears as it was 4.2 years ago. The farther away an object, the longer it takes for its light to reach Earth.

Measuring Brightness

When you look at stars, you can see that some are dim and some are bright. Astronomers measure the brightness of stars in two ways: by how bright they appear from Earth and by how bright they actually are.

Apparent Magnitude

Scientists measure how bright stars appear from Earth using a scale developed by the ancient Greek astronomer Hipparchus (hi PAR kus). Hipparchus assigned a number to every star he saw in the night sky, based on the star's brightness. Astronomers today call these numbers magnitudes. *The **apparent magnitude** of an object is a measure of how bright it appears from Earth.*

As shown in **Figure 5,** some objects have negative apparent magnitudes. That is because Hipparchus assigned a value of 1 to all of the brightest stars. He also did not assign values to the Sun, the Moon, or Venus. Astronomers later assigned negative numbers to the Sun, the Moon, Venus, and a few bright stars.

ACADEMIC VOCABULARY

apparent

(adjective) appearing to the eye or mind

Figure 5 The fainter a star or other object in the sky appears, the greater its apparent magnitude.

9. **Visual Check** Find Use the diagram below to determine the apparent magnitude of Sirius.

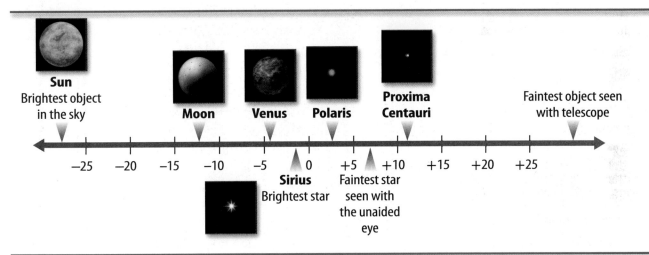

Sun Brightest object in the sky

Moon

Venus

Polaris

Proxima Centauri

Faintest object seen with telescope

−25   −20   −15   −10   −5   0   +5   +10   +15   +20   +25

Sirius Brightest star

Faintest star seen with the unaided eye

Absolute Magnitude

Stars can appear bright or dim depending on their distances from Earth. But stars also have actual, or absolute, magnitudes. **Luminosity** (lew muh NAH sih tee) *is the true brightness of an object.* The luminosity of a star, measured on an absolute magnitude scale, depends on the star's temperature and size, not its distance from Earth. A star's luminosity, apparent magnitude, and distance are related. If scientists know two of these factors, they can determine the third using mathematical formulas.

10. **NGSSS Check** Summarize How do scientists measure the brightness of stars? SC.8.E.5.5

Visual Summary

Greek astronomers identified dozens of star patterns and named those patterns after heroes and animals from Greek mythology.

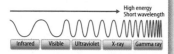

Different wavelengths of the electromagnetic spectrum carry different energies.

Astronomers measure distances within the solar system using astronomical units.

Inquiry SC.8.N.3.1

LAB STATION **Try It!**

Skill Lab *How can you use scientific illustrations to locate constellations?* at connectED.mcgraw-hill.com

Use Vocabulary

1. A device that spreads light into different wavelengths is a(n) _____ .

2. **Distinguish** between *apparent magnitude* and *luminosity*. SC.8.E.5.5

_____

_____

_____

Understand Key Concepts 🔑

3. Which does a light-year measure? SC.8.E.5.1
   - (A) brightness
   - (B) distance
   - (C) time
   - (D) wavelength

4. **Describe** how scientists divide the sky.

_____

_____

Interpret Graphics

5. **Organize Information** Fill in the graphic organizer below to list three things astronomers can learn from a star's light. SC.8.E.5.5

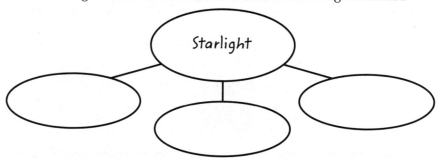

Starlight

Critical Thinking

6. **Evaluate** why astronomers use modern constellation regions instead of ancient constellation patterns to divide the sky.

_____

_____

Math Skills   MA.6.A.3.6

7. The Andromeda galaxy is about 25,000,000,000,000,000,000 km from Earth. How long does it take light to reach Earth from the Andromeda galaxy?

Explain facts associated with viewing the night sky.

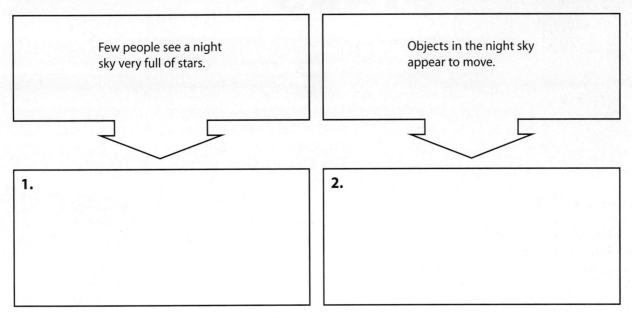

Characterize astronomy before the invention of the telescope.

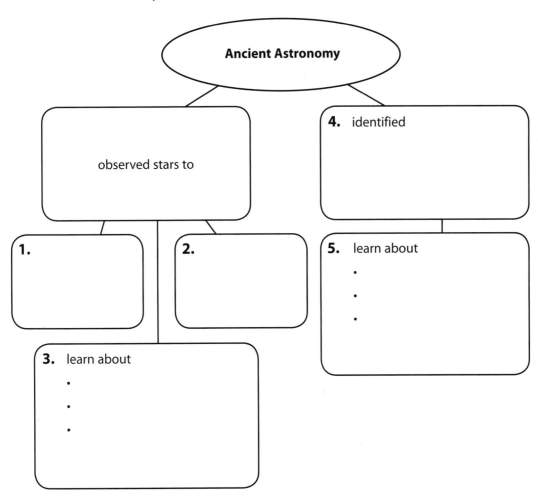

The Sun and Other STARS

ESSENTIAL QUESTIONS

How do stars shine?

How are stars layered?

How does the Sun change over short periods of time?

How do scientists classify stars?

Vocabulary

nuclear fusion p. 103

star p. 103

radiative zone p. 104

convection zone p. 104

photosphere p. 104

chromosphere p. 104

corona p. 104

Hertzsprung-Russell diagram p. 107

Florida NGSSS

LA.8.2.2.3 The student will organize information to show understanding or relationships among facts, ideas, and events (e.g., representing key points within text through charting, mapping, paraphrasing, summarizing, or comparing/contrasting);

LA.8.4.2.2 The student will record information (e.g., observations, notes, lists, charts, legends) related to a topic, including visual aids to organize and record information, as appropriate, and attribute sources of information;

SC.8.E.5.5 Describe and classify specific physical properties of stars: apparent magnitude (brightness), temperature (color), size, and luminosity (absolute brightness).

SC.8.E.5.6 Create models of solar properties including: rotation, structure of the Sun, convection, sunspots, solar flares, and prominences.

SC.8.N.1.1 Define a problem from the eighth grade curriculum using appropriate reference materials to support scientific understanding, plan and carry out scientific investigations of various types, such as systematic observations or experiments, identify variables, collect and organize data, interpret data in charts, tables, and graphics, analyze information, make predictions, and defend conclusions.

 SC.8.E.5.6

(Inquiry) Launch Lab

15 minutes

What are those spots on the Sun?

If you could see the Sun up close, what would it look like? Does it look the same all the time?

Procedure

1. Examine a **collage of Sun images**. Notice the dates on which the pictures were taken.

2. Discuss with a partner what the dark spots might be and why they change position.

3. Select one spot. Estimate how long it took the spot to move completely across the surface of the Sun. Record your estimate below.

Data and Observations

Think About This

1. What do you think the spots are?

2. Why do you think the spots move across the surface of the Sun

3. 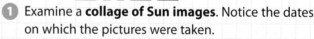 **Key Concept** How do you think the Sun changes over days, months, and years?

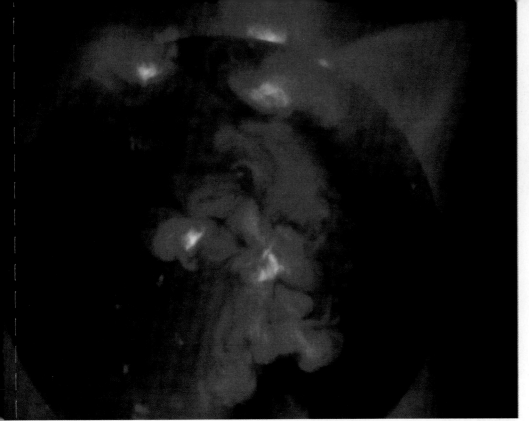

1. No, it's the Sun's atmosphere! The Sun's atmosphere can extend millions of kilometers into space. Solar storms have caused Earth to lose contact with orbiting satellites. How might this disruption impact people?

How Stars Shine

The hotter something is, the more quickly its atoms move. As atoms move, they collide. If a gas is hot enough and its atoms move quickly enough, the nuclei of some of the atoms stick together. **Nuclear fusion** *is a process that occurs when the nuclei of several atoms combine into one larger nucleus.*

Nuclear fusion releases a great amount of energy. This energy powers stars. A **star** *is a large ball of gas held together by gravity with a core so hot that nuclear fusion occurs.* A star's core can reach millions or hundreds of millions of degrees Celsius. When energy leaves a star's core, it travels throughout the star and radiates into space. As a result, the star shines.

 2. NGSSS Check **Summarize** How do stars shine? SC.8.E.5.5

Composition and Structure of Stars

The Sun is the closest star to Earth. Because it is so close, scientists easily can observe it. They can send probes to the Sun, and they can study its spectrum using spectroscopes on Earth-based telescopes. Spectra of the Sun and other stars provide information about **stellar** composition. The Sun and most stars are made almost entirely of hydrogen and helium gas. A star's composition changes slowly over time as hydrogen in its core fuses into more complex nuclei.

Active Reading

FOLDABLES LA.8.2.2.3

Make a vertical four-tab book. Label it as shown. Use it to organize your notes about the changing features of the Sun.

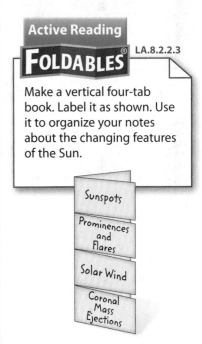

Sunspots

Prominences and Flares

Solar Wind

Coronal Mass Ejections

SCIENCE USE V. COMMON USE

stellar

Science Use anything related to stars

Common Use outstanding, exemplary

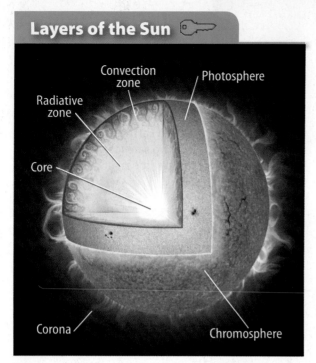

Layers of the Sun 🔑

- Convection zone
- Photosphere
- Radiative zone
- Core
- Corona
- Chromosphere

Figure 6 The Sun is divided into six layers.

✔ **3. Visual Check** **Describe** Where is the photosphere located in relation to the Sun's other layers?

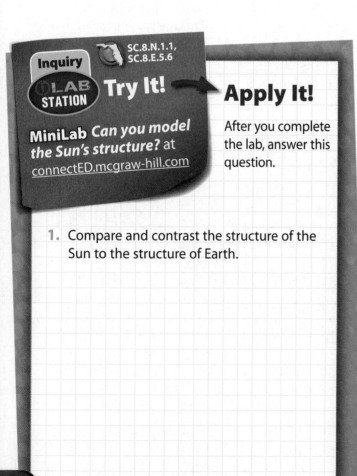

Inquiry
SC.8.N.1.1, SC.8.E.5.6

🔬**LAB STATION** **Try It!**

MiniLab *Can you model the Sun's structure?* at connectED.mcgraw-hill.com

Apply It!

After you complete the lab, answer this question.

1. Compare and contrast the structure of the Sun to the structure of Earth.

Interior of Stars

When first formed, all stars fuse hydrogen into helium in their cores. Helium is denser than hydrogen, so it sinks to the inner part of the core after it forms.

The core is one of three interior layers of a typical star, as shown in the drawing of the Sun in **Figure 6**. *The* **radiative zone** *is a shell of cooler hydrogen above a star's core.* Hydrogen in this layer is dense. Light energy bounces from atom to atom as it gradually makes its way upward, out of the radiative zone.

Above the radiative zone is the **convection zone**, *where hot gas moves up toward the surface and cooler gas moves deeper into the interior.* Light energy moves quickly upward in the convection zone.

 4. NGSSS Check **List** What are the interior layers of a star? SC.8.E.5.5, SC.8.E.5.6

Atmosphere of Stars

Beyond the convection zone are the three outer layers of a star. These layers make up a star's atmosphere. *The* **photosphere** *is the apparent surface of a star.* In the Sun, it is the dense, bright part you can see, where light energy radiates into space. From Earth, the Sun's photosphere looks smooth. But like the rest of the Sun, it is made of gas.

Above the photosphere are the two outer layers of a star's atmosphere. *The* **chromosphere** *is the orange-red layer above the photosphere,* as shown in **Figure 6**. *The* **corona** *is the wide, outermost layer of a star's atmosphere.* The temperature of the corona is higher than that of the photosphere or the chromosphere. It has an irregular shape and can extend for several million kilometers.

The Sun's Changing Features

The interior features of the Sun are stable over millions of years. But the Sun's atmosphere can change over years, months, or even minutes. Some of these features are illustrated in **Table 1** on the following page.

Table 1 The Sun is dynamic. It changes over years, months, hours, and minutes.

5. **NGSSS Check Explain** Which parts of the Sun's atmosphere change over short periods of time? SC.8.E.5.6

Table 1 Changing Features of the Sun

Sunspots
Regions of strong magnetic activity are called sunspots. Cooler than the rest of the photosphere, sunspots appear as dark splotches on the Sun. They seem to move across the Sun as the Sun rotates. The number of sunspots changes over time. They follow a cycle, peaking in number every 11 years. An average sunspot is about the size of Earth.

 6. **State** What causes sunspots?

_____

_____

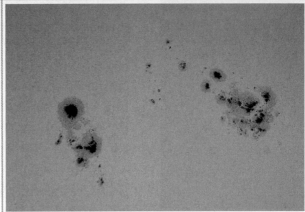

Prominences and Flares
The loop shown here is a prominence. Prominences are clouds of gas that make loops and jets extending into the corona. They can last for weeks. Flares are sudden increases in brightness often found near sunspots or prominences. They are violent eruptions that can last hours. Both features begin at or just above the photosphere.

 7. **Contrast** Differentiate between a flare and a prominence.

_____

_____

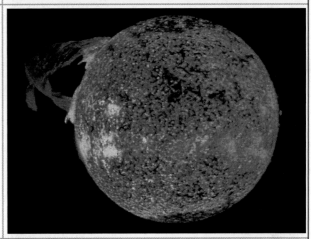

Coronal Mass Ejections (CMEs)
Huge bubbles of gas ejected from the corona are coronal mass ejections (CMEs). They are much larger than flares and occur over the course of several hours. Material from a CME can reach Earth, occasionally causing a radio blackout or a malfunction in an orbiting satellite.

The Solar Wind
Charged particles that stream continually away from the Sun create the solar wind. The solar wind passes Earth and extends to the edge of the solar system. The northern lights, or auroras, shown here, are curtains of light. They are created when particles from the solar wind or a CME interact with Earth's magnetic field.

Figure 7 Open clusters (top) contain fewer than 1,000 stars. Globular clusters (bottom) can contain hundreds of thousands of stars.

Active Reading 8. **Compare** How are stars in a star cluster related to each other?

_____

_____

_____

WORD ORIGIN

globular
from Latin *globus*, means "round mass, sphere"

Groups of Stars

The Sun has no stellar companion. The star closest to the Sun is 4.2 light-years away. Many stars are single stars, such as the Sun. But most stars exist in star systems bound by gravity.

The most common star system is a binary system, where two stars orbit each other. By studying the orbits of binary stars, astronomers can determine the stars' masses. Many stars exist in large groupings called clusters. Two types of star clusters—open clusters and **globular** clusters—are shown in **Figure 7**. Stars in a cluster all formed at about the same time and are the same distance from Earth. If astronomers determine the distance to or the age of one star in a cluster, they know the distance to or the age of every star in the cluster.

Classifying Stars

How do you classify a star? Which properties are important? Scientists classify stars according to their spectra. Recall that a star's spectrum is the light it emits spread out by wavelength. Stars have different spectra and different colors depending on their surface temperatures.

Temperature, Color, and Mass

Have you ever seen coals in a fire? Red coals are the coolest, and blue-white coals are the hottest. Stars are similar. Blue-white stars are hotter than red stars. Orange, yellow, and white stars are intermediate in temperature. Though there are exceptions, color in most stars is related to mass, as shown in **Figure 8**. Blue-white stars tend to have the most mass, followed by white stars, yellow stars, orange stars, and red stars.

Active Reading 9. **Explain** How does star color relate to mass?

_____

_____

As shown in **Figure 8**, the Sun is tiny compared to large, blue-white stars. However, scientists suspect that most stars—as many as 90 percent—are smaller than the Sun. These stars are called red dwarfs. The smallest star in **Figure 8** is a red dwarf.

Figure 8 The most massive stars are usually the hottest and are blue-white. The smallest stars tend to be cooler and red.

Sun

Hertzsprung-Russell Diagram 🗝

Figure 9 The H-R diagram plots luminosity against temperature. Most stars exist along the main sequence, the band that stretches from the upper left to the lower right.

Active Reading **10. Identify** (Circle) the location of the Sun on this diagram.

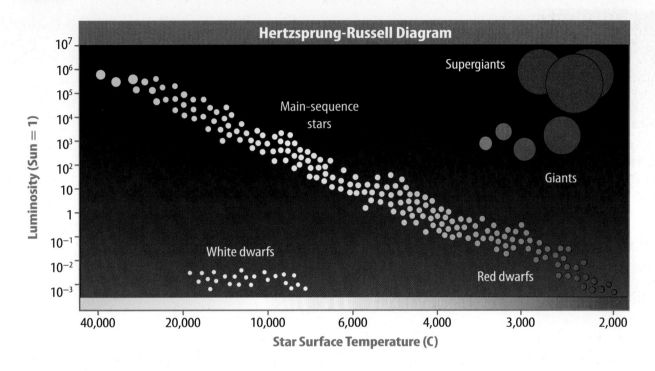

Hertzsprung-Russell Diagram

When scientists plot the temperatures of stars against their luminosities, the result is a graph like that shown in **Figure 9**. The **Hertzsprung-Russell diagram** (or H-R diagram) *is a graph that plots luminosity v. temperature of stars*. The *y*-axis of the H-R diagram displays increasing luminosity. The *x*-axis displays decreasing temperature.

The H-R diagram is named after two astronomers who developed it in the early 1900s. It is an important tool for categorizing stars. It also is an important tool for determining distances of some stars. If a star has the same temperature as a star on the H-R diagram, astronomers often can determine its luminosity. As you read earlier, if scientists know a star's luminosity, they can calculate its distance from Earth.

 11. NGSSS Check Cite Underline the purpose of the Hertzsprung-Russell diagram. SC.8.E.5.5

The Main Sequence

Most stars, including the stars shown in **Figure 8** on the previous page, exist along the main sequence. On the H-R diagram, main sequence stars form a line from the upper left corner to the lower right corner of the graph. The mass of a main-sequence star determines both its temperature and its luminosity. Because high-mass stars have more gravity pulling inward than low-mass stars, their cores have higher temperatures and produce more energy through fusion.

As shown in **Figure 9**, some groups of stars on the H-R diagram do not fit on the main sequence. Stars at the top right are cool yet luminous. This is because they are unusually large, not because they produce more energy. Massive stars are giants. The most massive are supergiants. The white dwarfs at the bottom of the H-R diagram are hot yet dim. This is because they are unusually small. You will read more about these stars in Lesson 3.

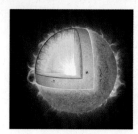

Hot gas moves up and cool gas moves down in the Sun's convection zone.

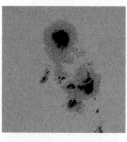

Sunspots are relatively dark areas on the Sun that have strong magnetic activity.

Globular clusters contain hundreds of thousands of stars.

Use Vocabulary

1 The _____ is a graph that plots luminosity v. temperature. SC.8.E.5.5

2 **Use the term** *photosphere* in a sentence. SC.8.E.5.6

_____

_____

3 **Define** *star* in your own words. SC.8.E.5.5

_____

_____

Understand Key Concepts

4 Which part of a star extends millions of kilometers into space? SC.8.E.5.5

   (A) chromosphere   (C) photosphere

   (B) corona   (D) radiative zone

5 **Explain** how stars produce and release energy. SC.8.E.5.5

_____

_____

_____

6 **Construct** an H-R diagram, and show the positions of the main sequence and the Sun. SC.8.E.5.5

Interpret Graphics

7 **Organize Information** Sequence the Sun's radiative zone, corona, convection zone, chromosphere, and photosphere in order outward from the core. SC.8.E.5.6

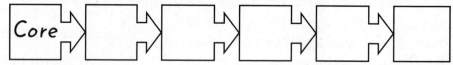

Core →

Critical Thinking

8 **Evaluate** In what way is the Sun an average star? In what way is it not an average star? SC.8.E.5.5

_____

_____

_____

Viewing the Sun in 3-D

NASA's Solar Terrestrial Relations Observatory

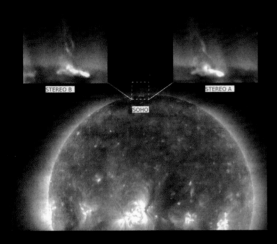

You might have used a telescope to look at objects far in the distance or to look at stars and planets. Although telescopes allow you to see a distant object in closer detail, you cannot see a three-dimensional view of objects in space. To get a three-dimensional view of the Sun, astronomers use two space telescopes. NASA's *Solar Terrestrial Relations Observatory* (STEREO) telescopes orbit the Sun in front of and behind Earth and give astronomers a 3-D view of the Sun. Why is this important?

If a coronal mass ejection (CME) erupts from the Sun, it can blast more than a billion tons of material into space. The powerful energy in a CME can damage satellites and power grids if Earth happens to be in its way. Before STEREO, scientists had only a straight-on view of CMEs approaching Earth. With STEREO, they have two different views. Each STEREO telescope carries several cameras that can detect many wavelengths. Scientists combine the pictures from each type of camera to make one 3-D image. In this way, they can track a CME from its emergence on the Sun all the way to its impact with Earth.

STEREO B is in orbit around the Sun behind Earth.

In January 2009, the telescopes were 90 degrees apart.

In February 2011, the craft will be 180 degrees apart.

Earth

STEREO A is in orbit around the Sun ahead of Earth.

It's Your Turn

RESEARCH AND REPORT How can power and satellite companies in Florida prepare for an approaching CME? Find out and write a short report on what you find. Share your findings with the class.

LA.8.4.2.2

Evolution of STARS

ESSENTIAL QUESTIONS

 How do stars form?

 How does a star's mass affect its evolution?

 How is star matter recycled in space?

Vocabulary

nebula p. 111

white dwarf p. 113

supernova p. 113

neutron star p. 114

black hole p. 114

Florida NGSSS

LA.8.2.2.3 The student will organize information to show understanding or relationships among facts, ideas, and events (e.g., representing key points within text through charting, mapping, paraphrasing, summarizing, or comparing/contrasting);

MA.6.A.3.6 Construct and analyze tables, graphs, and equations to describe linear functions and other simple relations using both common language and algebraic notation.

SC.8.E.5.4 Explore the Law of Universal Gravitation by explaining the role that gravity plays in the formation of planets, stars, and solar systems and in determining their motions.

SC.8.E.5.5 Describe and classify specific physical properties of stars: apparent magnitude (brightness), temperature (color), size, and luminosity (absolute brightness).

SC.8.N.1.1 Define a problem from the eighth grade curriculum using appropriate reference materials to support scientific understanding, plan and carry out scientific investigations of various types, such as systematic observations or experiments, identify variables, collect and organize data, interpret data in charts, tables, and graphics, analyze information, make predictions, and defend conclusions.

SC.8.N.3.1 Select models useful in relating the results of their own investigations.

 inquiry Launch Lab

SC.8.N.3.1

20 minutes

Do stars have life cycles?

You might have learned about the life cycles of plants or animals. Do stars, such as the Sun, have life cycles? Before you find out, review the life cycle of a sunflower.

Procedure

1. Read and complete a lab safety form.
2. Obtain an **envelope containing slips of paper** that explain the life cycle of a sunflower.
3. Use **colored pencils** to draw a sunflower in the middle of a piece of **paper**, or use a **glue stick** to glue a sunflower picture on the paper.
4. Using your knowledge of plant life cycles, arrange the slips of paper around the sunflower in the order in which the events listed on them occur. Draw arrows to show how the steps form a cycle.

Think About This

1. Does the life cycle of a sunflower have a beginning and an end? Explain your answer.

2. Do you think that every stage in the life cycle takes the same amount of time? Why or why not?

3. **Key Concept** How do you think the life cycle of a star compares to the life cycle of a sunflower? Do you think all stars have the same life cycle?

Inquiry Exploding Star?

1. No, this is a cloud of gas and dust where stars form. How do you think stars form? Do you think stars ever stop shining?

Life Cycle of a Star

Like living things, stars have life cycles. They are "born," and after millions or billions of years, they "die." Stars die in different ways, depending on their masses. But all stars—from white dwarfs to supergiants—form in the same way.

Nebulae and Protostars

Stars form deep inside clouds of gas and dust. *A cloud of gas and dust is a* **nebula** (plural, nebulae). Star-forming nebulae are cold, dense, and dark. Gravity causes the densest parts to collapse, forming regions called protostars. Protostars continue to contract, pulling in surrounding gas, until their cores are hot and dense enough for nuclear fusion to begin. As they contract, protostars produce enormous amounts of thermal energy.

Birth of a Star

Over many thousands of years, the energy produced by protostars heats the gas and dust surrounding them. Eventually, the surrounding gas and dust blows away, and the protostars become visible as stars. Some of this material might later become planets or other objects that orbit the star. During the star-formation process, nebulae glow brightly, as shown in the photograph on the previous page.

2. NGSSS Check **Describe** How do stars form? SC.8.E.5.4

WORD ORIGIN

nebula
from Latin *nebula*, means "mist" or "little cloud"

Active Reading

FOLDABLES LA.8.2.2.3

Make a vertical five-tab book. Label it as shown. Use it to organize your notes on the life cycle of a star.

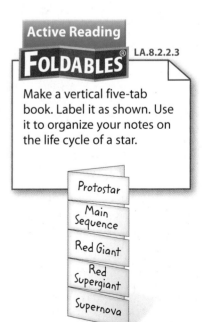

Protostar

Main Sequence

Red Giant

Red Supergiant

Supernova

Main-Sequence Stars

Recall the main sequence of the Hertzsprung-Russell diagram. Stars spend most of their lives on the main sequence. A star becomes a main-sequence star as soon as it begins to fuse hydrogen into helium. It remains on the main sequence for as long as it continues to fuse hydrogen into helium. Lower-mass stars such as the Sun stay on the main sequence for billions of years. High-mass stars stay on the main sequence for only a few million years. Even though massive stars have more hydrogen than lower-mass stars, they process it at a much faster rate.

When a star's hydrogen supply is nearly gone, the star leaves the main sequence. It begins the next stage of its life cycle, as shown in **Figure 10**. Not all stars go through all phases in **Figure 10**. Lower-mass stars, such as the Sun, do not have enough mass to become supergiants.

Figure 10 Massive stars become red giants, then larger red giants, then red supergiants.

✓ 3. **Visual Check**
Identify Which element forms in only the most massive stars?

A Massive Star's Life Cycle 🔑

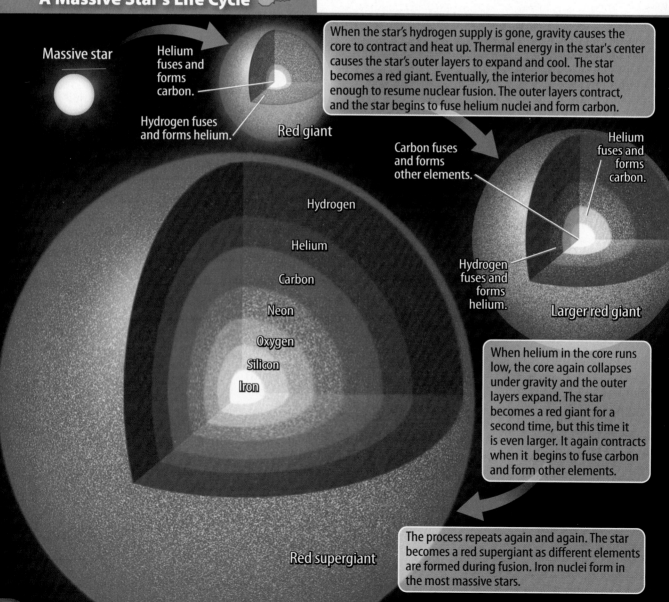

Massive star

Helium fuses and forms carbon.

Hydrogen fuses and forms helium.

Red giant

When the star's hydrogen supply is gone, gravity causes the core to contract and heat up. Thermal energy in the star's center causes the star's outer layers to expand and cool. The star becomes a red giant. Eventually, the interior becomes hot enough to resume nuclear fusion. The outer layers contract, and the star begins to fuse helium nuclei and form carbon.

Carbon fuses and forms other elements.

Helium fuses and forms carbon.

Hydrogen fuses and forms helium.

Larger red giant

Hydrogen
Helium
Carbon
Neon
Oxygen
Silicon
Iron

When helium in the core runs low, the core again collapses under gravity and the outer layers expand. The star becomes a red giant for a second time, but this time it is even larger. It again contracts when it begins to fuse carbon and form other elements.

Red supergiant

The process repeats again and again. The star becomes a red supergiant as different elements are formed during fusion. Iron nuclei form in the most massive stars.

End of a Star

All stars form in the same way. But stars die in different ways, depending on their masses. Massive stars collapse and explode. Lower-mass stars die more slowly.

White Dwarfs

Lower-mass stars, such as the Sun, do not have enough mass to fuse elements beyond helium. They do not get hot enough. After helium in their cores is gone, the stars cast off their gases, exposing their cores. The core becomes a **white dwarf**, *a hot, dense, slowly cooling sphere of carbon.*

What will happen to Earth and the solar system when the Sun runs out of fuel? When the Sun runs out of hydrogen, in about 5 billion years, it will become a red giant. Once helium fusion begins, the Sun will contract. When the helium is gone, the Sun will expand again, probably absorbing Mercury, Venus, and Earth and pushing Mars outward, as shown in **Figure 11.** Eventually, the Sun will become a white dwarf. Imagine the mass of the Sun squeezed a million times until it is the size of Earth. That's the size of a white dwarf. Scientists expect that all stars with masses less than 8–10 times that of the Sun will eventually become white dwarfs.

Active Reading 5. **Explain** <u>Underline</u> what will happen to Earth when the Sun runs out of fuel?

Supernovae

Stars with more than 10 times the mass of the Sun do not become white dwarfs. Instead, they explode. *A* **supernova** *(plural, supernovae) is an enormous explosion that destroys a star.* In the most massive stars, a supernova occurs when iron forms in the star's core. Iron is stable and does not fuse. After a star forms iron, it loses its internal energy source, and the core collapses quickly under the force of gravity. So much energy is released that the star explodes. When it explodes, a star can become one billion times brighter and form elements even heavier than iron.

Figure 11 In about 5 billion years, the Sun will become a red giant and then a white dwarf.

Active Reading 4. **Compose** Write captions under each image to describe what happens as the Sun becomes a white dwarf.

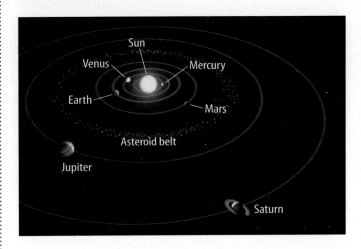

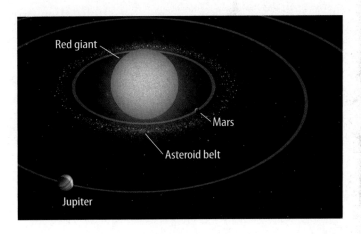

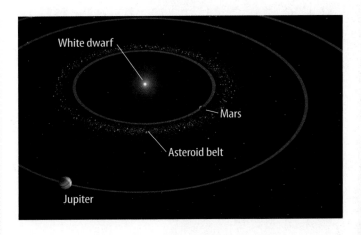

Neutron Stars

Have you ever eaten cotton candy? A bag of cotton candy is made from just a few spoonfuls of spun sugar. Cotton candy is mostly air. Similarly, atoms are mostly empty space. During a supernova, the collapse is so violent that the normal spaces inside atoms are eliminated, and a neutron star forms. *A* **neutron star** *is a dense core of neutrons that remains after a supernova.* Neutron stars are only about 20 km wide, with cores so dense that a teaspoonful would weigh more than 1 billion tons.

Black Holes

For the most massive stars, atomic forces holding neutrons together are not strong enough to overcome so much mass in such a small volume. Gravity is too strong, and the matter crushes into a black hole. *A* **black hole** *is an object whose gravity is so great that no light can escape.*

A black hole does not suck matter in like a vacuum cleaner. But its gravity is very strong because all of its mass is concentrated in a single point. Because astronomers cannot see a black hole, they only can infer its existence. For example, if they detect a star circling around something, but they cannot see what that something is, they suspect it is a black hole.

 6. NGSSS Check Summarize How does a star's mass determine if it will become a white dwarf, a neutron star, or a black hole? SC.8.E.5.4

_____

_____

_____

_____

Inquiry SC.8.N.3.1

LAB STATION **Try It!**

MiniLab *How do astronomers detect black holes?* at connectED.mcgraw-hill.com

Apply It! After you complete the lab, answer these questions.

1. What did the marble represent?

2. How could a space probe move safely past a black hole?

Recycling Matter

At the end of a star's life cycle, much of its gas escapes into space. This gas is recycled. It becomes the building blocks of future generations of stars and planets.

Planetary Nebulae

You read that lower-mass stars, such as the Sun, become white dwarfs. When a star becomes a white dwarf, it casts off hydrogen and helium gases in its outer layers, as shown in **Figure 12**. The expanding, cast-off matter of a white dwarf is a planetary nebula. Most of the star's carbon remains locked in the white dwarf. But the gases in the planetary nebula can be used to form new stars.

Planetary nebulae have nothing to do with planets. They are so named because early astronomers thought they were regions where planets were forming.

Supernova Remnants

During a supernova, a massive star comes apart. This sends a shock wave into space. The expanding cloud of dust and gas is called a supernova remnant. A supernova remnant is shown in **Figure 13.** Like a snowplow pushing snow in its path, a supernova remnant pushes on the gas and dust it encounters.

In a supernova, a star releases the elements that formed inside it during nuclear fusion. Almost all of the elements in the universe other than hydrogen and helium were created by nuclear reactions inside the cores of massive stars and released in supernovae. This includes the oxygen in air, the silicon in rocks, and the carbon in you.

8. **NGSSS Check Summarize** Underline how stars recycle matter. **SC.8.E.5.5**

Gravity causes recycled gases and other matter to clump together in nebulae and form new stars and planets. As you will read in the next lesson, gravity also causes stars to clump together into even larger structures called galaxies.

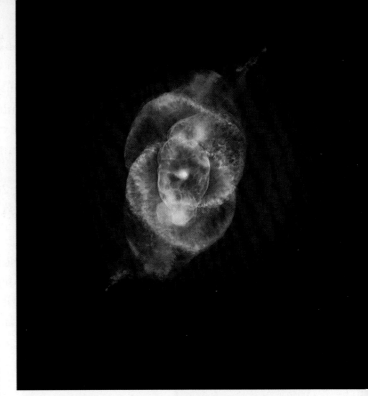

Figure 12 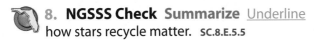 White dwarfs cast off helium and hydrogen as planetary nebulae.

7. **Visual Check Locate** Which part of the image represents a planetary nebula?

Figure 13 Many of the elements in you and in matter all around you were formed inside massive stars and released in supernovae.

Visual Summary

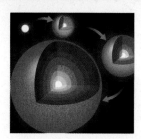

Iron is formed in the cores of the most massive stars.

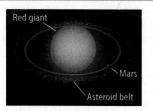

The Sun will become a red giant in about 5 billion years.

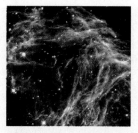

Matter is recycled in supernovae.

Use Vocabulary

1 Planetary nebulae are the expanding outer layers of a(n)

_____. SC.8.E.5.4

2 **Define** *supernova* in your own words.

_____

3 **Use the terms** *neutron star* and *black hole* in a sentence. SC.8.E.5.5

_____

Understand Key Concepts 🔑

4 Which type of star will the Sun eventually become? SC.8.E.5.5

(A) neutron star   (C) red supergiant

(B) red dwarf   (D) white dwarf

5 **Explain** how supernovae recycle matter. SC.8.E.5.5

_____

_____

6 **Rank** black holes, neutron stars, and white dwarfs from smallest to largest. Then rank them from most massive to least massive. SC.8.E.5.5

_____

_____

_____

Interpret Graphics

7 **Organize Information** Fill in the graphic organizer below to list what happens to a star following a supernova. LA.8.2.2.3

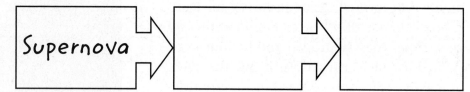

Supernova →   →

Critical Thinking

8 **Evaluate** why mass is so important in determining the evolution of a star. SC.8.E.5.4

_____

_____

_____

_____

Sequence what will happen to the solar system when the Sun runs out of fuel.

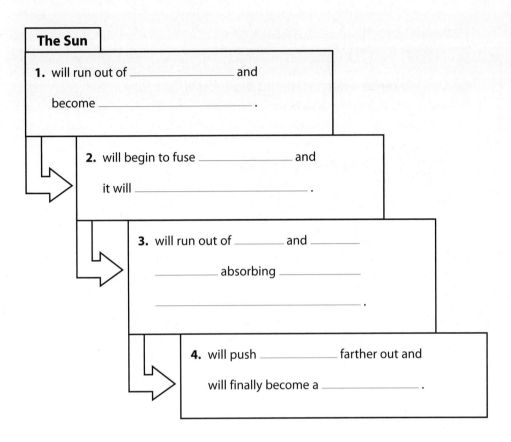

The Sun

1. will run out of _____ and

   become _____ .

2. will begin to fuse _____ and

   it will _____ .

3. will run out of _____ and _____

   _____ absorbing _____

   _____ .

4. will push _____ farther out and

   will finally become a _____ .

Sequence the change of a nebula to a visible star.

5. A nebula begins as a _____ , and _____ cloud of _____

   and _____ .

6. _____ causes the _____ parts to _____ forming

   _____ .

7. A protostar contracts until _____

   _____ .

8. The _____ around the protostar _____ and the protostar becomes

   _____ .

Galaxies and the UNIVERSE

ESSENTIAL QUESTIONS

 What are the major types of galaxies?

 What is the Milky Way, and how is it related to the solar system?

 What is the Big Bang theory?

Vocabulary

galaxy p. 119

dark matter p. 119

Big Bang theory p. 124

Doppler shift p. 124

Florida NGSSS

LA.8.2.2.3 The student will organize information to show understanding or relationships among facts, ideas, and events (e.g., representing key points within text through charting, mapping, paraphrasing, summarizing, or comparing/contrasting);

SC.8.E.5.1 Recognize that there are enormous distances between objects in space and apply our knowledge of light and space travel to understand this distance.

SC.8.E.5.2 Recognize that the universe contains many billions of galaxies and that each galaxy contains many billions of stars.

SC.8.N.1.1 Define a problem from the eighth grade curriculum using appropriate reference materials to support scientific understanding, plan and carry out scientific investigations of various types, such as systematic observations or experiments, identify variables, collect and organize data, interpret data in charts, tables, and graphics, analyze information, make predictions, and defend conclusions.

---

20 minutes

Inquiry Launch Lab

Does the universe move?

Procedure

1. Read and complete a lab safety form.

2. Use a **marker** to make three dots 5–7 cm apart on one side of a **large round balloon**. Label the dots *A*, *B*, and *C*. The dots represent galaxies.

3. Blow up the balloon to a diameter of about 8 cm. Hold the balloon closed as your partner uses a **measuring tape** to measure the distance between each galaxy on the balloon's surface. Record the distances.

4. Repeat step 3 two more times, blowing up the balloon a little more each time. Record your data.

Data and Observations

Balloon size	A–B (cm)	B–C (cm)	A–C (cm)
Small			
Medium			
Large			

Think About This

1. What happened to the distances between galaxies as the balloon expanded?

2. If you were standing in one of the galaxies, what would you observe about the other galaxies as the balloon expanded?

3. **Key Concept** If the balloon were a model of the universe, what do you think might have caused galaxies to move in this way?

Inquiry **Disk in Space?**

1. Yes, this is the disk of a galaxy—a huge collection of stars. You see this galaxy on its edge. If you were to look down on it from above, it would look like a two-armed spiral. Do you think all galaxies are shaped like spirals? What about the galaxy we live in?

Galaxies

Most people live in towns or cities where houses are close together. Not many houses are found in the wilderness. Similarly, most stars exist in galaxies. **Galaxies** *are huge collections of stars.* The universe contains hundreds of billions of galaxies, and each galaxy can contain hundreds of billions of stars.

Active Reading

2. Define What are galaxies?

Dark Matter

Gravity holds stars together. Gravity also holds galaxies together. When astronomers examine how galaxies, such as those in **Figure 14**, rotate and gravitationally interact, they find that most of the matter in galaxies is invisible. *Matter that emits no light at any wavelength is* **dark matter**. Scientists hypothesize that more than 90 percent of the universe's mass is dark matter. Scientists do not fully understand dark matter. They do not know what material it contains.

Figure 14 By examining interacting galaxies such as these, astronomers hypothesize that most mass in the universe is invisible dark matter.

Types of Galaxies

There are three major types of galaxies: spiral, elliptical, and irregular. **Table 2** gives a brief description of each type.

 3. NGSSS Check Name What are the major types of galaxies? SC.8.E.5.2

_____

_____

Table 2 Types of Galaxies

Spiral Galaxies

The stars, gas, and dust in a spiral galaxy exist in spiral arms that begin at a central disk. Some spiral arms are long and symmetrical; others are short and stubby. Spiral galaxies are thicker near the center, a region called the central bulge. A spherical halo of globular clusters and older, redder stars surrounds the disk. NGC 5679, shown here, contains a pair of spiral galaxies.

Active Reading 4. Describe What is the structure of a spiral galaxy?

_____

_____

Elliptical Galaxies

Unlike spiral galaxies, elliptical galaxies do not have internal structure. Some are spheres, like basketballs, while others resemble footballs. Elliptical galaxies have higher percentages of old, red stars than spiral galaxies do. They contain little or no gas and dust. Scientists suspect that many elliptical galaxies form by the gravitational merging of two or more spiral galaxies. The elliptical galaxy pictured here is NGC 5982, part of the Draco Group.

Active Reading 5. Summarize What type of stars do elliptical galaxies contain?

_____

Irregular Galaxies

Irregular galaxies are oddly shaped. Many form from the gravitational pull of neighboring galaxies. Irregular galaxies contain many young stars and have areas of intense star formation. Shown here is the irregular galaxy NGC 1427A.

Active Reading 6. Compare How is star formation activity different in irregular galaxies than the other galaxies?

_____

_____

Inquiry

LAB STATION SC.8.N.1.1, SC.8.E.5.2

Try It!

MiniLab *Can you identify a galaxy?* at connectED.mcgraw-hill.com

Apply It! After you complete the lab, answer these questions.

1. How did the structure of the galaxy affect your observations?

2. How can gravity affect the shape of galaxies?

Groups of Galaxies

Galaxies are not distributed evenly in the universe. Gravity holds them together in groups called clusters. Some clusters of galaxies are enormous. The Virgo Cluster is 60 million light-years from Earth. It contains about 2,000 galaxies. Most clusters exist in even larger structures called superclusters. Between superclusters are voids, which are regions of nearly empty space. Scientists hypothesize that the large-scale structure of the universe resembles a sponge.

Active Reading

7. **Identify** What holds clusters of galaxies together?

_____

_____

The Milky Way

The solar system is in the Milky Way, a spiral galaxy that contains gas, dust, and almost 200 billion stars. The Milky Way is a member of the Local Group, a cluster of about 30 galaxies. Scientists expect the Milky Way will begin to merge with the Andromeda Galaxy, the largest galaxy in the Local Group, in about 3 billion years. Because stars are far apart in galaxies, it is not likely that many stars will actually collide during this event.

Where is Earth in the Milky Way? **Figure 15** on the next two pages shows an artist's drawing of the Milky Way and Earth's place in it.

Active Reading

FOLDABLES LA.8.2.2.3

Make a horizontal single-tab matchbook. Label it as shown. Use it to describe the contents of the Milky Way.

Milky Way

Galaxy

WORD ORIGIN

galaxy
from Greek *galactos*, means "milk"

The Milky Way

Figure 15 The Milky Way is shown here in two separate views, from the top (left page) and on edge (right page). Because Earth is located inside the disk of the Milky Way, people cannot see beyond the central bulge to the other side.

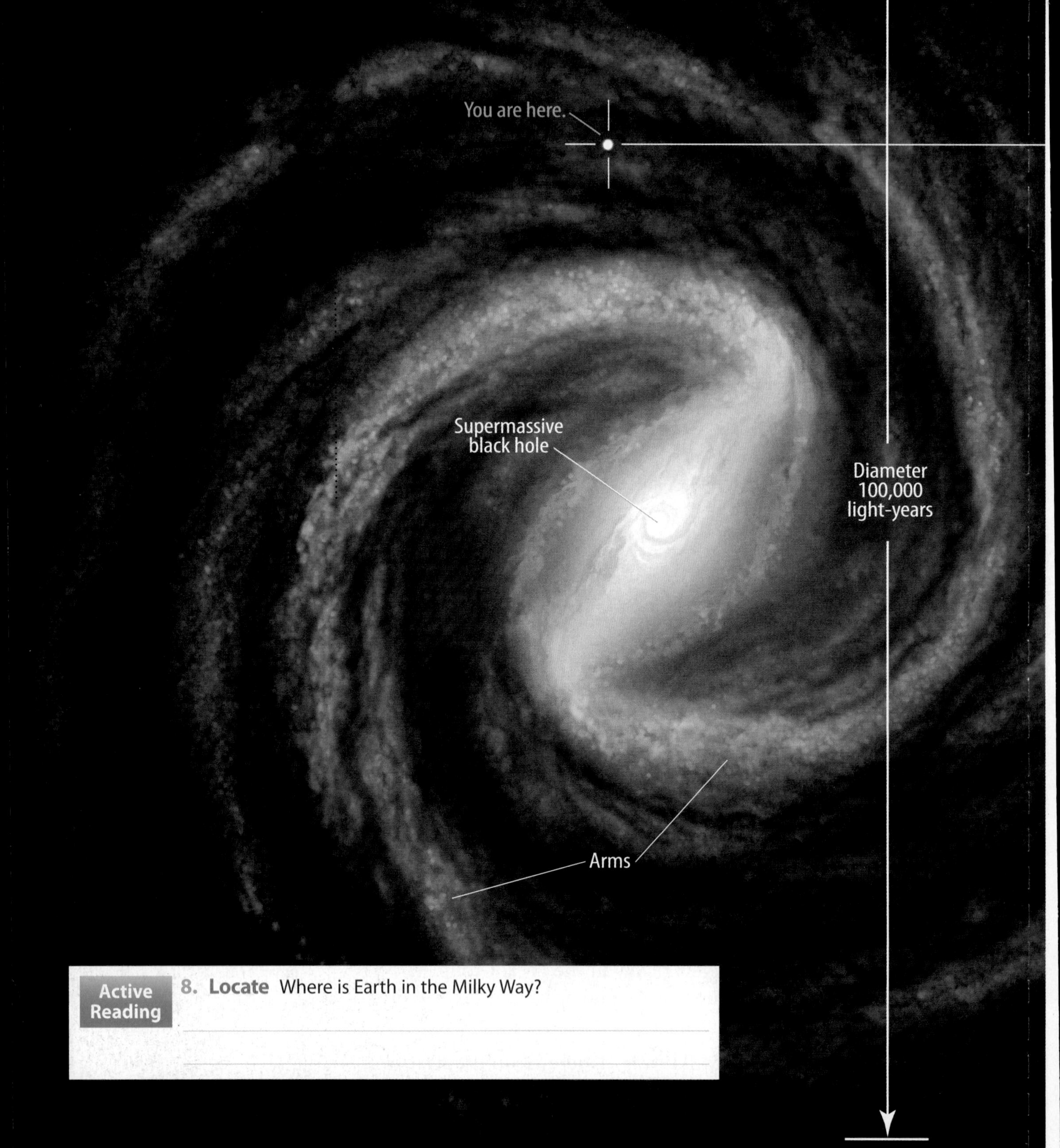

You are here.
Supermassive black hole
Diameter 100,000 light-years
Arms

Active Reading **8. Locate** Where is Earth in the Milky Way?

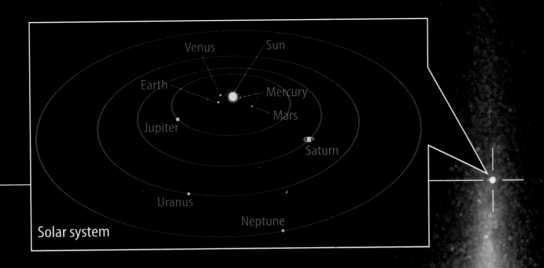

Solar system

Venus
Sun
Earth
Mercury
Jupiter
Mars
Saturn
Uranus
Neptune

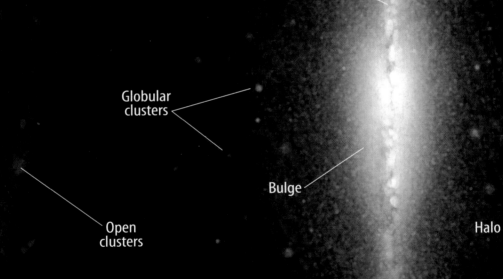

Disk

Globular
clusters

Open
clusters

Bulge

Halo

Arms

The Big Bang Theory

When astronomers look into space, they look back in time. Is there a beginning to time? According to the **Big Bang theory**, *the universe began from one point billions of years ago and has been expanding ever since.*

Origin and Expansion of the Universe

Most scientists agree that the universe is 13–14 billion years old. When the universe began, it was dense and hot—so hot that even atoms didn't exist. After a few hundred thousand years, the universe cooled enough for atoms to form. Eventually, stars formed, and gravity pulled them into galaxies.

As the universe expands, space stretches and galaxies move away from one another. The same thing happens in a loaf of unbaked raisin bread. As the dough rises, the raisins move apart. Scientists observe how space stretches by measuring the speed at which galaxies move away from Earth. As the galaxies move away, their wavelengths lengthen and stretch out. How does light stretch?

Doppler Shift

You have probably heard the siren of a speeding police car. As **Figure 16** illustrates, when the car moves toward you, the sound waves compress. As the car moves away, the sound waves spread out. Similarly, when visible light travels toward you, its wavelength compresses. When light travels away from you, its wavelength stretches out. It shifts to the red end of the electromagnetic spectrum. *The shift to a different wavelength is called the* **Doppler shift**. Because the universe is expanding, light from galaxies is red-shifted. The more distant a galaxy is, the faster it moves away from Earth, and the more it is red-shifted.

Dark Energy

Will the universe expand forever? Or will gravity cause the universe to contract? Scientists have observed that galaxies are moving away from Earth faster over time. To explain this, they suggest a force called dark energy is pushing the galaxies apart.

Dark energy, like dark matter, is an active area of research. There is still much to learn about the universe and all it contains.

Active Reading **9. Restate** What is the Big Bang theory?

_____

_____

Doppler Shift

Figure 16 The sound waves from an approaching police car are compressed. As the car speeds away, the sound waves are stretched out. Similarly, when an object is moving away, its light is stretched out. The light's wavelength shifts toward a longer wavelength.

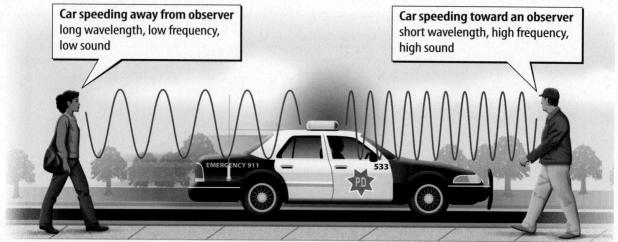

Car speeding away from observer
long wavelength, low frequency, low sound

Car speeding toward an observer
short wavelength, high frequency, high sound

Visual Summary

By studying interacting galaxies, scientists have determined that most mass in the universe is dark matter.

The Sun is one of billions of stars in the Milky Way.

When an object moves away, its light stretches out, just as a siren's sound waves stretch out as the siren moves away.

Inquiry SC.8.N.1.1, SC.8.E.5.1

iLAB STATION Try It!

Inquiry Lab *Describe a Trip Through Space* at connectED.mcgraw-hill.com

Use Vocabulary

1 Stars exist in huge collections called _____ .
SC.8.E.5.2

2 **Use the term** *dark matter* in a sentence.
_____

3 **Define** the *Big Bang theory*.
_____

Understand Key Concepts

4 Which is NOT a major galaxy type? SC.8.E.5.2
  (A) dark   (C) irregular
  (B) elliptical   (D) spiral

5 **Identify** Sketch the Milky Way, and identify the location of the solar system. SC.8.E.5.1

6 **Explain** how scientists know the universe is expanding.
_____
_____

Interpret Graphics

7 **Organize Information** Fill in the graphic organizer below. List the three major types of galaxies and characteristics of each. SC.8.E.5.2

Galaxy Type	Characteristics

Critical Thinking

8 **Predict** what the solar system and the universe might be like in 10 billion years. SC.8.E.5.2
_____
_____

Think About It! The universe is made up of stars, gas, and dust, as well as invisible dark matter. Material in the universe is pulled by gravity into galaxies, including our own Milky Way galaxy.

Key Concepts Summary

Vocabulary

LESSON 1 The View from Earth

- The sky is divided into 88 constellations.
- Astronomers learn about the energy, distance, temperature, and composition of stars by studying their light.
- Astronomers measure distances in space in **astronomical units** and in **light-years**. They measure star brightness as **apparent magnitude** and as **luminosity**.

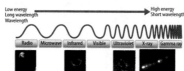

spectroscope p. 97
astronomical unit p. 98
light-year p. 98
apparent magnitude p. 99
luminosity p. 99

LESSON 2 The Sun and Other Stars

- **Stars** shine because of **nuclear fusion** in their cores.
- Stars have a layered structure—they conduct energy through their **radiative zones** and their **convection zones** and release the energy at their **photospheres**.
- Sunspots, prominences, flares, and coronal mass ejections are temporary phenomena on the Sun.
- Astronomers classify stars by their temperatures and luminosities.

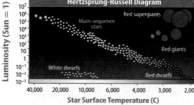

nuclear fusion p. 103
star p. 103
radiative zone p. 104
convection zone p. 104
photosphere p. 104
chromosphere p. 104
corona p. 104
Hertzsprung-Russell diagram p. 107

LESSON 3 Evolution of Stars

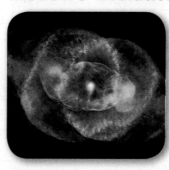

- Stars are born in clouds of gas and dust called **nebulae**.
- What happens to a star when it leaves the main sequence depends on its mass.
- Matter is recycled in the planetary nebulae of **white dwarfs** and the remnants of **supernovae**.

nebula p. 111
white dwarf p. 113
supernova p. 113
neutron star p. 114
black hole p. 114

LESSON 4 Galaxies and the Universe

- The three major types of **galaxies** are spiral, elliptical, and irregular.
- The Milky Way is the spiral galaxy that contains the solar system.
- The **Big Bang theory** explains the origin of the universe.

galaxy p. 119
dark matter p. 119
Big Bang theory p. 124
Doppler shift p. 124

FOLDABLES® **Chapter Project**

Assemble your lesson Foldables as shown to make a Chapter Project. Use the project to review what you have learned in this chapter.

Use Vocabulary

1 **Explain** how nebulae are related to stars.

2 **Define** *Doppler shift*.

3 **Compare** neutron stars and black holes.

4 **Explain** the role of white dwarfs in recycling matter.

5 **Distinguish** between an astronomical unit and a light-year.

6 **Explain** how a convection zone transfers energy.

7 **Use the term** *dark matter* in a sentence.

8 On what diagram would you find a plot of stellar luminosity v. temperature?

9 **Compare** photosphere and corona.

Link Vocabulary and Key Concepts

Use vocabulary terms from the previous page to complete the concept map.

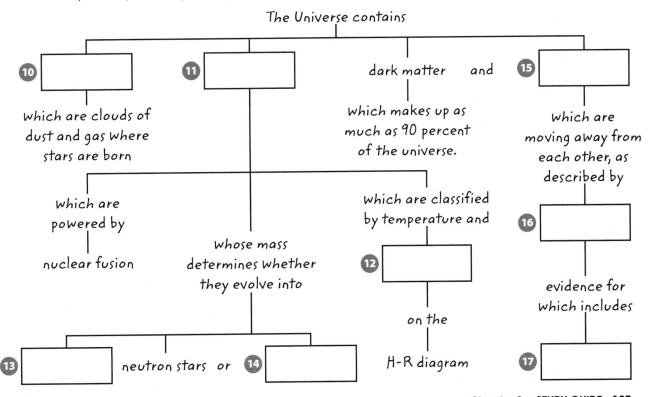

Fill in the correct answer choice.

🔑 Understand Key Concepts

1 Scientists divide the sky into SC.8.E.5.1
- (A) astronomical units.
- (B) clusters.
- (C) constellations.
- (D) light-years.

2 Which part of the Sun is marked with an *X* on the diagram below? SC.8.E.5.6

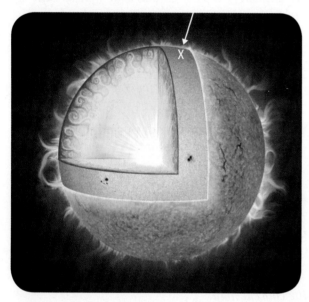

- (A) convection zone
- (B) corona
- (C) photosphere
- (D) radiative zone

3 Which might change, depending on the distance to a star? SC.8.E.5.5
- (A) absolute magnitude
- (B) apparent magnitude
- (C) composition
- (D) luminosity

4 Which is the average distance between Earth and the Sun? SC.8.E.5.1
- (A) 1 AU
- (B) 1 km
- (C) 1 light-year
- (D) 1 magnitude

Critical Thinking

5 **Explain** how energy is released in a star. SC.8.E.5.5

_____

_____

_____

_____

6 **Assess** how the invention of the telescope changed people's views of the universe. SC.8.E.5.1

_____

_____

_____

_____

7 **Imagine** you are asked to classify 10,000 stars. Which properties would you measure? SC.8.E.5.5

_____

_____

_____

8 **Deduce** why supernovae are needed for life on Earth. SC.8.E.5.5

_____

_____

_____

9 **Predict** how the Sun would be different if it were twice as massive. SC.8.E.5.5

_____

_____

_____

10 **Imagine** that you are writing to a friend who lives in the Virgo Cluster of galaxies. What would you write as your return address? Be specific. SC.8.E.5.1

_____

_____

_____

_____

11 **Interpret** The figure below shows part of the solar system. Explain what is happening. SC.8.E.5.4

_____

_____

_____

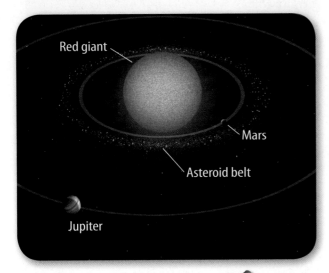

Red giant

Mars

Asteroid belt

Jupiter

Writing in Science

12 **Write** You are a scientist being interviewed by a magazine on the topic of black holes. On a separate sheet of paper write three questions an interviewer might ask, as well as your answers. SC.8.E.5.3

Big Idea Review

13 What makes up the universe, and how does gravity affect the universe? SC.8.E.5.4

_____

_____

_____

_____

Math Skills   MA.6.A.3.6

Use Proportions

14 The Milky Way galaxy is about 100,000 light-years across. What is this distance in kilometers?

15 Astronomers sometimes use a distance unit called a parsec. One parsec is 3.3 light-years. What is the distance, in parsecs, of a nebula that is 82.5 light-years away?

16 The distance to the Orion nebula is about 390 parsecs. What is this distance in light-years?

Fill in the correct answer choice.

Multiple Choice

1 Which characteristics can by studied by analyzing a star's spectrum? SC.8.E.5.5

Ⓐ absolute and apparent magnitudes

Ⓑ formation and evolution

Ⓒ movement and luminosity

Ⓓ temperature and composition

2 Which feature of the Sun appears in cycles of about 11 years? SC.8.E.5.6

Ⓕ coronal mass ejections

Ⓖ solar flares

Ⓗ solar wind

Ⓘ sunspots

Use the graph below to answer question 3.

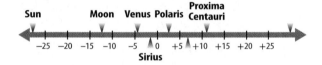

3 Which star on the graph has the greatest apparent magnitude? SC.8.E.5.5

Ⓐ Polaris

Ⓑ Proxima Centauri

Ⓒ Sirius

Ⓓ the Sun

4 What do astronomical units and light-years have in common? SC.8.E.5.1

Ⓕ Both measure the speed of light.

Ⓖ Both measure the speed of spacecraft.

Ⓗ Both measure the speed of galaxies.

Ⓘ Both measure the distance between objects in space.

Use the figure below to answer question 5.

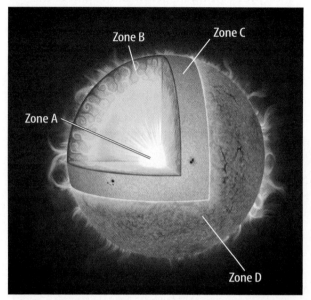

5 Which zone contains hot gas moving up toward the surface and cooler gas moving down toward the center of the Sun? SC.8.E.5.6

Ⓐ zone A

Ⓑ zone B

Ⓒ zone C

Ⓓ zone D

6 Which contains most of the mass of the universe? SC.8.E.5.5

Ⓕ black holes

Ⓖ dark matter

Ⓗ gas and dust

Ⓘ stars

7 Which stellar objects eventually form from the most massive stars? SC.8.E.5.5

Ⓐ black holes

Ⓑ diffuse nebulae

Ⓒ planetary nebulae

Ⓓ white dwarfs

Use the figure below to answer question 8.

8 Which is a characteristic for this type of galaxy? SC.8.E.5.2

   Ⓕ It contains no dust.

   Ⓖ It contains little gas.

   Ⓗ It contains many young stars.

   Ⓘ It contains mostly old stars.

9 Where do stars form? SC.8.E.5.5

   Ⓐ in black holes

   Ⓑ in constellations

   Ⓒ in nebulae

   Ⓓ in supernovae

10 What term describes the process that causes a star to shine? SC.8.E.5.6

   Ⓕ binary fission

   Ⓖ coronal mass ejection

   Ⓗ nuclear fusion

   Ⓘ stellar composition

11 What property of a star can be determined by its color? SC.8.E.5.5

   Ⓐ mass

   Ⓑ size

   Ⓒ distance from Earth

   Ⓓ surface temperature

Use the diagram below to answer question 12.

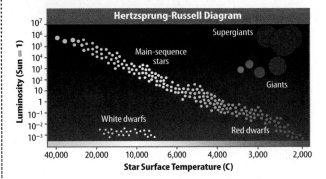

12 How is a red giant different from most main-sequence stars? SC.8.E.5.5

   Ⓕ It has a hotter surface temperature and less luminosity.

   Ⓖ It has a cooler surface temperature and greater luminosity.

   Ⓗ It has the same absolute magnitude and cooler surface temperature.

   Ⓘ It has the same absolute magnitude and less luminosity.

13 The Sun belongs to which class of star? SC.8.E.5.6

   Ⓐ black hole

   Ⓑ red giant

   Ⓒ white dwarf

   Ⓓ main-sequence

NEED EXTRA HELP?

If You Missed Question...	1	2	3	4	5	6	7	8	9	10	11	12	13
Go to Lesson...	1	2	1	1	2	4	3	4	3	2	2	2	2

Benchmark Mini-Assessment Chapter 3 • Lesson 1

mini BAT

Multiple Choice *Bubble the correct answer.*

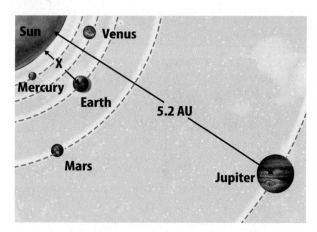

1. In the image above, which is X? SC.8.E.5.1

Ⓐ 1 AU

Ⓑ 1 km

Ⓒ 1 light-year

Ⓓ 1 year

2. Which helps scientists learn about the temperature and composition of stars? SC.8.E.5.5

Ⓕ apparent magnitude

Ⓖ astronomical unit

Ⓗ spectroscope

Ⓘ telescope

Use the image below to answer questions 3 and 4.

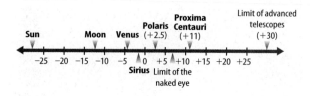

3. Based on the scale, which object has the lowest apparent magnitude? SC.8.E.5.5

Ⓐ Moon

Ⓑ Sirius

Ⓒ Sun

Ⓓ Venus

4. Based on the scale, which object appears to be the brightest? SC.8.E.5.5

Ⓕ Moon

Ⓖ Polaris

Ⓗ Sirius

Ⓘ Venus

Benchmark Mini-Assessment | Chapter 3 • Lesson 2

mini BAT

Multiple Choice *Bubble the correct answer.*

Use the image below to answer questions 1 and 2.

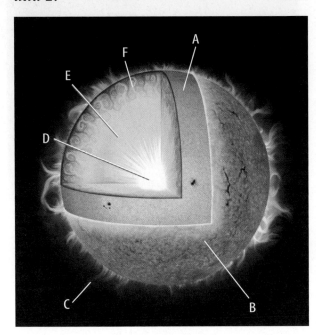

1. In which part of the Sun would the most helium be found? **SC.8.E.5.6**

- (A) A
- (B) B
- (C) D
- (D) E

2. Which correctly shows the temperature relationship among the layers of the Sun? **SC.8.E.5.5**

- (F) A > C
- (G) A = C
- (H) B < C
- (I) B = C

Use the image below to answer questions 3 and 4.

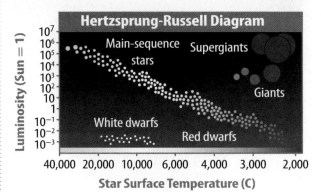

3. Which statement is true? **SC.8.E.5.5**

- (A) White dwarf stars are the hottest and most luminous stars.
- (B) In the main sequence, hotter stars are more luminous.
- (C) Red dwarfs, giants, and supergiants have similar luminosity.
- (D) Supergiants tend to be very hot but have low luminosity.

4. A scientist examines a newly discovered star. It has a luminosity of 10^{-3} and a temperature of 18,000 C. Based on the H-R diagram, which kind of star is the scientist most likely studying? **SC.8.E.5.5**

- (F) supergiants
- (G) main-sequence
- (H) red dwarf
- (I) white dwarf

Multiple Choice *Bubble the correct answer.*

Use the image below to answer questions 1 and 2.

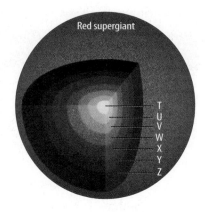

1. In which layer would you expect to find iron? **SC.8.E.5.4**

    (A) T

    (B) U

    (C) V

    (D) Z

2. In which layer would you expect to find helium? **SC.8.E.5.4**

    (F) W

    (G) X

    (H) Y

    (I) Z

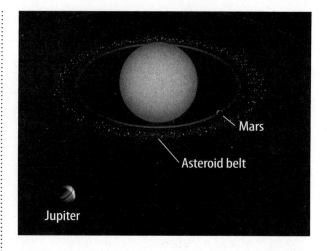

3. In the life cycle of the Sun, which stage is shown in the image above? **SC.8.N.1.1**

    (A) red dwarf

    (B) red giant

    (C) white dwarf

    (D) white giant

4. A scientist is studying a star system in which a star is orbiting another object. However, the scientist cannot see what the other object is. What is the visible star most likely orbiting? **SC.8.E.5.4**

    (F) black hole

    (G) neutron star

    (H) red giant

    (I) white dwarf

Benchmark Mini-Assessment   Chapter 3 • Lesson 4

mini BAT

Multiple Choice *Bubble the correct answer.*

Use the table below to answer questions 1 and 2.

Types of Galaxies

Galaxy	Description	Examples
A	Lack internal structure; spherical or football-shaped	Sextans A, Large Magellanic Cloud
B	Oddly shaped; areas of intense star formation; many young stars	Maffei 1, M32
C	Arms begin at a central disk of stars; thicker near the center	Andromeda, NGC 5679

1. Which statement correctly identifies the type of galaxy? **SC.8.E.5.2**

   (A) A: elliptical galaxy

   (B) A: irregular galaxy

   (C) C: elliptical galaxy

   (D) C: irregular galaxy

2. Which example is similar to the Milky Way? **SC.8.E.5.2**

   (F) M32

   (G) Maffei 1

   (H) NGC 5679

   (I) Sextans A

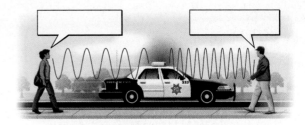

3. During science class, Sofia drew a picture similar to the one above. Which phrase would appear in the caption box on the right side of the figure? **SC.8.N.1.1**

   (A) long wavelength

   (B) low frequency

   (C) low sound

   (D) short wavelength

4. Which provided the most evidence in support of the Big Bang theory? **SC.8.E.5.1**

   (F) dark energy

   (G) dark matter

   (H) Doppler shift

   (I) spiral galaxies

Name _____ Date _____

How far have humans traveled?

In 1969, the Apollo 11 astronauts were the first humans to land on the Moon. More than 50 years later, astronauts continue to travel in space. What do you think is the farthest distance humans have traveled from Earth since astronauts landed on the Moon in 1969?

A. About 350 km above Earth
B. About halfway to the Moon (191,250 km)
C. To the Moon (about 382,500 km)
D. About 10,000 km past the Moon
E. Halfway to Mars (about 28,000,000 km)
F. To Mars (about 56,000,000 km)
G. Beyond Mars

Explain your thinking. What helped you decide how far humans have traveled in space?

Space Exploration and FLORIDA

FLORIDA BIG IDEAS

1 The Practice of Science

3 The Role of Theories, Laws, Hypotheses and Models

4 Science and Society

5 Earth in Space and Time

Think About It!

How do humans observe and explore space?

The satellite shown here is the *Hubble Space Telescope* which was launched from Florida's Kennedy Space Center. It collects light from distant objects in space. But, most satellites you might be familiar with point toward Earth. They provide navigation assistance, monitor weather, and bounce communication signals to and from Earth.

1 Why would scientists want to put a telescope in space?

_____

_____

2 In what other ways do you think scientists observe and explore space?

_____

_____

3 What do you think are goals of some current and future space missions?

_____

_____

Get Ready to Read

What do you think about space exploration?

Do you agree or disagree with each of these statements? As you read this chapter, see if you change your mind about any of the statements.

		AGREE	DISAGREE
1	Astronomers put telescopes in space to be closer to the stars.	☐	☐
2	Telescopes can work using only visible light.	☐	☐
3	Humans have walked on the Moon.	☐	☐
4	Some orthodontic braces were developed using space technology.	☐	☐
5	Humans have landed on Mars.	☐	☐
5	A solar eclipse happens when Earth moves between the Moon and the Sun.	☐	☐
6	Scientists have detected water on other bodies in the solar system.	☐	☐

Observing the UNIVERSE

 How do scientists use the electromagnetic spectrum to study the universe?

 What types of telescopes and technology are used to explore space?

Vocabulary

electromagnetic spectrum p. 142

refracting telescope p. 144

reflecting telescope p. 144

radio telescope p. 145

 Florida NGSSS

LA.8.2.2.3 The student will organize information to show understanding or relationships among facts, ideas, and events (e.g., representing key points within text through charting, mapping, paraphrasing, summarizing, or comparing/contrasting);

MA.6.A.3.6 Construct and analyze tables, graphs, and equations to describe linear functions and other simple relations using both common language and algebraic notation.

SC.8.E.5.1 Recognize that there are enormous distances between objects in space and apply our knowledge of light and space travel to understand this distance.

SC.8.E.5.10 Assess how technology is essential to science for such purposes as access to outer space and other remote locations, sample collection, measurement, data collection and storage, computation, and communication of information.

SC.8.E.5.11 Identify and compare characteristics of the electromagnetic spectrum such as wavelength, frequency, use, and hazards and recognize its application to an understanding of planetary images and satellite photographs.

SC.8.E.5.10

 Launch Lab

15 minutes

Do you see what I see?

Your eyes have lenses. Eyeglasses, cameras, telescopes, and many other tools involving light also have lenses. Lenses are transparent materials that refract light, or cause light to change direction. Lenses can cause light rays to form images as they come together or move apart.

Procedure

1. Read and complete a lab safety form.
2. Place each of the lenses on the words of this sentence.
3. Slowly move each lens up and down over the words to observe if or how the words change. Record your observations below.
4. Hold each lens at arm's length and focus on an object a few meters away. Observe how the object looks through each lens. Make simple drawings to illustrate what you observe below.

Data and Observations

Think About This

1. What happened to the words as you moved the lenses toward and away from the sentence?

2. What did the distant object look like through each lens?

3. **Key Concept** How do you think lenses are used in telescopes to explore space?

How can you see this?

1. This is an expanding halo of dust in space, illuminated by the light from the star in the center. This photo was taken with a telescope. How do you think telescopes obtain such clear images?

_____

_____

_____

_____

_____

Observing the Sky

If you look up at the sky on a clear night in Florida, you might be able to see the Moon, planets, and stars. These objects have not changed much since people first turned their gaze skyward. People in the past spent a lot of time observing the sky. They told stories about the stars, and they used stars to tell time. Most people thought Earth was the center of the universe.

Astronomers today know that Earth is part of a system of eight planets revolving around the Sun. The Sun, in turn, is part of a larger system called the Milky Way galaxy that contains billions of other stars. And the Milky Way is one of billions of other galaxies in the universe. As small as Earth might seem in the universe, it could be unique. Scientists have not found life anywhere else.

One advantage astronomers have over people in the past is the **telescope**. Telescopes enable astronomers to observe many more stars than they could with their eyes alone. Telescopes gather and focus light from objects in space. The photo above was taken with a telescope that orbits Earth. Astronomers use many kinds of telescopes to study the light energy emitted by stars and other objects in space.

WORD ORIGIN

telescope
from Greek _tele_, means "far"; and Greek _skopos_, means "seeing"

Active Reading 2. **Explain** What is the purpose of telescopes?

_____

_____

_____

_____

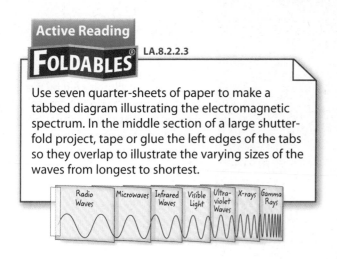
Electromagnetic Waves

Stars produce energy that radiates into space, traveling in waves called electromagnetic (ih lek troh mag NEH tik) waves. Electromagnetic waves are different from mechanical waves, such as sound waves. Sound waves can transfer energy through solids, liquids, or gases. Electromagnetic waves can transfer energy through matter or through a vacuum, such as space. The energy they carry is called radiant energy.

The Electromagnetic Spectrum

The entire range of radiant energy carried by electromagnetic waves is the **electromagnetic spectrum.** As shown in **Figure 1,** electromagnetic-spectrum waves are continuous. They range from gamma rays with short wavelengths to radio waves with long wavelengths. Electromagnetic waves with shorter wavelengths have higher frequencies. Waves with longer wavelengths have lower frequencies.

4. NGSSS Check Describe How is the electromagnetic spectrum used? SC.8.E.5.11

_____

_____

Figure 1 🗝️ Objects emit radiation in continuous wavelengths. Most wavelengths are not visible to the human eye.

Active Reading **3. Identify** (Circle) the wavelengths of microwaves. Highlight the waves that can cause sunburns.

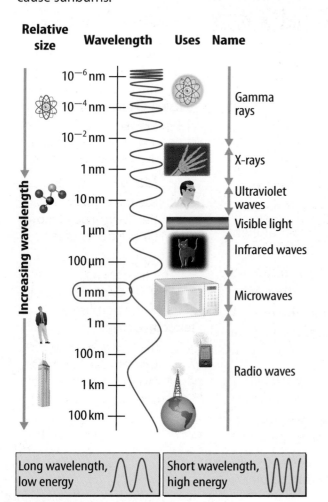

Relative size | Wavelength | Uses | Name

10^{-6} nm
10^{-4} nm — Gamma rays
10^{-2} nm
1 nm — X-rays
10 nm — Ultraviolet waves
1 μm — Visible light
100 μm — Infrared waves
1 mm — Microwaves
1 m
100 m — Radio waves
1 km
100 km

Increasing wavelength →

Long wavelength, low energy

Short wavelength, high energy

Humans observe only the visible light of the electromagnetic spectrum. The electromagnetic waves humans cannot see can be dangerous. Ultraviolet waves in sunlight help your body make vitamin D. But too much exposure to ultraviolet waves can cause sunburn or skin cancer. X-rays create images of broken bones. Overexposure to X-rays can also cause cancer.

Radiant Energy and Stars

Most stars emit energy in all wavelengths. But the wavelengths they emit depend on their temperatures. Hot stars emit mostly shorter waves with higher energy, such as X-rays, gamma rays, and ultraviolet waves. Cool stars emit mostly longer waves with lower energy, such as infrared waves and radio waves. The Sun emits much of its energy as visible light.

Apply It! After you complete the lab, answer these questions.

1. The prism and flashlight produced the entire light spectrum. Where in nature does this happen?

2. How do you think this occurs in nature?

3. Using the electromagnetic spectrum on the previous page, what are the two types of waves that border visible light?

Why You See Planets and Moons

Planets and moons are much cooler than even the coolest stars. They do not make their own energy and, therefore, do not emit light. However, you can see the Moon and the planets because they reflect light from the Sun.

Light from the Past

All light waves, from radio waves to gamma rays, travel through space at a constant speed of 300,000 km/s. This is the speed of light. The speed of light might seem incredibly fast, but the universe is very large. Even moving at the speed of light, it can take millions or billions of years for some light waves to reach Earth.

Because it takes time for light to travel, you see planets and stars as they were when their light started its journey to Earth. It takes very little time for light to travel within the solar system. Reflected light from the Moon reaches Earth in about 1 second. Light from the Sun reaches Earth in about 8 minutes. It reaches Jupiter in about 40 minutes.

Light from stars is much older. Some stars are so far away that it can take millions or billions of years for their radiant energy to reach Earth. Therefore, by studying energy from stars, astronomers can learn what the universe was like millions or billions of years ago.

Active Reading 5. **Restate** How is looking at stars like looking at the past?

_____

_____

Math Skills MA.6.A.3.6

Scientific Notation

Scientists use scientific notation to work with large numbers. Express the speed of light in scientific notation using the following process.

1. Move the decimal point until only one nonzero digit remains on the left.

   $300,000 \rightarrow 3.00000$

2. Use the number of places the decimal point moved (5) as a power of ten.

   300,000 km/s = 3.0×10^5 km/s

6. **Practice**
   The Sun is 150,000,000 km from Earth. Express this distance in scientific notation.

Refracting telescope

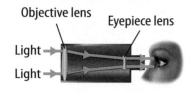

Objective lens
Eyepiece lens
Light →
Light →

Reflecting telescope

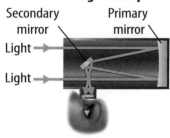

Secondary mirror
Primary mirror
Light →
Light →

Figure 2 🔑 Optical telescopes collect visible light in two different ways.

Active Reading **7. Contrast** What are the main differences between the two telescope types?

_____

_____

_____

_____

Figure 3 Each 10-m primary mirror in the twin Keck Telescopes consists of 36 small mirrors.

Earth-Based Telescopes

Telescopes are designed to collect a certain type of electromagnetic wave. Some telescopes detect visible light, and others detect radio waves and microwaves.

Optical Telescopes

There are two kinds of optical telescopes—refracting telescopes and reflecting telescopes, illustrated in **Figure 2.**

Refracting Telescopes Have you ever used a magnifying lens? You might have noticed that the lens was curved and thick in the middle. This is a convex lens. _A telescope that uses a convex lens to concentrate light from a distant object is a_ **refracting telescope**. As shown at the top of **Figure 2,** the objective lens in a refracting telescope is the lens closest to the object being viewed. The light goes through the objective lens and refracts, forming a small, bright image. The eyepiece is a second lens that magnifies the image.

Active Reading **8. Identify** Which electromagnetic waves do refracting telescopes collect? _____

Reflecting Telescopes Most large telescopes use curved mirrors instead of curved lenses. _A telescope that uses a curved mirror to concentrate light from a distant object is a_ **reflecting telescope**. As shown at the bottom of **Figure 2,** light is reflected from a primary mirror to a secondary mirror. The secondary mirror is tilted to allow the viewer to see the image. Generally, larger primary mirrors produce clearer images than smaller mirrors. However, there is a limit to mirror size. The largest reflecting telescopes, such as the Keck Telescopes on Hawaii's Mauna Kea, shown in **Figure 3,** have many small mirrors linked together. These small mirrors act as one large primary mirror.

Radio Telescope 🔑

Radio image

Radio Telescopes

Unlike a telescope that collects visible light waves, a **radio telescope** *is a telescope that collects radio waves and some microwaves using an antenna that looks like a TV satellite dish.* Because these waves have long wavelengths and carry little energy, radio antennae must be large to collect them. Radio telescopes are often built together and used as if they were one telescope. The telescopes shown in **Figure 4** are part of the Very Large Array in New Mexico. The 27 instruments in this array act as a single telescope with a 36-km diameter.

Distortion and Interference

Moisture in Earth's atmosphere can absorb and distort radio waves. Therefore, most radio telescopes are located in remote deserts, which have dry environments. Remote deserts also tend to be far from radio stations, which emit radio waves that interfere with radio waves from space.

Water vapor and other gases in Earth's atmosphere also distort visible light. Stars seem to twinkle because gases in the atmosphere move, refracting the light and causing the star's image to change slightly. At high elevations, the atmosphere is thin and produces less distortion than it does at low elevations. That is why most optical telescopes are built on mountains. New technology called adaptive optics lessens the effects of atmospheric distortion even more, as shown in **Figure 5.**

Active Reading 9. **Explain** Why would Florida not be a suitable location for a radio telescope array?

Figure 4 Radio telescopes are often built in large arrays. Computers convert radio data into images.

Figure 5 These are two images of the same celestial object. Notice the difference when adaptive optics are used.

Before Adaptive Optics

After Adaptive Optics

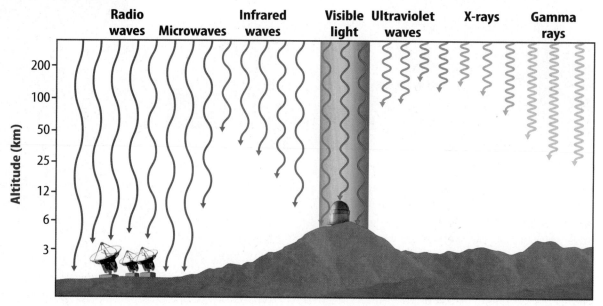

Radio waves Microwaves Infrared waves Visible light Ultraviolet waves X-rays Gamma rays

Altitude (km)

200
100
50
25
12
6
3

Figure 6 🔑 Most electro-magnetic waves do not penetrate Earth's atmosphere. Even though the atmosphere blocks most UV rays, some still reach Earth's surface.

Active Reading **10. Locate** (Circle) the altitude that gamma rays reach above Earth's surface.

Figure 7 The *Hubble Space Telescope* is controlled by astronomers on Earth.

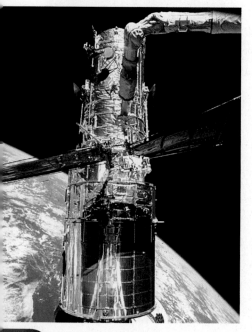

Space Telescopes

Why would astronomers want to put a telescope in space? The reason is Earth's atmosphere. Earth's atmosphere absorbs some types of electromagnetic radiation. As shown in **Figure 6,** visible light, radio waves, and some microwaves reach Earth's surface. Other types of electromagnetic waves do not. Telescopes on Earth can collect only the electromagnetic waves that are not absorbed by Earth's atmosphere. Telescopes in space can collect energy at all wavelengths, including those that Earth's atmosphere would absorb, such as most infrared light, most ultraviolet light, and X-rays.

11. NGSSS Check **Summarize** Why do astronomers put some telescopes in space? SC.8.E.5.11

_____

_____

Optical Space Telescopes

Optical telescopes collect visible light on Earth's surface, but optical telescopes work better in space. The reason, again, is Earth's atmosphere. As you read earlier, gases in the atmosphere can absorb some wavelengths. In space, there are no atmospheric gases. The sky is darker, and there is no weather.

The first optical space telescope was launched from Kennedy Space Center in Florida. The *Hubble Space Telescope,* shown in **Figure 7,** is a reflecting telescope that orbits Earth. Its primary mirror is 2.4 m in diameter. At first the *Hubble* images were blurred because of a flaw in the mirror. In 1993, astronauts repaired the telescope. Since then, *Hubble* has routinely sent to Earth spectacular images of distant objects.

Using Other Wavelengths

The *Hubble Space Telescope* is the only space telescope that collects visible light. Dozens of other space telescopes, operated by many different countries, gather ultraviolet, X-ray, gamma ray, and infrared light. Each type of telescope can point at the same region of sky and produce a different image. The image of the star Cassiopeia A (ka see uh PEE uh • AY) in **Figure 8** was made with a combination of optical, X-ray, and infrared data. The colors represent different kinds of material left over from the star's explosion many years ago.

Spitzer Space Telescope Young stars and planets hidden by dust and gas cannot be viewed in visible light. However, wavelengths of infrared light can penetrate the dust and reveal what is inside. Infrared light can also be used to observe objects that are too old and too cold to emit visible light. In 2003, the *Spitzer Space Telescope* was launched from Cape Canaveral at Kennedy Space Center in Florida to collect infrared waves while orbiting the Sun.

Active Reading
12. **Name** Which type of radiant energy does the *Spitzer Space Telescope* collect?

James Webb Space Telescope A larger space telescope, scheduled for launch in 2013, is also designed to collect infrared radiation as it orbits the Sun. The *James Webb Space Telescope*, illustrated in **Figure 9,** will have a mirror with an area 50 times larger than *Spitzer*'s mirror and seven times larger than *Hubble*'s mirror. Astronomers plan to use the telescope to detect galaxies that formed very early in the history of the universe.

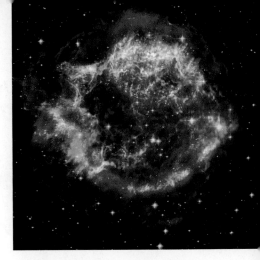

Figure 8 Each color in this image of Cassiopeia A is derived from a different wavelength—yellow: visible; pink/red: infrared; green and blue: X-ray.

Figure 9 The advanced technology of the *James Webb Space Telescope* will help astronomers study the origin of the universe.

Active Reading
13. **Recall** What type of radiant energy will this telescope collect?

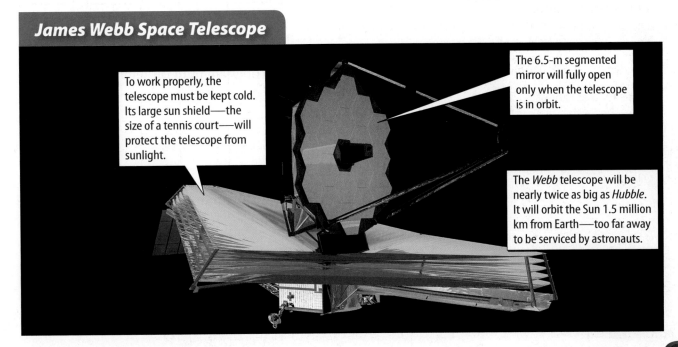

James Webb Space Telescope

To work properly, the telescope must be kept cold. Its large sun shield—the size of a tennis court—will protect the telescope from sunlight.

The 6.5-m segmented mirror will fully open only when the telescope is in orbit.

The *Webb* telescope will be nearly twice as big as *Hubble*. It will orbit the Sun 1.5 million km from Earth—too far away to be serviced by astronauts.

Reflecting telescopes use mirrors to concentrate light.

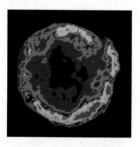

Earth-based telescopes can collect energy in the visible, radio, and microwavelengths.

Space-based telescopes can collect wavelengths of energy that cannot penetrate Earth's atmosphere.

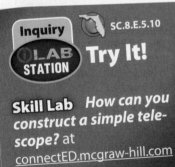

Inquiry SC.8.E.5.10
LAB STATION **Try It!**

Skill Lab *How can you construct a simple telescope?* at connectED.mcgraw-hill.com

Use Vocabulary

1. **Distinguish** between a *reflecting telescope* and a *refracting telescope*.

2. **Use the term** *electromagnetic spectrum* in a sentence. SC.8.E.5.11

3. **Define** *radio telescope* in your own words.

Understand Key Concepts

4. Which emits visible light?
   - (A) moon
   - (C) satellite
   - (B) planet
   - (D) star

5. **Compare** the *Hubble Space Telescope* and the *James Webb Space Telescope*. SC.8.E.5.10

Interpret Graphics

6. **Organize Information** List the wavelengths collected by space telescopes from the longest to the shortest. SC.8.E.5.11

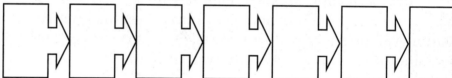

Critical Thinking

7. **Suggest** a reason—besides the lessening of atmospheric distortion—why optical telescopes are built on remote mountains.

Math Skills   MA.6.A.3.6

8. Light travels 9,460,000,000,000 km in 1 year. Express this number in scientific notation.

One Giant Leap
FOR FLORIDA

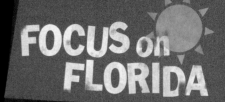

From Citrus Groves to Space Shuttles

FOR YEARS, Florida's Kennedy Space Center has been the place where NASA's historic space flights begin. The first Americans in space, the only humans to walk on the Moon, and all of the space-shuttle crews left Earth from Florida. These missions have enabled scientists to learn more about the Moon, the solar system, and the universe than they can learn from Earth.

Why Florida?

When President Harry S. Truman needed a place to test missiles in 1949, Cape Canaveral, Florida, was selected. Its mild climate allows for year-round missile tests. Also, the Cape's location near the Atlantic Ocean meant missiles could be launched over water and not over communities.

Cape Canaveral served as a major launch base for the Air Force for over ten years. In 1961, President John F. Kennedy announced that the United States would send a human to the Moon. NASA selected Cape Canaveral as the location for its new facility.

In 1962, NASA purchased land on nearby Merritt Island that would become the John F. Kennedy Space Center (KSC). Since then, KSC has become the launch site for most of NASA's space missions.

This Delta rocket blasted into space carrying the *Spitzer Space Telescope* in 2003 from Kennedy Space Center in Cape Canaveral, Florida.

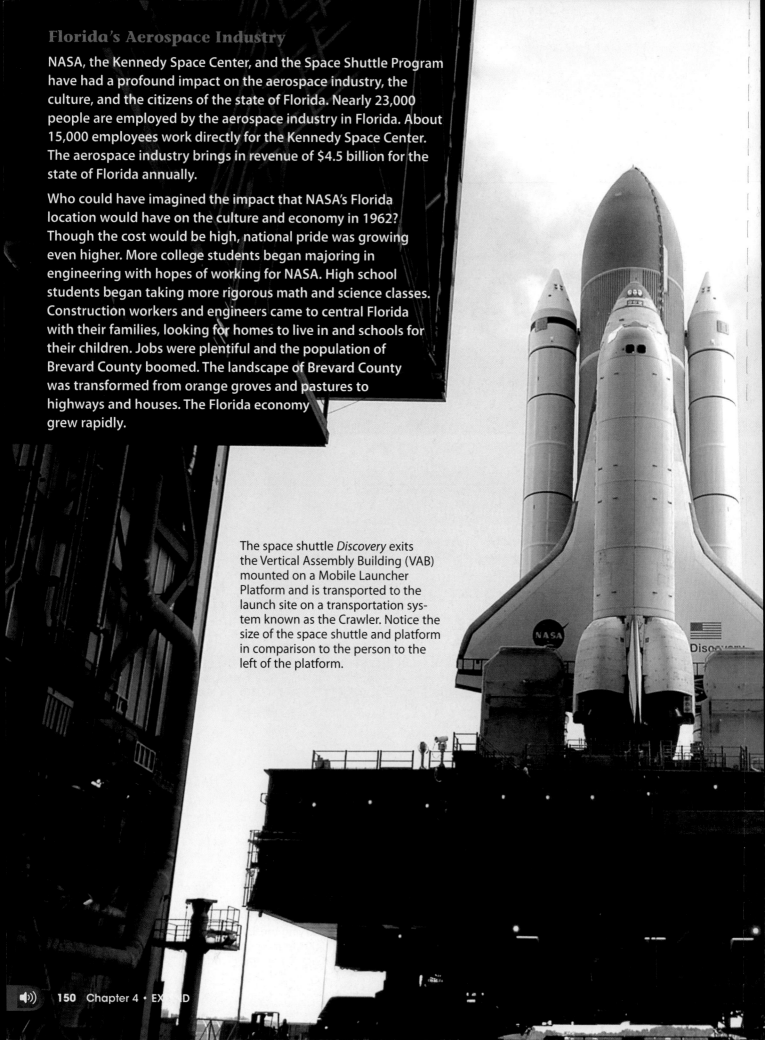

Florida's Aerospace Industry

NASA, the Kennedy Space Center, and the Space Shuttle Program have had a profound impact on the aerospace industry, the culture, and the citizens of the state of Florida. Nearly 23,000 people are employed by the aerospace industry in Florida. About 15,000 employees work directly for the Kennedy Space Center. The aerospace industry brings in revenue of $4.5 billion for the state of Florida annually.

Who could have imagined the impact that NASA's Florida location would have on the culture and economy in 1962? Though the cost would be high, national pride was growing even higher. More college students began majoring in engineering with hopes of working for NASA. High school students began taking more rigorous math and science classes. Construction workers and engineers came to central Florida with their families, looking for homes to live in and schools for their children. Jobs were plentiful and the population of Brevard County boomed. The landscape of Brevard County was transformed from orange groves and pastures to highways and houses. The Florida economy grew rapidly.

The space shuttle *Discovery* exits the Vertical Assembly Building (VAB) mounted on a Mobile Launcher Platform and is transported to the launch site on a transportation system known as the Crawler. Notice the size of the space shuttle and platform in comparison to the person to the left of the platform.

Florida Technology

Florida is home to more than 24,000 companies in the Information Technology (IT) field alone. These companies employ more than 250,000 workers in a variety of IT job fields including modeling, simulation and training, optics, digital media, software, microelectronics, and telecommunications. All of these technologies are a direct result of NASA scientists and experiments.

Science and experimental technologies have contributed many new products and technologies to the public. These products resulted from the race to space. Such products include invisible braces; protective sports equipment, such as helmets and shin guards; satellite television; medical imaging; home insulation; and smoke detectors.

Tourism

Today, the Kennedy Space Center is one of the top four tourist attractions in Florida. Tourists visiting the Kennedy Space Center contribute over $20 million annually to the Florida economy.

At the space center, you can see the Vertical Assembly Building (VAB). It is 160 m tall and is the largest one-story building in the world. When the doors of this building open, a fully assembled rocket or space shuttle can be rolled out on a track and transported to a launchpad.

This satellite image shows the proximity of the launch site at Cape Canaveral to water. This is one of the reasons Cape Canaveral was chosen as the headquarters for NASA.

How You Can Help

As a future Florida voter, it is important that you learn about science. One day, you might be called upon to make decisions at the community, state, and national levels. You also might be called upon to cast votes that affect the scientific progress of our country. Your understanding of the process and benefits of science and scientific research can help you make more informed decisions.

Assess Your Understanding

1. **Identify** <u>Underline</u> the effects of space exploration on the economy and the culture of Florida.

2. **Explain** Why were Merritt Island and Cape Canaveral chosen as the site for NASA?

3. **Describe** What are some products we use in our everyday lives that are a result of the space program?

Early History of Space EXPLORATION

ESSENTIAL QUESTIONS

 How are rockets and artificial satellites used?

 Why do scientists send both crewed and uncrewed missions into space?

 What are some ways that people use space technology to improve life on Earth?

Vocabulary

rocket p. 153

satellite p. 154

space probe p. 155

lunar p. 155

Project Apollo p. 156

space shuttle p. 156

 Florida NGSSS

LA.8.2.2.3 The student will organize information to show understanding or relationships among facts, ideas, and events (e.g., representing key points within text through charting, mapping, paraphrasing, summarizing, or comparing/contrasting);

SC.8.E.5.1 Recognize that there are enormous distances between objects in space and apply our knowledge of light and space travel to understand this distance.

SC.8.E.5.10 Assess how technology is essential to science for such purposes as access to outer space and other remote locations, sample collection, measurement, data collection and storage, computation, and communication of information.

SC.8.E.5.12 Summarize the effects of space exploration on the economy and culture of Florida.

SC.8.N.4.2 Explain how political, social, and economic concerns can affect science, and vice versa.

Inquiry Launch Lab

20 minutes

How do rockets work?

Procedure

1. Read and complete a lab safety form.

2. Use **scissors** to carefully cut a 5-m piece of **string.**

3. Insert the string into a **drinking straw.** Tie each end of the string to a stationary object. Make sure the string is taut. Slide the drinking straw to one end of the string.

4. Blow up a **balloon.** Do not tie it. Instead, twist the neck and clamp it with a **clothespin** or a **paper clip. Tape** the balloon to the straw.

5. Remove the clothespin or paper clip to launch your rocket. Observe how the rocket moves. Record your observations below.

Data and Observations

Think About This

1. Describe how your rocket moved along the string.

2. How might you get your rocket to go farther or faster?

3. [Key Concept] How do you think rockets are used in space exploration?

 Inquiry **Where has it been?**

1. The crew of *Apollo 11* is inside the capsule waiting to be retrieved after splashing down in the ocean. Explain how this method of return to Earth would have been the best method at the time. How have the methods of return changed for crewed missions?

_____

_____

_____

_____

_____

Figure 10 🔑 Rockets are used once and then fall safely back to Earth and land in the ocean. Space shuttle *Atlantis* launches safely into orbit from Kennedy Space Center in Florida.

Rockets

Think about listening to a recording of your favorite music. Now think about how different it is to experience the same music at a live performance. This is like the difference between exploring space from a distance with a telescope and actually going there.

A big problem in launching an object into space is overcoming the force of Earth's gravity. This is accomplished with rockets. A **rocket** *is a vehicle designed to propel itself by ejecting exhaust gas from one end.* Fuel burned inside the rocket builds up pressure. The force from the exhaust thrusts the rocket forward, as shown in **Figure 10.** Rocket engines do not draw in oxygen from the surrounding air to burn their fuel, as jet engines do. They carry their oxygen with them. As a result, rockets can operate in space where there is very little oxygen.

2. NGSSS Check **Describe** How are rockets used in space exploration? SC.8.E.5.10

_____

_____

_____

Active Reading

FOLDABLES® LA.8.2.2.3

Make a vertical two-tab book. Record what you learn about crewed and uncrewed space missions under the tabs.

Crewed Missions

Uncrewed Missions

Artificial Satellites

Any small object that orbits a larger object is a **satellite**. The Moon is a natural satellite of Earth. Artificial satellites are made by people and launched by rockets. They orbit Earth or other bodies in space, transmitting radio signals back to Earth.

The First Satellites—*Sputnik* and *Explorer*

The first artificial satellite sent into Earth's orbit was *Sputnik 1*. Many people think this satellite, launched in 1957 by the former Soviet Union, represents the beginning of the space age. In 1958, the United States launched its first Earth-orbiting satellite, *Explorer I*. Today, thousands of satellites orbit Earth.

How Satellites Are Used

The earliest satellites were developed by the military for navigation and to gather information. Today, Earth-orbiting satellites are also used to transmit television and telephone signals and to monitor weather and climate. An array of satellites called the Global Positioning System (GPS) is used for navigation in cars, boats, airplanes, and even for hiking.

3. NGSSS Check Summarize How are Earth-orbiting satellites used? SC.8.E.5.10

_____

Figure 11 Space exploration began with the first rocket launch in 1926.

Active Reading 4. **Identify** (Circle) the information about the first U.S. satellite.

Early Exploration and Florida

In 1958, the National Aeronautics and Space Administration (NASA) was established. NASA oversees all U.S. space missions—often controlling space missions from NASA centers in Florida. Some early steps in U.S. space exploration are shown in **Figure 11**.

Early Space Exploration 🔑

1926 First rocket: Robert Goddard's liquid-fueled rocket rose 12 m into the air.

1958 First U.S. satellite: In the same year NASA was founded, *Explorer 1* was launched from Kennedy Space Center in Florida. It orbited Earth 58,000 times before burning up in Earth's atmosphere in 1970.

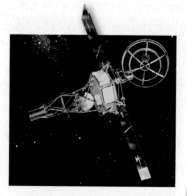

1962 First planetary probe: *Mariner 2* traveled to Venus and collected data for 3 months. The craft now orbits the Sun.

1972 First probe to outer solar system: After flying past Jupiter, *Pioneer 10* is still traveling onward, someday to exit the solar system.

Figure 12 Scientists use space probes to explore the planets and some moons in the solar system.

Active Reading

5. State Which type of probe uses a parachute? _____

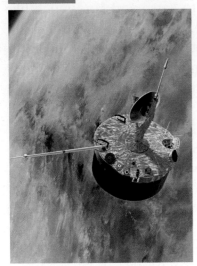

Once orbiters reach their destinations, they use rockets to slow down enough to be captured in a planet's orbit. How long they orbit depends on their fuel supply. The orbiter probe here, *Pioneer,* orbited Venus.

Landers touch down on surfaces. Sometimes they release rovers. Landers use rockets and parachutes to slow their descent. The lander probe here, *Phoenix,* analyzed the Martian surface for evidence of water.

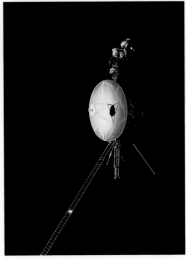

Flybys do not orbit or land. When its mission is complete, a flyby continues through space, eventually leaving the solar system. *Voyager 1,* shown here, explored Jupiter and Saturn and will soon leave the solar system.

Space Probes

Some spacecraft have human crews, but most do not. *A* **space probe** *is an uncrewed spacecraft sent from Earth to explore objects in space.* Space probes are robots that work automatically or by remote control. They take pictures and gather data. Probes can make trips that would be too long or too dangerous for humans. Probes are essential to science because they can gather information about space that humans are not able to collect. The data they gather are relayed to Earth via radio waves. **Figure 12** shows three major types of space probes.

Lunar and Planetary Probes

The first probes to the Moon were sent by the United States and the former Soviet Union in 1959. Probes to the Moon are called lunar probes. *The term* **lunar** *refers to anything related to the Moon.* The first spacecraft to gather information from another planet was the flyby *Mariner 2,* sent to Venus in 1962. Since then, space probes have been sent to all the planets.

SCIENCE USE V. COMMON USE

probe

Science Use an uncrewed spacecraft

Common Use to question or examine closely

6. NGSSS Check
Summarize Why do scientists send uncrewed missions to space? SC.8.E.5.10

_____

_____

Human Spaceflight from Florida

Sending humans into space was a major goal of the early space program. In 1962, the first American astronaut, John Glenn, orbited Earth in the spacecraft *Friendship 7*. Some highlights of the early U.S. human spaceflight program are shown in **Figure 13.** All crewed spaceflights in the U.S. are launched from Cape Canaveral at the Kennedy Space Center in Florida.

The Apollo Program

In 1961, U.S. President John F. Kennedy challenged the American people to place a person on the Moon by the end of the decade. The result was **Project Apollo**—*a series of space missions designed to send people to the Moon.* In 1969, Neil Armstrong and Buzz Aldrin, *Apollo 11* astronauts, were the first people to walk on the Moon.

Active Reading **7. Identify** <u>Underline</u> the goal of Project Apollo.

Space Transportation Systems

Early spacecraft and the rockets used to launch them were used only once. **Space shuttles** *are reusable spacecraft that transport people and materials to and from space.* Space shuttles return to Earth and land much like airplanes. NASA's fleet of space shuttles began operating in 1981. As the shuttles aged, NASA began developing a new transportation system, *Orion,* to replace them.

The *International Space Station*

The United States space program cooperates with the space programs of other countries. In 1998, it joined 15 other nations to begin building the *International Space Station.* Occupied since 2000, this Earth-orbiting satellite is a research laboratory where astronauts from many countries work and live.

Space is a challenging environment. Objects in space are nearly weightless. They must be tied down or they will float away. There is little friction to stop them.

U.S. Human Spaceflight 🔑

Figure 13 Forty years after human spaceflight began in Florida, people were living and working in space.

Active Reading **8. Compose** Write your own captions for each image.

Inquiry LAB STATION **Try It!**

MiniLab *How does lack of friction in space affect simple tasks?* at connectED.mcgraw-hill.com

Apply It! After you complete the lab, answer these questions.

1. Why would it be easier to move a table on the *International Space Station* than in your classroom?

2. There is a lack of friction in space. As the space shuttle reenters Earth's atmosphere, it encounters a large amount of friction. How do you think the increased friction affects the space shuttle?

Space Technology from Florida

Space exploration programs based in Florida influence everyone on Earth. The space program requires materials that can withstand the extreme temperatures and pressures of space. These materials are used by people every day and influence culture. NASA scientists test new technology for space, like in **Figure 14**, and then pass this new technology on to the public.

Space materials must be protective while being flexible and strong. Materials developed for space suits are now used to make racing suits for swimmers, lightweight firefighting gear, running shoes, and other sports clothing. Sports protective gear, including helmets and shin guards, are a result of space materials. Polarized, scratch-resistent sunglasses also had a technology start in the space industry.

NASA developed a strong, fibrous material to make parachute cords for spacecraft that land on planets and moons. This material, five times stronger than steel, is used to make radial tires.

Prosthetic limbs, infrared ear thermometers, and robotic and laser surgery all have roots in the space program. Orthodontic braces contain ceramic material originally developed to strengthen the heat resistance of space shuttles.

9. **NGSSS Check** **Summarize** What are some ways that space exploration has impacted Florida? SC.8.E.5.12

Figure 14 This scientist at the Marshall Space Flight Center is working on a mirror that will be used in the *James Webb Telescope*. The technology used to create this mirror could be used in everyday items in the future.

Exhaust from burned fuel accelerates a rocket.

Some space probes can land on the surface of a planet or a moon.

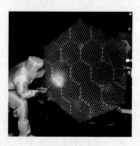

Technologies developed for the space program have been applied to everyday life on Earth.

Use Vocabulary

1 **Define** *space shuttle* in your own words.

2 **Use the term** *satellite* in a sentence.

3 The mission that sent people to the Moon was

Understand Key Concepts 🔑

4 What are rockets used for? SC.8.E.5.10

  (A) carrying people   (C) observing planets

  (B) launching satellites   (D) transmitting signals

5 **Explain** why *Sputnik 1* is considered the beginning of the space age. SC.8.E.5.10

6 **Compare and contrast** crewed and uncrewed space missons.

Interpret Graphics

7 **Organize Information** Fill in the graphic organizer with the following in the correct order: *first human in space, invention of rockets, first human on the Moon, first artificial satellite.* LA.8.2.2.3

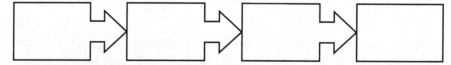

Critical Thinking

8 **Predict** how your life would be different if all artificial satellites stopped working. SC.8.E.5.10

9 **Evaluate** the benefits and drawbacks of international cooperation in space exploration.

Going Up

Could a space elevator make space travel easier?

If you wanted to travel into space, the first thing you would have to do is overcome the force of Earth's gravity. So far, the only way to do that has been to use rockets. Rockets are expensive, however. They are used only once, and they require a lot of fuel. It takes a lot of resources to build and power a rocket. But what if you could take an elevator into space instead?

Space elevators were once science fiction, but scientists are now taking the possibility seriously. With the lightweight but strong materials under development today, experts say it could take only 10 years to build a space elevator. The image here shows how it might work.

It generally costs more than $100 million to place a 12,000-kg spacecraft into orbit using a rocket. Some people estimate that a space elevator could place the same craft into orbit for less than $120,000. A human passenger with luggage, together totaling 150 kg, might be able to ride the elevator to space for less than $1,500.

Counterweight: The spaceward end of the cable would attach to a captured asteroid or an artificial satellite. The asteroid or satellite would stabilize the cable and act as a counterweight.

Cable: Made of super-strong but thin materials, the cable would be the first part of the elevator to be built. A rocket-launched spacecraft would carry reels of cable into orbit. From there the cable would be unwound until one end reached Earth's surface.

Anchor Station: The cable's Earthward end would be attached here. A movable platform would allow operators to move the cable away from space debris in Earth's orbit that could collide with it. The platform would be movable because it would float on the ocean.

Climber: The "elevator car" would carry human passengers and objects into space. It could be powered by Earth-based laser beams, which would activate solar-cell "ears" on the outside of the car.

It's Your Turn

DEBATE Form an opinion about the space elevator and debate with a classmate. Could a space elevator become a reality in the near future? Would a space elevator benefit ordinary people? Should the space elevator be used for space tourism?

Recent and Future Space MISSIONS

ESSENTIAL QUESTIONS

What are goals for future space exploration?

What conditions are required for the existence of life on Earth?

How can exploring space help scientists learn about Earth?

Vocabulary

extraterrestrial life p. 165

astrobiology p. 165

Florida NGSSS

LA.8.2.2.3 The student will organize information to show understanding or relationships among facts, ideas, and events (e.g., representing key points within text through charting, mapping, paraphrasing, summarizing, or comparing/contrasting);

LA.8.4.2.2 The student will record information (e.g., observations, notes, lists, charts, legends) related to a topic, including visual aids to organize and record information, as appropriate, and attribute sources of information;

SC.8.E.5.1 Recognize that there are enormous distances between objects in space and apply our knowledge of light and space travel to understand this distance.

SC.8.E.5.10 Assess how technology is essential to science for such purposes as access to outer space and other remote locations, sample collection, measurement, data collection and storage, computation, and communication of information.

SC.8.N.1.1 Define a problem from the eighth grade curriculum using appropriate reference materials to support scientific understanding, plan and carry out scientific investigations of various types, such as systematic observations or experiments, identify variables, collect and organize data, interpret data in charts, tables, and graphics, analyze information, make predictions, and defend conclusions.

SC.8.N.3.1 Select models useful in relating the results of their own investigations.

SC.8.N.4.2 Explain how political, social, and economic concerns can affect science, and vice versa.

 Inquiry Launch Lab SC.8.N.3.1

15 minutes

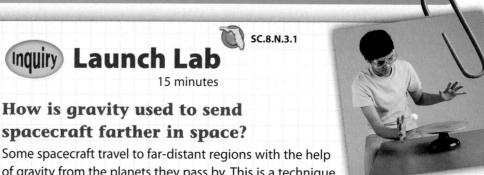

How is gravity used to send spacecraft farther in space?

Some spacecraft travel to far-distant regions with the help of gravity from the planets they pass by. This is a technique called gravity assist. You can model gravity assist using a simple table-tennis ball.

Procedure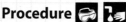

1. Read and complete a lab safety form.

2. Set a **turntable** in motion.

3. Gently throw a **table-tennis ball** so that it just skims the top of the spinning surface. You might have to practice before you're able to get the ball to glide over the surface.

4. Describe or draw a picture of what you observed below.

Data and Observations

Think About This

1. Use your observations to describe how this activity is similar to gravity assist.

2. **Key Concept** How do you think gravity assist helps scientists learn about the solar system?

Inquiry Views from above?

1. This image was taken as the space shuttle orbits over southern Florida. How might this image have been captured? What does the future of space flight hold?

Missions to the Sun and the Moon

What is the future for space exploration? Scientists at NASA and other space agencies around the world have cooperatively developed goals for future space exploration. One goal is to expand human space travel within the solar system. Two steps leading to this goal are sending probes to the Sun and sending probes to the Moon.

Active Reading

2. Summarize What is a goal of future space exploration?

Solar Probes

The Sun emits high-energy radiation and charged particles. Storms on the Sun can eject powerful jets of gas and charged particles into space, as shown in **Figure 15.** The Sun's high-energy radiation and charged particles can harm astronauts and damage spacecraft. To better understand these hazards, scientists study data collected by solar probes that orbit the Sun. The solar probe *Ulysses,* launched in 1990, orbited the Sun and gathered data for 19 years.

Lunar Probes

NASA and other space agencies also plan to send several probes to the Moon. The *Lunar Reconnaissance Orbiter*, launched in 2009, collects data to learn more about the Moon's environment.

Figure 15 Storms on the Sun send charged particles far into space. These solar storms can interfere with communication satellites and impact communication systems on Earth.

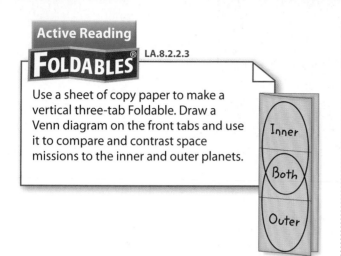

Inner

Both

Outer

Missions to the Inner Planets

The inner planets are the four rocky planets closest to the Sun: Mercury, Venus, Earth, and Mars. Scientists have sent many probes to the inner planets, and more are planned. These probes help scientists learn how the inner planets formed, what geologic forces are active on them, and whether any of them could support life. Some recent and current missions to the inner planets are described in **Figure 16.**

Active Reading

3. Describe What do scientists want to learn about the inner planets?

_____

_____

Planetary Missions

Figure 16 Studying the solar system remains a major goal of space exploration.

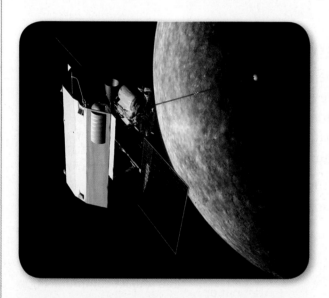

Messenger The first probe to visit Mercury—the planet closest to the Sun—since *Mariner 10* flew by the planet in 1975 is *Messenger*. After a 2004 launch and two passes of Venus, *Messenger* will fly past Mercury several times before entering Mercury's orbit in 2011. *Messenger* will study Mercury's geology and chemistry. It will send images and data back to Earth for one Earth year. On its first pass by Mercury, in 2008, *Messenger* returned more than 1,000 images in many wavelengths.

Spirit **and** *Opportunity* Since the first flyby reached Mars in 1964, many probes have been sent to Mars. In 2003, two robotic rovers, *Spirit* and *Opportunity,* began exploring the Martian surface for the first time. These solar-powered rovers traveled more than 20 km and relayed data for more than 5 years. They have sent thousands of images to Earth.

Active Reading

4. Recall What type of space probe are the rovers?

_____

_____

Missions to the Outer Planets and Beyond

The outer planets are the four large planets farthest from the Sun: Jupiter, Saturn, Uranus, and Neptune. Pluto was once considered an outer planet, but it is now included with other small, icy **dwarf planets** observed orbiting the Sun outside the orbit of Neptune. Missions to outer planets are long and difficult because the planets are so far from Earth. Some missions to the outer planets and beyond are described in **Figure 16** below. The next major mission to the outer planets will be an international mission to Jupiter and its four largest moons.

Active Reading

5. **Explain** Why are missions to the outer planets difficult?

Active Reading

6. **Name** Which planet has been explored by rovers?

Cassini The first orbiter sent to Saturn, *Cassini* was launched in 1997 as part of an international effort involving 19 countries. *Cassini* traveled for 7 years before entering Saturn's orbit in 2004. When it arrived, it sent a smaller probe to the surface of Saturn's largest moon, Titan, as shown in an artist's depiction at left. *Cassini* is so large—6,000 kg—that no rocket was powerful enough to send it directly to Saturn. Scientists used gravity from closer planets—Venus, Earth, and Jupiter—to help power the trip. The gravity from each planet gave the spacecraft a boost toward Saturn.

Active Reading

7. **Restate** What is the name given to the gravity used to help propel *Cassini* to its destination?

New Horizons A much smaller spacecraft, *New Horizons,* is speeding toward Pluto. *New Horizons* is also using gravity assist for its journey, with a swing past Jupiter. Launched in 2006, *New Horizons* won't reach Pluto until 2015. It will leave the solar system in 2029. Without a gravity assist from Jupiter, it would take *New Horizons* 5 years longer to reach Pluto.

Figure 17 This inflatable structure could serve as housing for astronauts. It has been tested in the harsh environment of Antarctica.

Human Space Missions and Florida

Do you think there will ever be cities or communities built outside Earth? That is looking very far ahead. No person has ever been farther than the Moon. But human space travel remains a goal of NASA and other space agencies around the world. Every crewed U.S. space flight launches at the Kennedy Space Center in Cape Canaveral, Florida.

Visiting Mars

A visit to Mars will probably not occur for several more decades. To prepare for a visit to Mars, NASA plans to send additional probes to the planet. The probes will explore sites on Mars that might have resources that can support life. NASA is studying different housing structure **options** for future planetary outposts like the one seen in **Figure 17.**

Active Reading 8. **Differentiate** Contrast the human lunar landings of Project Apollo and the next human visits to the solar system.

Apollo lunar landings	
Future missions	

Inquiry

LAB STATION Try It!

MiniLab *What conditions are required for life on Earth?* at connectED.mcgraw-hill.com

Apply It! After you complete the lab, answer these questions.

1. What living necessities did the environments have in common?

2. How are scientists at NASA preparing for travel to other planets and the living conditions on those planets?

The Search for Life

No one knows if life exists beyond Earth, but people have thought about the possibility for a long time. It even has a name. *Life that originates outside Earth is* **extraterrestrial** (ek struh tuh RES tree ul) **life.**

Conditions Needed for Life

Astrobiology *is the study of life in the universe, including life on Earth and the possibility of extraterrestrial life.* Investigating the conditions for life on Earth helps scientists predict where they might find life elsewhere in the solar system. Astrobiology also can help scientists locate environments in space where humans and other Earth life might be able to survive.

Life exists in a wide range of environments on Earth. Life-forms survive on dark ocean floors, deep in solid rocks, and in scorching water, such as the hot spring shown in **Figure 18.** No matter how extreme their environments, all known life-forms on Earth need liquid water, organic molecules, and some source of energy to survive. Scientists assume that if life exists elsewhere in space, it would have the same requirements.

Active Reading 9. **Describe** What is required for life on Earth?

_____

_____

Water in the Solar System

Evidence from space probes suggests that water vapor or ice exists on many planets and moons in the solar system. Photographs of Mars suggest that liquid water once existed on the Martian surface and might still be present below the surface. NASA plans to launch the *Mars Science Laboratory* in 2011 to sample a variety of soils and rocks on Mars. This mission will investigate the possibility that life exists or once existed on the planet.

Some of the moons in the outer solar system, such as Jupiter's moon Europa, shown in **Figure 19,** might also have large amounts of liquid water beneath their surfaces.

Figure 18 Bacteria live in the boiling water of this hot spring in Yellowstone National Park.

WORD ORIGIN

astrobiology
from Greek *astron,* means "star"; Greek *bios,* means "life"; and Greek *logia,* means "study"

Figure 19 The dark patches in the inset photo might represent areas where water from an underground ocean has seeped to Europa's surface.

Active Reading 10. **Infer** What might the existence of water on the surface of Europa suggest?

_____

_____

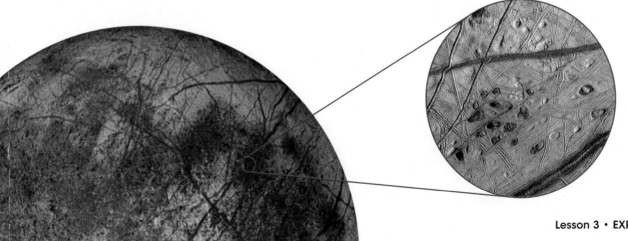

Understanding Earth by Exploring Space

Space provides frontiers for the human spirit of exploration and discovery. The exploration of space also provides insight into planet Earth. Information gathered in space helps scientists understand how the Sun and other bodies in the solar system influence Earth, how Earth formed, and how Earth supports life. Looking for Earthlike planets outside the solar system helps scientists learn if Earth is unique in the universe.

Searching for Other Planets

Astronomers have detected more than 300 planets outside the solar system. Most of these planets are much bigger than Earth and probably could not support liquid water—or life. To search for Earthlike planets, NASA launched the *Kepler telescope* in 2009. The *Kepler telescope*, illustrated in **Figure 20,** focuses on a single area of sky containing about 100,000 stars. However, though it might detect Earthlike planets orbiting other stars, *Kepler* will not be able to detect life on any planet.

Understanding Our Home Planet

Not all of NASA's missions are to other planets, to other moons, or to look at stars and galaxies. NASA and space agencies around the world also launch and maintain Earth-observing satellites. Satellites that orbit Earth provide large-scale images of Earth's surface. These images help scientists understand Earth's climate and weather. **Figure 21** is a 2005 satellite image showing changes in ocean temperature associated with Hurricane Katrina, one of the deadliest storms in U.S. history.

Figure 20 *Kepler* is orbiting the Sun, searching a single area of sky for Earthlike planets.

11. NGSSS Check
Describe How can exploring space help scientists learn about Earth? SC.8.E.5.10

Figure 21 Earth-orbiting satellites collect data in many wavelengths. This satellite image of Hurricane Katrina was made with a microwave sensor.

Active Reading **12. Identify** Which states did Hurricane Katrina affect?

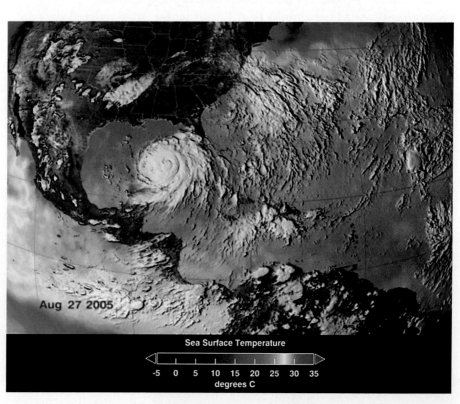

Aug 27 2005

Sea Surface Temperature

-5   0   5   10   15   20   25   30   35
degrees C

The *New Horizons* space-craft will reach Pluto in 2015.

Scientists think liquid water might exist on or below the surfaces of Mars and some moons.

Earth-orbiting satellites help scientists under-stand weather and cli-mate patterns on Earth.

Inquiry
SC.8.N.1.1,
SC.8.N.3.1,
SC.8.E.5.10,
LA.8.4.2.2

LAB STATION Try It!

Inquiry Lab *Design and Construct a Moon Habitat* at
connectED.mcgraw-hill.com

Use Vocabulary

1 **Use the term** *extraterrestrial life* in a sentence.

_____

2 The study of life in the universe is

_____ .

Understand Key Concepts

3 Which gave *Cassini* a boost toward Saturn? SC.8.E.5.10

(A) buoyancy   (C) magnetism

(B) gravity   (D) wind

4 **Explain** why bodies that have liquid water are the best candidates for supporting life.

_____

_____

5 **Identify** some phenomena on Earth best viewed by artificial satellites. SC.8.E.5.10

_____

_____

Interpret Graphics

6 **Organize** Fill in the graphic organizer below to list requirements for life on Earth. LA.8.2.2.3

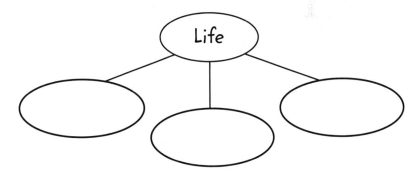

Critical Thinking

7 **Predict** some of the challenges people might face living in a planetary outpost. SC.8.E.5.10

_____

_____

_____

_____

Think About It! Humans develop knowledge and understanding of the universe with Earth-based and space-based telescopes. They explore the solar system with crewed and uncrewed space probes.

Key Concepts Summary

Vocabulary

LESSON 1 Observing the Universe

- Scientists use different parts of the **electromagnetic spectrum** to study stars and other objects in space and to learn what the universe was like many millions of years ago.
- Telescopes in space can collect radiant energy that Earth's atmosphere would absorb or refract.

electromagnetic spectrum p. 142

refracting telescope p. 144

reflecting telescope p. 144

radio telescope p. 145

LESSON 2 Early History of Space Exploration

- **Rockets** are used to overcome the force of Earth's gravity when sending **satellites**, **space probes**, and other spacecraft into space.
- Uncrewed missions can make trips that are too long or too dangerous for humans.
- Materials and technologies from the space program have been applied to everyday life.

rocket p. 153

satellite p. 154

space probe p. 155

lunar p. 155

Project Apollo p. 156

space shuttle p. 156

LESSON 3 Recent and Future Space Missions

- A goal of the space program is to expand human space travel within the solar system and develop lunar and Martian outposts.
- All known life-forms need liquid water, energy, and organic molecules.
- Information gathered in space helps scientists understand how the Sun influences Earth, how Earth formed, whether life exists outside of Earth, and how weather and climate affect Earth.

extraterrestrial life p. 165

astrobiology p. 165

FOLDABLES® Chapter Project

Assemble your lesson Foldables as shown to make a Chapter Project. Use the project to review what you have learned in this chapter.

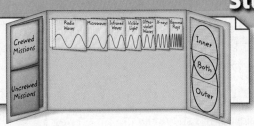

Use Vocabulary

1 All radiation is classified by wavelength in the

_____.

2 Two types of telescopes that collect visible light

are _____ and

_____.

3 The space mission that sent the first humans to

the Moon was _____.

4 An example of a human space transportation

system is a(n) _____.

5 An uncrewed spacecraft is a(n) _____

_____.

6 The discipline that investigates life in the universe

is _____.

7 The best place to find _____

is on solar system bodies containing water.

Link Vocabulary and Key Concepts

Use vocabulary terms from the previous page to complete the concept map.

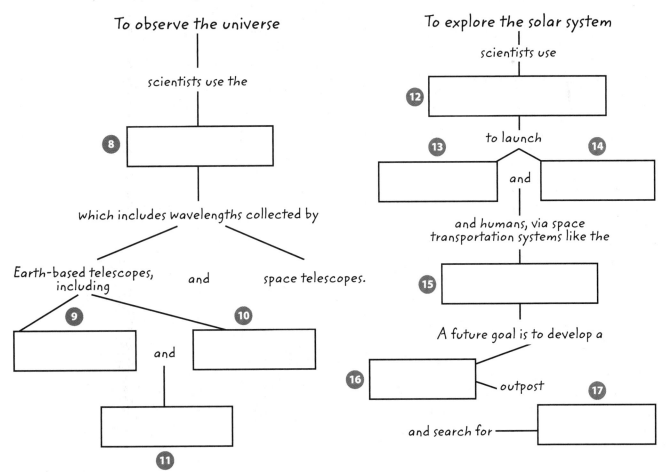

Fill in the correct answer choice.

🔑 Understand Key Concepts

1. Which type of telescope is shown in the figure below? SC.8.E.5.11

Reflecting Telescope

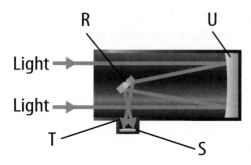

Ⓐ infrared telescope
Ⓑ radio telescope
Ⓒ reflecting telescope
Ⓓ refracting telescope

2. In which wavelength would you expect the hottest stars to emit most of their energy? SC.8.E.5.11
Ⓐ gamma rays
Ⓑ microwaves
Ⓒ radio waves
Ⓓ visible light

3. Which best describes *Hubble?* SC.8.E.5.11
Ⓐ infrared telescope
Ⓑ radio telescope
Ⓒ refracting telescope
Ⓓ space telescope

4. What is special about the *Kepler* mission? SC.8.E.5.10
Ⓐ *Kepler* can detect objects at all wavelengths.
Ⓑ *Kepler* has found the most distant objects in the universe.
Ⓒ *Kepler* is dedicated to finding Earthlike planets.
Ⓓ *Kepler* is the first telescope to orbit the Sun.

Critical Thinking

5. **Contrast** waves in the electromagnetic spectrum with water waves in the ocean. SC.8.E.5.11

6. **Differentiate** If you wanted to study new stars forming inside a huge dust cloud, which wavelength might you use? Explain. SC.8.E.5.11

7. **Deduce** Why do optical telescopes only work at night, while radio telescopes work all day and all night long? SC.8.E.5.11

8. **Analyze** Why it is more challenging to send space probes to the outer solar system than to the inner solar system? SC.8.E.5.10

9. **Create** a list of requirements that must be satisfied before humans can live on the Moon. SC.8.E.5.10

10. **Choose** a body in the solar system that you think would be a good place to look for life. Explain. SC.8.E.5.10

11 **Interpret Graphics** Label the relative positions of ultraviolet waves, X-rays, visible light, infrared waves, microwaves, gamma rays, and radio waves on the diagram of electromagnetic waves. SC.8.E.5.11

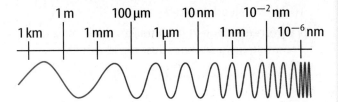

Writing in Science

12 On a separate sheet of paper, write a paragraph comparing colonizing North America and colonizing the Moon. Include a main idea, supporting details, and a concluding sentence. LA.8.2.2.3

Big Idea Review

13 In what different ways do humans observe and explore space? SC.8.E.5.10

_____

_____

_____

_____

14 The *Hubble Space Telescope* orbits Earth. What are advantages of space-based telescopes? What are disadvantages? SC.8.E.5.11

_____

_____

_____

_____

_____

Math Skills   MA.6.A.3.6

Use Scientific Notation

15 The distance from Saturn to the Sun averages 1,430,000,000 km. Express this distance in scientific notation.

16 The nearest star outside our solar system is Proxima Centauri, which is about 39,900,000,000,000 km from Earth. What is this distance in scientific notation?

17 The *Hubble Space Telescope* has taken pictures of an object that is 1,400,000,000,000,000,000,000 km away from Earth. Express this number in scientific notation.

Fill in the correct answer choice.

Multiple Choice

1 Why are radio telescopes built in remote deserts? SC.8.E.5.11

  Ⓐ to avoid interference with water vapor in the atmosphere

  Ⓑ to avoid interference with people

  Ⓒ to power the telescopes with heat

  Ⓓ to use the interference with water vapor in the atmosphere

2 Which technology does NOT leave Earth to collect data? SC.8.E.5.10

  Ⓕ a probe

  Ⓖ a rocket

  Ⓗ a satellite

  Ⓘ a radio telescope

Use the figure below to answer question 3.

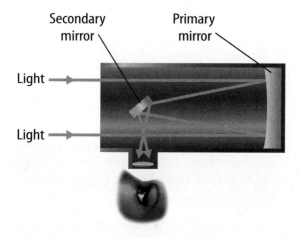

3 Which piece of space technology, shown above, uses mirrors to create images of planets and stars? SC.8.E.5.10

  Ⓐ lunar probe

  Ⓑ refracting telescope

  Ⓒ reflecting telescope

  Ⓓ orbiter

4 Which NASA location in Florida launches rockets for space exploration programs? SC.8.E.5.12

  Ⓕ Johnson Space Center

  Ⓖ Kennedy Space Center

  Ⓗ NASA Headquarters

  Ⓘ Jet Propulsion Laboratory

Use the table below to answer question 5.

Planet	Average Distance from Sun (in millions of kilometers)
Earth	150
Mars	228
Saturn	1,434

5 It takes about 8.3 min for light to travel from the Sun to Earth. It takes about 12.6 min for light to travel from the Sun to Mars. How long would you expect it to take light to travel from the Sun to Saturn? SC.8.E.5.1

  Ⓐ 8.5 min

  Ⓑ 1.3 h

  Ⓒ 13.5 h

  Ⓓ 26.3 h

6 Why does light from some stars take millions of years to reach Earth? SC.8.E.5.1

  Ⓕ because light does not travel in a straight line

  Ⓖ because light moves at 0.000003 km/s

  Ⓗ because the stars are extremely close to Earth

  Ⓘ because the stars are extremely far from Earth

7 What is the advantage of using gravity assist for a mission to Saturn? **SC.8.E.5.1, SC.8.E.5.10**

Ⓐ The spacecraft can be made of a nonmagnetic material.

Ⓑ The spacecraft can travel at the speed of light.

Ⓒ The spacecraft needs less fuel.

Ⓓ The spacecraft needs more weight.

8 The space probe *Phoenix*, which analyzed the martian surface for evidence of water, is what type of space probe? **SC.8.E.5.10**

Ⓕ orbiter

Ⓖ lander

Ⓗ flyby

Ⓘ rocket

Use the figure below to answer question 9.

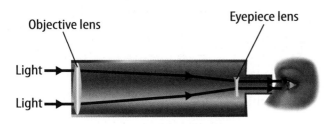

9 What type of light does the telescope shown above collect? **SC.8.E.5.10**

Ⓐ infrared waves

Ⓑ X-rays

Ⓒ visible light waves

Ⓓ ultraviolet waves

Use the figure below to answer questions 10 and 11.

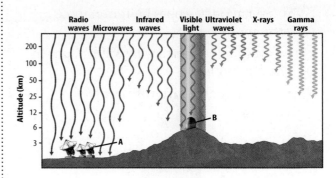

10 How do the electromagnetic waves collected by telescope A compare to those collected by telescope B? **SC.8.E.5.11**

Ⓕ Telescope A collects shorter wavelengths.

Ⓖ Telescope B collects longer wavelengths.

Ⓗ Telescope A and telescope B collect waves with the same energy.

Ⓘ Telescope A collects waves with more energy.

11 How do the frequencies of the electromagnetic waves collected by telescope A compare to those collected by telescope B? **SC.8.E.5.11**

Ⓐ Telescope A collects waves with higher frequencies.

Ⓑ Telescope B collects waves with higher frequencies.

Ⓒ Telescope A collects waves with no frequencies.

Ⓓ Telescope A and telescope B collect waves with the same frequencies.

NEED EXTRA HELP?

If You Missed Question...	1	2	3	4	5	6	7	8	9	10	11
Go to Lesson...	1	2	1	2	1	1	3	2	1	1	1

Benchmark Mini-Assessment — Chapter 4 • Lesson 1

mini BAT

Multiple Choice *Bubble the correct answer.*

Uses   Name

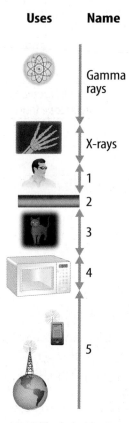

Gamma rays

X-rays

1

2

3

4

5

Reflecting Telescope

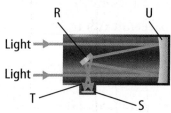

3. Which letter in the image above indicates the part in the reflecting telescope that concentrates light from a distant object? **SC.8.E.5.10**

  (A) R

  (B) S

  (C) T

  (D) U

4. Star A mostly emits microwaves. Star B mostly emits X-rays. The energy from Star B reaches Earth sooner than the energy from Star A because **SC.8.E.5.1**

  (F) Star B is brighter than Star A is.

  (G) Star B is closer to Earth than Star A is.

  (H) X-rays have more energy than microwaves.

  (I) X-rays move more quickly than microwaves.

1. Which number in the image above indicates the part of the electromagnetic spectrum used by a refracting telescope? **SC.8.E.5.11**

  (A) 2

  (B) 3

  (C) 4

  (D) 5

2. Infrared waves **SC.8.E.5.11**

  (F) are brighter than visible light waves.

  (G) are longer than visible light waves.

  (H) have more energy than visible light waves.

  (I) move more quickly than visible light waves.

Benchmark Mini-Assessment   Chapter 4 • Lesson 2

Multiple Choice *Bubble the correct answer.*

Use the time line below to answer questions 1 through 4.

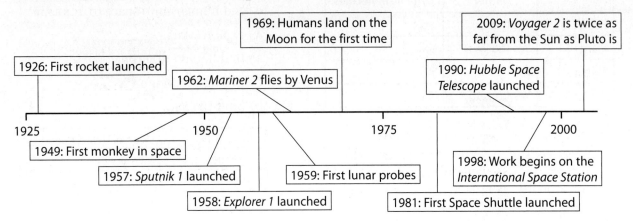

1926: First rocket launched

1969: Humans land on the Moon for the first time

1962: *Mariner 2* flies by Venus

2009: *Voyager 2* is twice as far from the Sun as Pluto is

1990: *Hubble Space Telescope* launched

1925   1950   1975   2000

1949: First monkey in space

1957: *Sputnik 1* launched

1958: *Explorer 1* launched

1959: First lunar probes

1998: Work begins on the *International Space Station*

1981: First Space Shuttle launched

1. Which year is most commonly considered to be the beginning of the space age? **SC.8.E.5.12**

   Ⓐ 1926
   Ⓑ 1949
   Ⓒ 1957
   Ⓓ 1969

2. According to the events shown in the time line, when did humans travel farthest from Earth? **SC.8.E.5.10**

   Ⓕ 1969
   Ⓖ 1981
   Ⓗ 1998
   Ⓘ 2009

3. When was the first reusable spacecraft launched? **SC.8.E.5.10**

   Ⓐ 1959
   Ⓑ 1969
   Ⓒ 1981
   Ⓓ 1998

4. When did the United States join other nations to create a research laboratory in space? **SC.8.E.5.10**

   Ⓕ 1949
   Ⓖ 1959
   Ⓗ 1981
   Ⓘ 1998

Benchmark Mini-Assessment   **Chapter 4 • Lesson 3**   mini BAT

Multiple Choice *Bubble the correct answer.*

	Water	**Organic Molecules**
Planet R	Yes (vapor)	No
Planet S	Yes (boiling)	Yes
Planet T	No	Yes
Planet U	Yes (frozen)	Yes

1. The chart above shows important characteristics of four newly discovered planets. Which planet is the best candidate for supporting extraterrestrial life? **SC.8.E.5.10**

   Ⓐ Planet R

   Ⓑ Planet S

   Ⓒ Planet T

   Ⓓ Planet U

2. NASA would most likely use gravity assist when sending a probe to which object in our solar system? **SC.8.E.5.10**

   Ⓕ Moon

   Ⓖ Sun

   Ⓗ Neptune

   Ⓘ Venus

3. Sometime in the next 20 years, NASA plans to send humans into space to **SC.8.E.5.10**

   Ⓐ explore Mars.

   Ⓑ find liquid on Europa.

   Ⓒ live on the Moon.

   Ⓓ orbit the Sun.

Probe	**Planet/Dwarf Planet**
Messenger	Mercury
Spirit	Mars
Cassini	Z
New Horizons	Pluto

4. Which planet does Z refer to in the table above? **SC.8.E.5.10**

   Ⓕ Jupiter

   Ⓖ Saturn

   Ⓗ Uranus

   Ⓘ Venus

Notes

Unit 2
Properties of Matter

1000 B.C. **1700** **1800**

350 B.C.
Greek philosopher Aristotle defines an element as "one of those bodies into which other bodies can decompose, and that itself is not capable of being divided into another."

1704
Isaac Newton proposes that atoms attach to each other by some type of force.

1869
Dmitri Mendeleev publishes the first version of the periodic table.

1874
G. Johnstone Stoney proposes the existence of the electron, a subatomic particle that carries a negative electric charge, after experiments in electrochemistry.

1897
J.J. Thompson demonstrates the existence of the electron, proving Stoney's claim.

1900

1907
Physicists Hans Geiger and Ernest Marsden, under the direction of Ernest Rutherford, conduct the famous gold foil experiment. Rutherford concludes that the atom is mostly empty space and that most of the mass is concentrated in the atomic nucleus.

1918
Ernest Rutherford reports that the hydrogen nucleus has a positive charge, and he names it the proton.

1950

1932
James Chadwick discovers the neutron, a subatomic particle with no electric charge and a mass slightly larger than a proton.

? Inquiry
Visit ConnectED for this unit's **STEM** activity.

Technology

SC.8.N.4.1, SC.8.N.4.2, SC.8.E.5.10

Scientists use technology to develop materials with desirable properties. **Technology** is the practical use of scientific knowledge, especially for industrial or commercial use. In the late 1800s, scientists developed the first plastic material, called celluloid, from cotton. In the 1900s, scientists developed other plastic materials, such as polystyrene, rayon, and nylon. These new materials were inexpensive, durable, lightweight, and could be molded into any shape.

New technologies often come with problems. For example, many plastics are made from petroleum and contain harmful chemicals. The high pressures and temperatures needed to produce plastics require large amounts of energy. Bacteria and fungi that easily break down natural materials do not easily decompose plastics. Often, plastics accumulate in landfills where they can remain for hundreds, or even thousands, of years, as shown in **Figure 1.**

Figure 1 Nature cannot easily recycle many human-made materials. Much of our trash remains in landfills for years. Scientists are developing materials that degrade quickly. This will help decrease the amount of pollution.

Active Reading

1. Identify What are three negative aspects of producing and using plastics?

_____

_____

_____

Types of Materials

Figure 2 Some organisms produce materials with properties that are useful to people. Scientists are trying to replicate these materials for new technologies.

Most human-made adhesives attach to some surfaces, but not others. Mussels, which are similar to clams, produce a "superglue" that is stronger than anything people can make. It also works on wet or dry surfaces. Chemists are trying to develop a technology that will replicate the mussel glue. This glue would provide solutions to difficult problems. Ships could be repaired under water, chipped teeth could be repaired, and broken bones could be more accurately set in place.

Abalone and other mollusks construct a protective shell from proteins and seawater. The material is four times stronger than human-made metal alloys and ceramics. Using technology, scientists are working to duplicate this material. They hope to use the new product in many ways, including for hip and elbow replacements. Automakers could use these strong, lightweight materials for automobile body panels.

Consider the Possibilities!

Chemists are looking to nature for ideas for new materials. For example, some sea sponges have skeletons that beam light deep inside the animal, similar to the way fiber-optic cables work. A bacterium from a snail-like nudibranch contains compounds that stop other sea creatures from growing on the nudibranch's back. These compounds could be used in paints to stop creatures from forming a harmful crust on submerged parts of boats and docks. Chrysanthemum flowers produce a product that keeps ticks and mosquitoes away. **Figure 2** includes other organisms that produce materials with remarkable properties.

Chemists and biologists are teaming up to understand and replicate the processes that organisms use to survive. Hopefully, these processes will lead to technologies and materials with unique properties that are helpful to people.

SC.8.N.1.6, SC.8.N.3.1

Inquiry

LAB STATION **Try It!**

MiniLab *How would you use it?* at connectED.mcgraw-hill.com

Apply It!

After you complete the lab, answer these questions.

1. **Estimate** How would your invention save money, resources, time, and effort as designed from an organism?

2. **Recognize** What are some problems that might be solved using technology spinoffs from organisms?

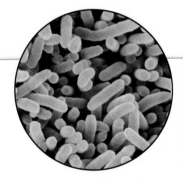

A British company has developed bacteria that produce large amounts of hydrogen gas when fed a diet of sugar. Chemists are working to produce tanks of these microorganisms that produce enough hydrogen to replace other fuels used to heat homes. Bacteria may become the power plants of the future.

Under a microscope, the horn of a rhinoceros looks much like the material used to make the wings of a Stealth aircraft. However, the rhino horn is self-healing. Picture a car with technologically advanced fenders similar to the horn of a rhinoceros; such a car could repair itself if it were in a fender-bender!

Spider silk begins as a liquid inside the spider's body. When ejected through openings, called spinnerets, it becomes similar to a plastic thread. However, its properties include strength five times greater than steel, stretchability greater than nylon, and toughness better than the material in bulletproof vests! Chemists are using technology to make a synthetic spider silk. They hope to someday use the material for cables strong enough to support a bridge or as reinforcing fibers in aircraft bodies.

Name _____ Date _____

Ball of Clay

Jenna placed a ball of clay on a table. She flattened it into the shape of a pancake. Which row in the chart best describes the properties of the clay after it was flattened?

	Weight	Mass
A	Stays the same	Increases
B	Stays the same	Decreases
C	Stays the same	Stays the same
D	Increases	Increases
E	Increases	Decreases
F	Increases	Stays the same
G	Decreases	Increases
H	Decreases	Decreases
I	Decreases	Stays the same

Explain your thinking about what happens to the mass and weight and why.

Matter: Properties and CHANGES

FLORIDA BIG IDEAS

1 **The Practice of Science**

8 **Properties of Matter**

9 **Changes in Matter**

Think About It!

What gives a substance its unique identity?

When designing a safe airplane, choosing materials with specific properties is important. Notice how the metal used in the outer shell of this airplane is curved, yet strong enough to hold its shape. Think about how the properties of the airplane's materials are important to the conditions in which it flies.

1 What properties do you think would be important to consider when constructing the outer shell of an airplane?

_____

_____

_____

2 Why do you think metal is used for electric wiring and plastic is used for the interior walls of the airplane?

_____

_____

_____

3 Why do you think different substances have different properties?

_____

_____

Get Ready to Read

What do you think about matter?

Before you read, decide if you agree or disagree with each of these statements. As you read this chapter, see if you change your mind about any of the statements.

	AGREE	DISAGREE
1 The particles in a solid object do not move.	☐	☐
2 Your weight depends on your location.	☐	☐
3 The particles in ice are the same as the particles in liquid water.	☐	☐
4 Mixing powdered drink mix with water causes a new substance to form.	☐	☐
5 If you combine two substances, bubbling is a sign that a new type of substance might be forming.	☐	☐
6 If you stir salt into water, the total amount of matter decreases.	☐	☐

There's More Online!
Video • Audio • Review • ⓘLab Station • WebQuest • Assessment • Concepts in Motion • Multilingual eGlossary

185

Lesson 1

Matter and Its PROPERTIES

ESSENTIAL QUESTIONS

 How do particles move in solids, liquids, and gases?

 How are physical properties different from chemical properties?

 How are properties used to identify a substance?

Vocabulary

volume p. 188

solid p. 188

liquid p. 188

gas p. 188

physical property p. 190

mass p. 190

density p. 191

solubility p. 192

chemical property p. 195

 Florida NGSSS

SC.8.P.8.1 Explore the scientific theory of atoms (also known as atomic theory) by using models to explain the motion of particles in solids, liquids, and gases.

SC.8.P.8.2 Differentiate between weight and mass recognizing that weight is the amount of gravitational pull on an object and is distinct from, though proportional to, mass.

SC.8.P.8.3 Explore and describe the densities of various materials through measurement of their masses and volumes.

SC.8.P.8.4 Classify and compare substances on the basis of characteristic physical properties that can be demonstrated or measured; for example, density, thermal or electrical conductivity, solubility, magnetic properties, melting and boiling points, and know that these properties are independent of the amount of the sample.

SC.8.N.1.1 Define a problem from the eighth grade curriculum using appropriate reference materials to support scientific understanding, plan and carry out scientific investigations of various types, such as systematic observations or experiments, identify variables, collect and organize data, interpret data in charts, tables, and graphics, analyze information, make predictions, and defend conclusions.

Also covers: LA.8.2.2.3, MA.6.A.3.6

 Launch Lab

SC.8.P.8.4

15 minutes

How can you describe a substance?

Think about the different ways you can describe a type of matter. Is it hard? Can you pour it? What color is it? Answering questions like these can help you describe the properties of a substance. In this lab, you will observe how the properties of a mixture can be very different from the properties of the substances it is made from.

Procedure

1 Read and complete a lab safety form.

2 Using a **small plastic spoon,** measure two spoonfuls of **cornstarch** into a **clear plastic cup.** What does the cornstarch look like? What does it feel like?

3 Slowly stir one spoonful of **water** into the cup containing the cornstarch. Gently roll the new substance around in the cup with your finger.

Think About This

1. **Identify** What were some properties of the cornstarch and water before they were mixed?

2.  **Differentiate** How were the properties of the mixture different from the original properties of the cornstarch and water?

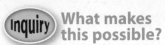 **What makes this possible?**

1. Scuba diving in the Florida Keys National Marine Sanctuary is a lot of fun, but you have to be prepared. Exploring below the water's surface has its dangers, and you need good equipment. What properties must a diver's wet suit, mask, flippers, and breathing apparatus have to make a safe dive possible?

_____

_____

_____

REVIEW VOCABULARY

matter
anything that has mass and takes up space

What is matter?

Have you ever experienced the excitement of scuba diving through a coral reef? The warm, pristine water can be absolutely breathtaking! As you drift through the quiet water, you are awestruck by the beautiful coral and colorful sea creatures. You may even discover an amazing shipwreck. Except for your air bubbles rising to the surface, there is not a sound.

Imagine looking around and asking yourself, "What is matter?" Sand, coral, fish, and all the things you might see during your dive are **matter** because they have mass and take up space. Air, even though you can't see it, is also matter because it too has mass and takes up space. Light from the Sun is not matter because it does not have mass and does not take up space. Sounds, forces, and energy also are not matter.

Think about the properties of matter around you below the water's surface. The face mask you wear is hard and clear. The hose from your air tank is soft and flexible. The colorful coral around you is alive. Matter has many properties. In this chapter, you will read about some of the properties of matter and how these properties help to identify many types of matter.

Active Reading **2. List** What are two examples of matter you can see in your classroom?

_____

_____

What are two examples of matter you cannot see?

_____

_____

What are two examples of things that are not matter?

_____

_____

States of Matter

One property that is useful when you describe different materials is state of matter. Three familiar states of matter are solids, liquids, and gases. A fourth state of matter is plasma, which is made of positive and negative particles. You can determine a material's state of matter by answering the following questions: Does it have a definite shape? Does it have a definite volume?

Volume *is the amount of space a sample of matter occupies.* As shown in **Table 1,** a material's state of matter determines whether its shape and its volume change when it is moved from one container to another.

Solids, Liquids, and Gases

Notice in **Table 1** that a **solid** *is a state of matter with a definite shape and volume.* The shape and volume of a solid do not change regardless of whether it is inside or outside a container. A **liquid** *is a state of matter with a definite volume but not a definite shape.* A liquid changes shape if it is moved to another container, but its volume does not change. *A state of matter without a definite shape or a definite volume is a* **gas.** A gas changes both shape and volume depending on the size and shape of its container.

Active Reading

3. Recognize Think of two examples each of solids, liquids, and gases. Write them in the table below. Do not use examples already listed in **Table 1.**

Table 1 Solids, Liquids, and Gases	
Solid Solids, such as rocks, do not change shape or volume regardless of whether they are inside or outside a container.	
Liquid A liquid, such as fruit juice, changes shape if it is moved from one container to another. Its volume does not change.	
Gas A gas, such as nitrogen dioxide, changes both shape and volume if it is moved from one container to another. If the container is not closed, the gas spreads out of the container.	

- no definite shape
- no definite volume
- particles very far apart
- very weak attractive forces between particles
- particles move freely

- a definite shape
- a definite volume
- particles close together
- strong attractive forces between particles
- particles vibrate in all directions

- no definite shape; takes the shape of its container
- definite volume
- particles close together
- weaker attractive forces between particles than in solids
- particles free to move past neighboring particles

Figure 1 The movement of particles and the attraction between them are different in solids, liquids, and gases.

Moving Particles

All matter is made of tiny particles that are constantly moving. Notice in **Figure 1** how the movement of particles is different in each state of matter. In solids, particles vibrate back and forth in all directions. However, particles in a solid cannot move from place to place. In liquids, the distance between particles is greater. Particles in liquids can slide past one another, similar to the way marbles in a box slide around. In a gas, particles move freely rather than staying close together.

🐟 **4. NGSSS Check Find** Highlight explanations of how particles move in solids, liquids, and gases.

Attraction Between Particles

Particles of matter that are close together exert an attractive force, or pull, on each other. The strength of the attraction depends on the distance between particles. Think about how this attraction affects the properties of the objects in **Figure 1**. A strong attraction holds particles of a solid close together in the same position. Liquids can flow because forces between the particles are weaker. Particles of a gas are so spread apart that they are not held together by attractive forces.

Active Reading 5. Label On the lines in each box in **Figure 1,** indicate whether the box is describing a solid, a liquid, or a gas. Then, in the circles, draw the correct particles in motion from the choices below.

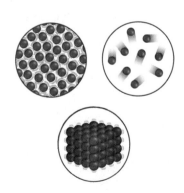

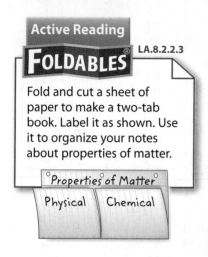

Active Reading

FOLDABLES® LA.8.2.2.3

Fold and cut a sheet of paper to make a two-tab book. Label it as shown. Use it to organize your notes about properties of matter.

Properties of Matter
Physical | Chemical

What are physical properties?

Think again about the properties of matter you might observe on a rafting trip. The water feels cold. The raft is heavy. The helmets are hard. The properties of all materials, or types of matter, depend on the substances that make them up. Recall that a substance is a type of matter with a composition that is always the same. *Any characteristic of matter that you can observe without changing the identity of the substances that make it up is a* **physical property**. **State** of matter, temperature, and the size of an object are all examples of physical properties.

Mass and Weight

Some physical properties of matter, such as mass and weight, depend on the size of the sample. **Mass** *is the amount of matter in an object*. Weight is the gravitational pull on the mass of an object. To measure the mass of a rock, you can use a balance, as shown in **Figure 2.** If more particles were added to the rock, its mass would increase, and the reading on the balance would increase. The weight of the rock would also increase.

Weight depends on the location of an object, but its mass does not. For example, the mass of an object is the same on Earth as it is on the Moon. The object's weight, however, is greater on Earth because the gravitational pull on the object is greater on Earth than on the Moon.

 6. NGSSS Check Differentiate <u>Underline</u> the definition of *mass* and highlight the explanation of *weight*. SC.8.P.8.2

Figure 2 You can measure a material's mass and volume and then calculate its density.

Mass, Volume, and Density

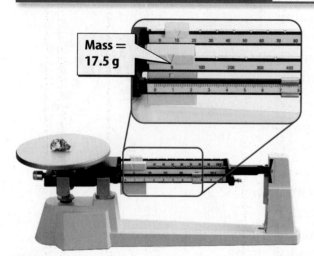

Mass = 17.5 g

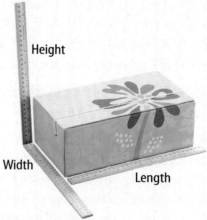

Volume = length × width × height

Height

Width

Length

Mass
A balance measures an object's mass by comparing it to the mass of the slides on the balance. Common units of mass are the kilogram (kg) and the gram (g).

Volume of a Rectangular-Shaped Solid
You find the volume of a rectangular solid by multiplying its length, its width, and its height together. A common unit of volume for a solid is the cubic centimeter (cm^3).

Volume

Another physical property of matter that depends on the amount or size of the sample is volume. You can measure the volume of a liquid by pouring it into a graduated cylinder or a measuring cup and reading the volume mark. Two ways to measure the volume of a solid are shown in **Figure 2.** If a solid has a regular geometric shape, you can calculate its volume by using the correct formula. If a solid has an irregular shape, you can use the displacement method to measure its volume.

Density

Density is a physical property of matter that does not depend on the size or amount of the sample. **Density** *is the mass per unit volume of a substance.* Density is useful when identifying unknown substances because it is constant for a given substance, regardless of the size of the sample. For example, imagine hiking in the mountains and finding a shiny yellow rock. Is it gold? Suppose you calculate that the density of the rock is 5.0 g/cm³. This rock cannot be gold because the density of gold is 19.3 g/cm³. A sample of pure gold, regardless of the size, will always have a density of 19.3 g/cm³.

Inquiry   SC.8.N.1.1, SC.8.P.8.3

LAB STATION Try It!

MiniLab *How can you find an object's mass and volume?* at connectED.mcgraw-hill.com

Apply It!

After you complete the lab, write a response to the following situation.

1. Suppose you have a bag of small glass beads. You want to measure the volume of one of the beads. However, when you place a bead in the water in the graduated cylinder, the water level does not rise enough for you to accurately read the volume. Using what you learned in the MiniLab, design a procedure you could follow to determine the volume of a single bead.

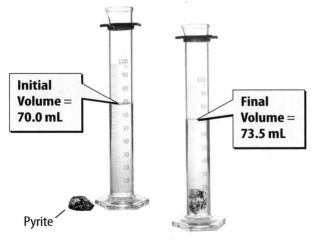

Initial Volume = 70.0 mL

Final Volume = 73.5 mL

Pyrite

Volume of an Irregular-Shaped Solid
The volume of an irregular-shaped object can be measured by displacement. The volume of the object is the difference between the water level before and after placing the object in the water. The common unit for liquid volume is the milliliter (mL).

Density Equation
$$\text{Density (in g/mL)} = \frac{\text{mass (in g)}}{\text{volume (in mL)}}$$

$$D = \frac{m}{v}$$

To find the density of the rock, first determine the mass and the volume of the rock:

mass: $m = 17.5$ g
volume: $V = 73.5$ mL -70.0 mL $= 3.5$ mL

Then, divide the mass by the volume:

$$D = \frac{17.5 \text{ g}}{3.5 \text{ mL}} = 5.0 \text{ g/mL}$$

Density Calculation
Density can be calculated using the density equation. The common units of density are grams per milliliter (g/mL) or grams per cubic centimeter (g/cm³). 1 mL = 1 cm³.

Conductivity

Another property that is independent of the sample size is conductivity. Electric conductivity is the ability to conduct an electric current. Copper often is used for electric wiring because it has high electric conductivity. Thermal conductivity is the ability of a material to conduct thermal energy. Metals tend to have high electric and thermal conductivity.

Solubility

You might have made lemonade by stirring a powdered drink mix into water. As you stir, the powder dissolves in the water. What would happen if you tried to dissolve sand in water? No matter how much you stir, the sand would remain solid. **Solubility** *is the ability of one substance to dissolve in another.* The powder is soluble in water, but sand is not.

Melting and Boiling Point

Melting point and boiling point also are physical properties. The melting point is the temperature at which a solid changes to a liquid. Ice cream melts when it warms enough to reach its melting point. The boiling point is the temperature at which a liquid changes to a gas. If you heat a pan of water, the water will boil at its boiling point. These temperatures do not depend on the size or the amount of the material.

SCIENCE USE v. COMMON USE

bond

Science Use a force between atoms or groups of atoms

Common Use a monetary certificate issued by a government or a business that earns interest

Table 2 🔑 This table contains the descriptions of several physical properties.

Active Reading **7. Summarize** Review the physical properties of matter by completing the table below.

Table 2 Physical Properties			
	Mass	**Conductivity**	**Volume**
Property			
Description of property	_____ _____ _____	The ability of matter to conduct, or carry along, electricity or heat	_____ _____ _____
Size-dependent or size-independent	_____	Size-independent	Size-dependent
How the property is used to separate a mixture (example)	Mass typically is not used to separate a mixture.	Conductivity typically is not used to separate a mixture.	Volume could be used to separate mixtures whose parts can be separated by filtration.

Magnetism

Some matter can be described by the specific way it behaves. For example, some materials pull iron toward them. These materials are said to be magnetic. Attraction to iron is a physical property of these substances. The ability to attract a magnet is independent of the sample size.

Separating Mixtures

Substances that make up mixtures are not held together by chemical **bonds**. When substances form a mixture, the properties of the individual substances do not change. One way a mixture and a compound differ is that the parts of a mixture often can be separated by physical properties. When salt and water form a solution, the salt and the water do not lose any of their individual properties. Therefore, you can separate the salt from the water by using differences in their boiling points. Other physical properties that can be used to separate different mixtures are described in **Table 2**.

Physical properties cannot be used to separate a compound into the elements it contains. The atoms that make up a compound are bonded together and cannot be separated by physical means. For example, you cannot separate the hydrogen atoms from the oxygen atoms in water by boiling water.

Active Reading

9. **Identify** What happens to the properties of individual substances when they are combined as a mixture? Find and highlight your answer in the text.

Math Skills   MA.6.A.3.6

Solve a One-Step Equation

A statement that two expressions are equal is an equation. For example, examine the density equation:

$$D = \frac{m}{V}$$

To solve an equation, place the variables you know into the equation. Then solve for the unknown variable. For example, if an object has a mass of **52 g** and a volume of **4 cm³**, calculate its density:

$$D = \frac{52\ g}{4\ cm^3} = 13\ g/cm^3$$

8. **Practice**
A cube of metal measures 3 cm on each side. It has a mass of 216 g. What is the density of the metal?

Boiling/Melting Points	State of Matter	Density	Solubility	Magnetism
_____ _____ _____	Whether something is a solid, a liquid, or a gas	The amount of mass per unit of volume	_____ _____	Attractive force for some metals, especially iron
Size-independent	Size-independent	_____	Size-independent	_____
Each part of a mixture will boil or melt at a different temperature.	A liquid can be poured off a solid.	Objects with greater density sink in objects with less density.	Dissolve a soluble material to separate it from a material with less solubility.	Attract iron from a mixture of materials.

Table 3 Identifying an Unknown Material by Its Physical Properties 🔑

Substance	Color	Mass (g)	Melting Point (°C)	Density (g/cm³)
Table salt	white	14.5	801	2.17
Sugar	white	11.5	148	1.53
Baking soda	white	16.0	50	2.16
Unknown	white	16.0	801	2.17

Identifying Matter Using Physical Properties

Physical properties are useful for describing types of matter, but they are also useful for identifying unknown substances. For example, look at the substances in **Table 3.** Notice how their physical properties are alike and how they are different. How can you use these properties to identify the unknown substance?

You cannot identify the unknown substance by its color. All of the substances are white. You also cannot identify the unknown substance by its mass or volume. Mass and volume are properties of matter that change with the amount of the sample present.

Recall that melting point and density are properties of matter that do not depend on the size or the amount of the sample. They are more reliable for identifying an unknown substance. Notice that both the melting point and the density of the unknown substance match those of table salt. The unknown substance must be table salt.

When you identify matter using physical properties, consider how the properties are alike and how they are different from known types of matter. It is important that the physical properties you use to identify an unknown type of matter are properties that do not change for any sample size. A cup of salt and a spoonful of salt will have the same melting point and density even though the mass and volume for each will be different. Therefore, melting point and density are physical properties that are reliable when identifying an unknown substance.

Active Reading 10. **Evaluate** In the table below, indicate if the property can be useful in identifying an unknown substance.

Physical property	Useful in identifying a substance? (Y/N)	Why or why not?
boiling point		
volume		
solubility		
color		

Figure 3 Flammability and the ability to rust are examples of chemical properties.

Flammability
In 1937 the airship *Hindenburg* caught fire and crashed. It was filled with hydrogen, a highly flammable gas.

Ability to rust
The metal parts of this beached shipwreck soon rust because the metal contains iron. The ability to rust is a chemical property of iron.

What are chemical properties?

Have you ever seen an apple turn brown? When you bite into or cut open apples or other fruits, substances that make up the fruit react with oxygen in the air. When substances react with each other, their particles combine to form a new, different substance. The ability of substances in fruit to react with oxygen is a chemical property of the substances. *A **chemical property** is the ability or inability of a substance to combine with or change into one or more new substances.* A chemical property is a characteristic of matter that you observe as it reacts with or changes into a different substance. For example, copper on the roof of a building turns green as it reacts with oxygen in the air. The ability to react with oxygen is a chemical property of copper. Two other chemical properties—flammability and the ability to rust—are shown in **Figure 3.**

 **11. Give examples** List two objects in your room that have the same chemical properties as those shown in **Figure 3.**

Flammability

Flammability is the ability of a type of matter to burn easily. Suppose you are on a camping trip and want to light a campfire. You see rocks, sand, and wood. Which would you choose for your fire? Wood is a good choice because it is flammable. Rocks and sand are not flammable.

Materials are often chosen for certain uses based on flammability. For example, gasoline is used in cars because it burns easily in engines. Materials that are used for cooking pans must not be flammable. The tragedy shown in **Figure 3** resulted when hydrogen, a highly flammable gas, was used in the airship *Hindenburg.* Today, airships are filled with helium, a nonflammable gas.

Ability to Rust

You have probably seen old boats that have begun to rust like the one in **Figure 3.** You might also have seen rust on bicycles or tools left outside. Rust is a substance that forms when iron reacts with water and oxygen in the air. The ability to rust is a chemical property of iron or metals that contain iron.

Visual Summary

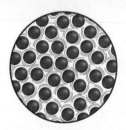

The movement of particles is different in a solid, a liquid, and a gas.

Physical properties and chemical properties are used to describe types of matter.

Physical properties such as magnetism can be used to separate mixtures.

Use Vocabulary

1. A state of matter that has a definite volume but not a definite shape is a _____.

2. **Distinguish** between a physical property and a chemical property.

Understand Key Concepts

3. **Analyze** Which can be used to identify an unknown substance: mass, melting point, density, volume, state of matter? SC.8.P.8.4

4. **Contrast** the movement of particles in a solid, a liquid, and a gas. SC.8.P.8.1

5. Which property is independent of sample amount? SC.8.P.8.4
   - (A) volume
   - (B) density
   - (C) mass
   - (D) weight

Interpret Graphics

6. **Calculate** the density of each object. SC.8.P.8.3

Object	Mass	Volume	Density
1	6.50 g	1.25 cm³	
2	8.65 g	2.50 mL	

Critical Thinking

7. **Design** an investigation you could use to find the density of a penny. SC.8.P.8.3

Inquiry SC.8.P.8.3, MA.6.A.3.6

LAB STATION **Try It!**

Skill Lab *How can you calculate density?* at connectED.mcgraw-hill.com

Math Skills   MA.6.A.3.6

8. The mass of a mineral is 9.6 g. The mineral is placed in a graduated cylinder containing 8.0 mL of water. The water level rises to 16.0 mL. What is the mineral's density?

Describe Complete the spider map to summarize the physical properties of matter.

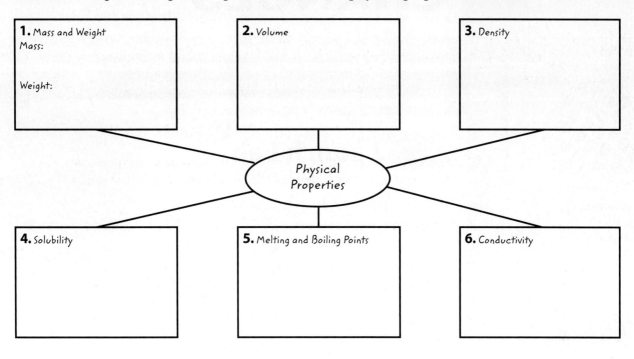

1. Mass and Weight
Mass:

Weight:

2. Volume

3. Density

Physical Properties

4. Solubility

5. Melting and Boiling Points

6. Conductivity

7. **Distinguish** properties as chemical or physical. Circle the chemical properties. Underline the physical properties. Then tell how physical properties are different from chemical properties.

ability to be bent or rolled   ability to be attracted to a magnet

ability to rust   ability to burn

ability to conduct electricity   ability to react with oxygen

You can observe _____ without changing the identity of the substances that make up a substance. You can observe _____ only as a substance reacts with or changes into a different substance.

Matter and Its CHANGES

Vocabulary

physical change p. 200

chemical change p. 202

law of conservation of mass p. 205

 Florida NGSSS

LA.8.2.2.3 The student will organize information to show understanding or relationships among facts, ideas, and events (e.g., representing key points within text through charting, mapping, paraphrasing, summarizing, or comparing/contrasting);

SC.8.P.8.3 Explore and describe the densities of various materials through measurement of their masses and volumes.

SC.8.P.8.4 Classify and compare substances on the basis of characteristic physical properties that can be demonstrated or measured; for example, density, thermal or electrical conductivity, solubility, magnetic properties, melting and boiling points, and know that these properties are independent of the amount of the sample.

SC.8.P.9.1 Explore the Law of Conservation of Mass by demonstrating and concluding that mass is conserved when substances undergo physical and chemical changes.

SC.8.P.9.2 Differentiate between physical changes and chemical changes.

SC.8.N.1.1 Define a problem from the eighth grade curriculum using appropriate reference materials to support scientific understanding, plan and carry out scientific investigations of various types, such as systematic observations or experiments, identify variables, collect and organize data, interpret data in charts, tables, and graphics, analyze information, make predictions, and defend conclusions.

Inquiry Launch Lab SC.8.P.9.2

10 minutes

What does a change in the color of matter show?

Matter has many different properties. Chemical properties can be observed only if the matter changes from one type to another. How can you tell if a chemical property has changed? Sometimes a change in the color of matter shows that its chemical properties have changed.

Procedure

1. Read and complete a lab safety form.

2. Obtain the **red indicator sponge** and the **red acid solution** from your teacher. Predict what will happen if the red acid solution touches the red sponge.

3. Use a **dropper** to remove a few drops of acid solution from the **beaker**. Place the drops on the sponge. ⚠ *Be careful not to splash the liquid onto yourself or your clothing.*

4. Record your observations.

Data and Observations

Think About This

1. **Compare** the properties of the sponge before and after you placed the acid solution on the sponge. Was your prediction correct?

2.  **Key Concept** **Identify** How do you know that physical properties and chemical properties changed?

 **Why is it orange?**

1. Streams are usually filled with clear freshwater. What happened to this water? Chemicals from a nearby mine seeped through rocks before flowing into the stream. How does the orange color form as these chemicals and metals from the rocks combine?

_____

_____

_____

_____

Changes of Matter

Imagine going to a park in the spring and then going back to the same spot in the fall. What changes do you think you might see? The changes would depend on where you live. An example of what a park in the fall might look like in many places is shown in **Figure 4.** Leaves that are soft and green in the spring might turn red, yellow, or brown in the fall. The air that was warm in the spring might be cooler in the fall. If you visit the park early on a fall morning, you might notice a thin layer of frost on the leaves. Matter, such as the things you see at a park, can change in many ways. These changes can be either physical or chemical.

Active Reading

2. Describe What are some changes that occur around you? How do the physical and chemical properties change?

_____

_____

_____

Figure 4 The physical and chemical properties of matter change in a park throughout the year.

What are physical changes?

A change in the size, shape, form, or state of matter that does not change the matter's identity is a **physical change**. You can see an example of a physical change in **Figure 5.** Recall that mass is an example of a physical property. Notice that the mass of the modeling clay is the same before and after its shape is changed. When a physical change occurs, the chemical properties of the matter stay the same. The substances that make up matter are exactly the same before and after a physical change.

Dissolving

One of the physical properties you read about in Lesson 1 was solubility—the ability of one material to dissolve, or mix evenly, in another. Dissolving is a physical change because the identities of the substances do not change when they are mixed. As shown in **Figure 6,** the identities of the water molecules and the sugar molecules do not change when sugar crystals dissolve in water.

Active Reading

3. **Defend** In the text, find and <u>underline</u> the explanation of why dissolving is classified as a physical change.

Figure 5 Changing the shape of the modeling clay does not change its mass.

Dissolving—A Physical Change

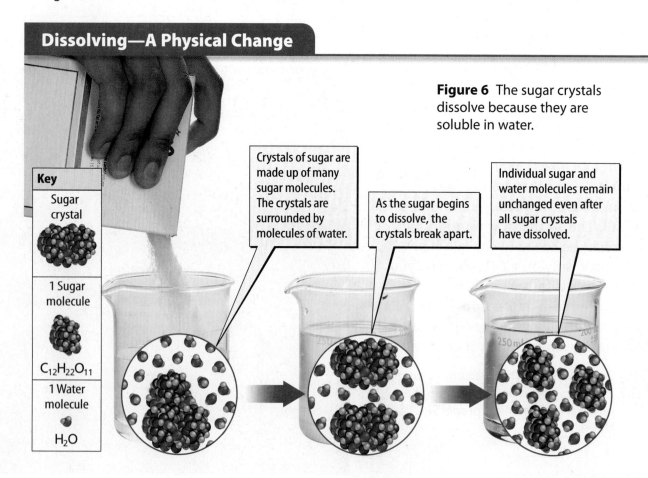

Figure 6 The sugar crystals dissolve because they are soluble in water.

Crystals of sugar are made up of many sugar molecules. The crystals are surrounded by molecules of water.

As the sugar begins to dissolve, the crystals break apart.

Individual sugar and water molecules remain unchanged even after all sugar crystals have dissolved.

Key

Sugar crystal

1 Sugar molecule

$C_{12}H_{22}O_{11}$

1 Water molecule

H_2O

Changing State

In Lesson 1 you read about three states of matter—solid, liquid, and gas. Can you think of examples of matter changing from one state to another? Changes in the state of matter are physical changes.

Melting and Boiling If you heat ice in a pot on the stove, the ice will melt, forming water that soon begins to boil. When a material melts, it changes from a solid to a liquid. When it boils, it changes from a liquid to a gas. The substances that make up the material do not change during a change in the state of matter, as shown in **Figure 7.** The particles that make up ice are the same as the particles that make up water as a liquid or gas.

Energy and Change in State The energy of the particles and the distances between the particles are different for a solid, a liquid, and a gas. Changes in energy cause changes in the state of matter. For example, energy must be added to a substance to change it from a solid to a liquid or from a liquid to a gas. Adding energy to a substance can increase its temperature. When the temperature reaches the substance's melting point, the solid changes to a liquid. At the boiling point, the liquid changes to a gas.

What would happen if you changed the rate at which you add energy to a substance? For example, what would happen if you heated an ice cube in your hand instead of in a pot on the stove? The ice would reach its melting point more slowly in your hand. The rate at which one state of matter changes to another depends on the rate at which energy is added to or taken away from the substance.

Figure 7 The particles that make up ice (solid water), liquid water, and water vapor (water in the gaseous state) are the same.

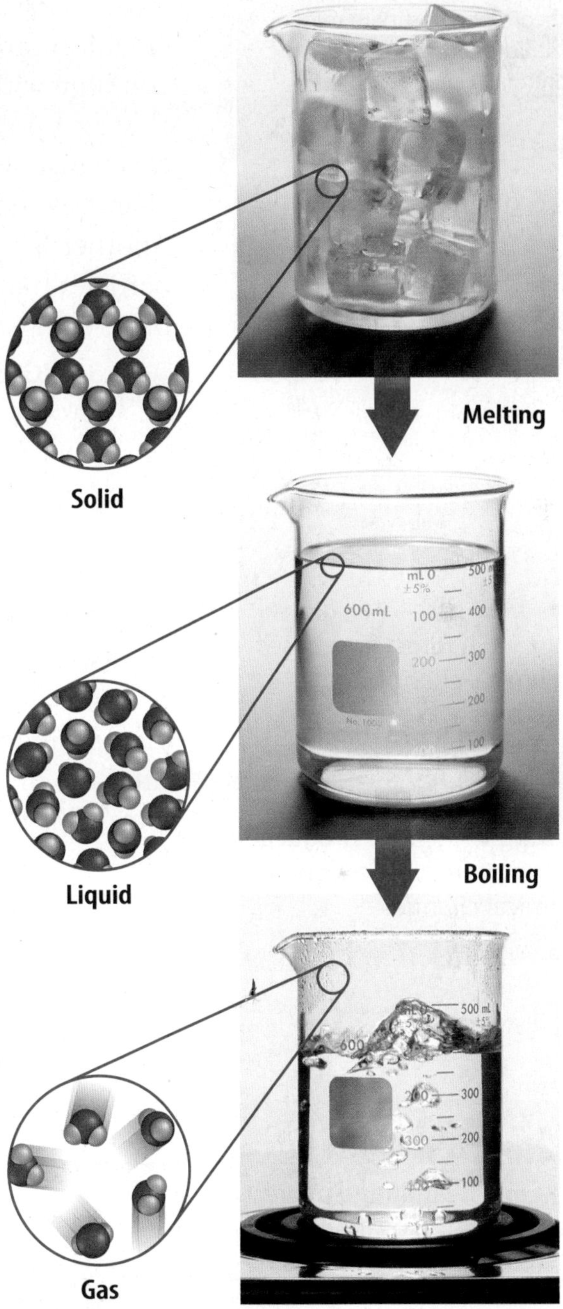

Solid

Melting

Liquid

Boiling

Gas

Active Reading 4. **Assess** What changes occur to the particles of a material during a change in state?

_____

_____

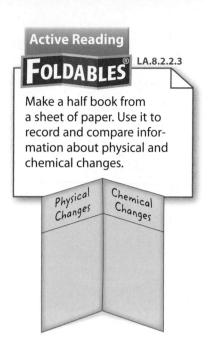
What are chemical changes?

Some changes in matter involve more than just changing physical properties. *A **chemical change** is a change in matter in which the substances that make up the matter change into other substances with different chemical and physical properties.* Recall that a chemical property is the ability or inability of a substance to combine with or change into one or more new substances. During a physical change, only the physical properties of matter change. However, the new substance produced during a chemical change has different chemical and physical properties. Another name for a chemical change is a chemical reaction. The particles that make up two or more substances react, or combine, with each other and form a new substance.

 5. NGSSS Check Describe How do chemical changes and physical changes differ? SC.8.P.9.2

_____

_____

_____

Signs of a Chemical Change

How can you tell that the burning of the trees in **Figure 8** is a chemical change? The reaction produces two gases—carbon dioxide and water vapor—even though you cannot see them. After the fire, you can see that any part of the trees that remains is black, and you can see ash—another new substance. But with some changes, the only new substance formed is a gas you cannot see. As trees burn in a forest fire, light and heat are signs of a chemical change. For many reactions, changes in physical properties, such as color or state of matter, are signs that a chemical change has occurred. However, the only sure sign of a chemical change is the formation of a new substance.

Figure 8 A forest fire causes a chemical change in the trees, producing new substances.

Active Reading **6. Explain** Why is the smoke produced during a forest fire a sign of a chemical change?

_____

_____

Chemical Change

Light and heat during a forest fire are signs that a chemical change is occurring.

After the fire, the formation of new substances shows that a chemical change has taken place.

Formation of Gas Bubbles of gas can form during both a physical change and a chemical change. When you heat a substance to its boiling point, the bubbles show that a liquid is changing to a gas—a physical change. When you combine substances, such as the medicine tablet and the water in **Figure 9,** gas bubbles show that a chemical change is occurring. Sometimes you cannot see the gas produced, but you might be able to smell it. The aroma of freshly baked bread, for example, is a sign that baking bread causes a chemical reaction that produces a gas.

Formation of a Precipitate Some chemical reactions result in the formation of a precipitate (prih SIH puh tut). As shown in the middle photo in **Figure 9,** a precipitate is a solid that sometimes forms when two liquids combine. When a liquid freezes, the solid formed is not a precipitate. A precipitate is not a state change from a liquid to a solid. Instead, the particles that make up two liquids react and form the particles that make up the solid precipitate, a new substance.

Color Change Suppose you want your room to be a different color. You would simply apply paint to the walls. The change in color is a physical change because you have only covered the wall. A new substance does not form. But notice the color of the precipitate in the middle photo of **Figure 9.** In this case, the change in color is a sign of a chemical change. The photo in the bottom of the figure shows that marshmallows change from white to brown when they are toasted. The change in the color of the marshmallows is also a sign of a chemical change.

Figure 9 The photographs below show several signs of chemical changes.

Formation of gas bubbles

Formation of a precipitate

Color change

Active Reading

7. **Identify** Fill in the graphic organizer with five types of chemical changes.

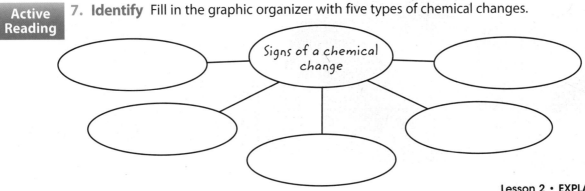

Signs of a chemical change

Figure 10 The flames, the lights, and the sounds of a fireworks display over Miami are signs of chemical changes.

Energy and Chemical Change

Think about a fireworks show. Again and again, you hear loud bangs as the fireworks burst into a display of colors, as in **Figure 10.** The release of thermal energy, light, and sound are signs that the fireworks result from chemical changes. All chemical reactions involve energy changes.

Figure 11 Thermal energy is needed for this chemical reaction.

Suppose you want to bake pretzels, as shown in **Figure 11.** What would happen if you placed one pan of unbaked pretzel dough in the oven and another pan of unbaked pretzel dough on the kitchen counter? Only the dough in the hot oven would become pretzels. Thermal energy is needed for the chemical reactions to occur that bake the pretzels.

Energy in the form of light is needed for other chemical reactions. Photosynthesis is a chemical reaction by which plants and some unicellular organisms produce sugar and oxygen. This process only occurs if the organisms are exposed to light. Many medicines also undergo chemical reactions when exposed to light. You might have seen some medicines stored in orange bottles. If the medicines are not stored in these light-resistant bottles, the ingredients can change into other substances.

Active Reading 8. **Select** On the lines below, list two familiar chemical changes, such as "burning a candle." In the boxes, describe the matter before and after adding energy.

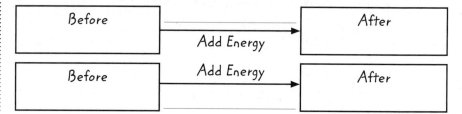

Before		After
	Add Energy →	

Before	Add Energy	After
	→	

Can changes be reversed?

Think again about the way matter changes form during a fireworks display. Once the chemicals combine and cause the explosions, you cannot get back the original chemicals. Like most chemical changes, the fireworks display cannot be reversed.

Grating a carrot and cutting an apple are physical changes, but you cannot reverse these changes either. Making a mixture by dissolving salt in a pan of water is also a physical change. You can reverse this change by boiling the mixture. The water will change to a gas, leaving the salt behind in the pan. Some physical changes can be easily reversed, but others cannot.

Active Reading

9. Identify In the text, find and (circle) one physical change that can be reversed and one physical change that cannot be reversed.

Conservation of Mass

Physical changes do not affect the mass of substances. When ice melts, for example, the mass of the ice equals the mass of the resulting liquid water. If you cut a piece of paper into strips, the total mass of the paper remains the same. Mass is conserved, or unchanged, during a physical change.

Mass is also conserved during a chemical change. Antoine Lavoisier (AN twon · luh VWAH zee ay) (1743–1794), a French chemist, made this discovery. Lavoisier carefully measured the masses of materials before and after chemical reactions. His discovery is now a scientific law. *The **law of conservation of mass** states that the total mass before a chemical reaction is the same as the total mass after the chemical reaction.* Weight also is the same because it depends on mass. For example, the mass of an unburned match plus the mass of the oxygen it reacts with equals the mass of the ashes plus the masses of all the gases given off when the match burns.

Inquiry

LAB STATION **Try It!**

SC.8.N.1.1, SC.8.P.9.1

MiniLab *Is mass conserved during a chemical reaction?* at connectED.mcgraw-hill.com

Apply It!

After you complete the lab, write a response to the following situation.

1. Suppose you measure the mass of a candle. Then, you burn it for several minutes. When you measure the candle's mass again, you find that it has lost mass. If the law of conservation of mass is true, explain where the missing mass might have gone.

Active Reading

10. Describe During a physical change, does the total mass of the matter decrease, increase, or stay the same? What about during a chemical change?

Physical change: _____

Chemical change: _____

11. Write What are some clues you can use to decide if a change is physical or chemical?

_____

_____

_____

_____

_____

_____

_____

Table 4 Chemical changes produce a new substance, but physical changes do not.

Comparing Physical and Chemical Changes

Suppose you want to explain to a friend the difference between a physical change and a chemical change. What would you say? You could explain that the identity of matter does not change during a physical change, but the identity of matter does change during a chemical change. However, you might not be able to tell just by looking at a substance whether its identity changed. You cannot tell whether the particles that make up the matter are the same or different.

Sometimes deciding if a change is physical or chemical is easy. Often, however, identifying the type of change is like being a detective. You have to look for clues that will help you figure out whether the identity of the substance has changed. For example, look at the summary of physical changes and chemical changes in **Table 4**. A change in color can occur during a chemical change or when substances are mixed (a physical change). Bubbles might indicate the formation of gas (a chemical change) or boiling (a physical change). You must consider many factors when comparing physical and chemical changes.

Table 4 Comparing Physical and Chemical Changes 🗝

Type of Change	Examples	Characteristics
Physical change	• melting • boiling • changing shape • mixing • dissolving • increasing or decreasing in temperature	• Substance is the same before and after the change. • Only physical properties change.
Chemical change	• changing color • burning • rusting • formation of gas • formation of a precipitate • spoiling food • tarnishing silver • digesting food	• Substance is different after the change. • Both physical and chemical properties change.

Physical change

Chemical change

Lesson Review 2

Visual Summary

The identity of a substance does not change during a physical change such as a change in the state of matter.

A new substance is produced during a chemical change.

The law of conservation of mass states that the mass of a material does not change during a chemical change.

Inquiry

Try It!

SC.8.N.1.1, SC.8.P.8.3, SC.8.P.8.4

Inquiry Lab *Identifying Unknown Minerals* at connectED.mcgraw-hill.com

Use Vocabulary

1. The particles that make up matter do not change during a(n) _____.

Understand Key Concepts

2. **Explain** how physical and chemical changes affect the mass of a material. SC.8.P.9.1

3. Which is a physical change? SC.8.P.9.2
   - (A) burning wood
   - (B) melting ice
   - (C) rusting iron
   - (D) spoiling food

Interpret Graphics

4. **Analyze** Suppose you mix 12.8 g of one substance with 11.4 g of another. The picture shows the mass you measure for the mixture. Is this reasonable? Explain.

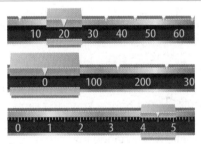

5. **Organize Information** List an example of each type of change. LA.8.2.2.3

Type of Change	Examples
Physical change with formation of bubbles	
Chemical change with formation of bubbles	

Critical Thinking

6. **Consider** Suppose you mix baking soda and white vinegar. What signs might indicate that a chemical change is occurring? SC.8.P.9.2

Chapter 5 Study Guide

Think About It! Physical and chemical properties give a substance its unique identity. The matter that makes up a substance can change physically or chemically.

 ## Key Concepts Summary

Vocabulary

LESSON 1 Matter and Its Properties

- Particles of a **solid** vibrate about a definite position. Particles of a **liquid** can slide past one another. Particles of a **gas** move freely within their container.

- A **physical property** is a characteristic of matter that you can observe without changing the identity of the substances that make it up. A **chemical property** is the ability or inability of a substance to combine with or change into one or more new substances.

- Some properties of matter do not depend on size or amount of the sample. You can identify a substance by comparing these properties to those of other known substances.

volume p. 188
solid p. 188
liquid p. 188
gas p. 188
physical property p. 190
mass p. 190
density p. 191
solubility p. 192
chemical property p. 195

LESSON 2 Matter and Its Changes

- A change in the size, shape, form, or state of matter in which the identity of the matter stays the same is a **physical change**. A change in matter in which the substances that make it up change into other substances with different chemical and physical properties is a **chemical change**.

- The **law of conservation of mass** states that the total mass before a chemical reaction is the same as the total mass after the reaction.

physical change p. 200
chemical change p. 202
law of conservation of mass p. 205

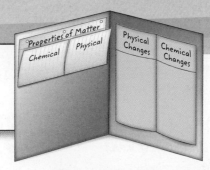

Active Reading

FOLDABLES® **Chapter Project**

Assemble your lesson Foldables as shown to make a Chapter Project. Use the project to review what you have learned in this chapter.

Use Vocabulary

1 A state of matter with a definite volume and a definite shape is a _____.

2 Flammability is an example of a _____ _____ of wood because when wood burns, it changes to different materials.

3 A drink mix dissolves in water because of its _____ in water.

4 The rusting of a metal tool left in the rain is an example of a _____.

5 According to the _____, the mass of an untoasted marshmallow equals its mass after it is toasted plus the mass of any gases produced as it was toasting.

6 Slicing an apple into sections is an example of a _____ that cannot be reversed.

Link Vocabulary and Key Concepts

Use vocabulary terms from the previous page to complete the concept map.

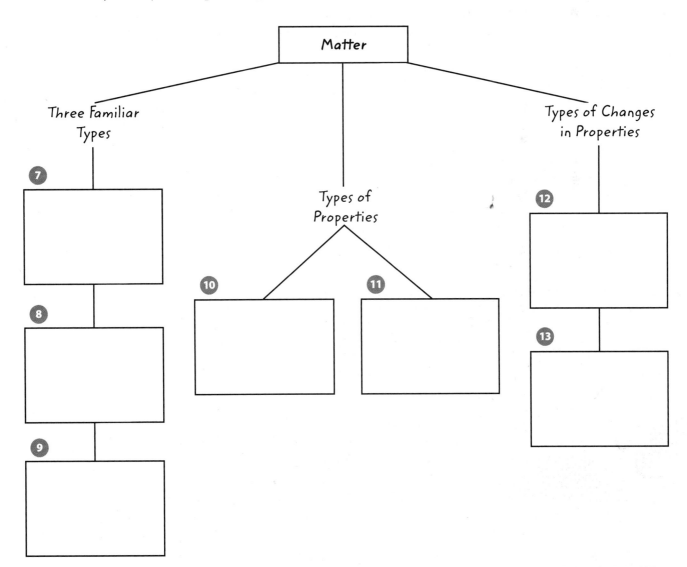

Fill in the correct answer choice.

🔑 Understand Key Concepts

1. Which is a property of all solids? SC.8.P.8.1
   - Ⓐ Particles are far apart.
   - Ⓑ Particles vibrate in all directions.
   - Ⓒ Volume and shape can easily change.
   - Ⓓ Weak forces exist between particles.

2. Which characteristic is a chemical property? SC.8.P.9.2
   - Ⓐ highly flammable
   - Ⓑ mass of 15 kg
   - Ⓒ woolly texture
   - Ⓓ golden color

3. Which property of an object depends on its location? SC.8.P.8.2
   - Ⓐ density
   - Ⓑ mass
   - Ⓒ volume
   - Ⓓ weight

4. How are the particles of a gas different from the particles of a liquid shown here? SC.8.P.8.1

   - Ⓐ They move more slowly.
   - Ⓑ They are farther apart.
   - Ⓒ They have less energy.
   - Ⓓ They have stronger attractions.

5. Which is a physical change? SC.8.P.9.2
   - Ⓐ burning natural gas
   - Ⓑ chopping onions
   - Ⓒ digesting food
   - Ⓓ exploding dynamite

6. Which stays the same when a substance changes from a liquid to a gas? SC.8.P.9.1
   - Ⓐ density
   - Ⓑ mass
   - Ⓒ forces between particles
   - Ⓓ distance between particles

Critical Thinking

7. **Apply** Suppose you find a gold-colored ring. Explain why you could use some physical properties but not others to determine whether the ring is actually made of gold. SC.8.P.8.4

8. **Reason** You make lemonade by mixing lemon juice, sugar, and water. Is this a physical change or a chemical change? Explain. SC.8.P.9.2

9. Give an example of a physical change you might observe at your school that is reversible and a physical change that is not reversible. SC.8.P.9.2

10. **Defend** A classmate defines a liquid as any substance that can be poured. Use the picture below to explain why this is not an acceptable definition. SC.8.P.8.1

11 **Suggest** a way that you could use displacement to determine the volume of a rock that is too large to fit into a graduated cylinder. LA.8.2.2.3

12 **Hypothesize** A scientist measures the mass of two liquids before and after combining them. The mass after combining the liquids is less than the sum of the masses before. Where is the missing mass? SC.8.P.9.1

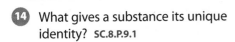

Writing in Science

13 **Write** a four-sentence description on a separate sheet of paper of an object in your home or classroom. Be sure to identify both physical properties and chemical properties of the object. LA.8.2.2.3

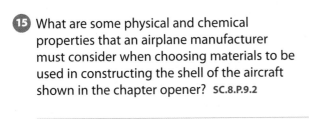

Big Idea Review

14 What gives a substance its unique identity? SC.8.P.9.1

15 What are some physical and chemical properties that an airplane manufacturer must consider when choosing materials to be used in constructing the shell of the aircraft shown in the chapter opener? SC.8.P.9.2

Math Skills MA.6.A.3.6

16 Use what you have learned about density to complete the table below. Then, determine the identities of the two unknown metals.

Metal	Mass (g)	Volume (cm³)	Density (g/cm³)
Iron	42.5	5.40	
Lead	28.8	2.55	
Tungsten	69.5	3.60	
Zinc	46.4	6.50	
	61.0	5.40	
	46.4	2.40	

Record your answers on the answer sheet provided by your teacher or on a sheet of paper.

Multiple Choice

1 Which describes the particles in a substance with no definite volume or shape? SC.8.P.8.1

  (A) Particles are close but can move freely.

  (B) Particles are close but can vibrate in all directions.

  (C) Particles are far apart and cannot move.

  (D) Particles are far apart and move freely.

2 Which diagram shows a chemical change? SC.8.P.9.2

  (F)

  (G)

  (H)

  (I)

3 Which is NOT true about firewood that burns completely? SC.8.P.9.1

  (A) Ashes and gases form from the substances in the wood.

  (B) Oxygen from the air combines with substances in the wood.

  (C) The total mass of substances in this process decreases.

  (D) The wood gives off thermal energy and light.

4 How do weight and mass differ? SC.8.P.8.2

  (F) Weight depends on the location of an object, but mass does not.

  (G) Weight does not depend on the location of an object, but mass does.

  (H) Weight and mass depend on the location of an object.

  (I) Weight and mass do not depend on the location of an object.

5 Which is true when an ice cube melts? SC.8.P.8.1

  (A) Volume and mass increase.

  (B) Volume and mass do not change.

  (C) Volume decreases, but mass does not change.

  (D) Volume increases, but mass decreases.

Use the table below to answer question 6.

Volume (mL)	Mass (g)
200	180
300	270
400	360

6 A liquid of unknown density is studied. Based on the data, what is the density of the liquid? SC.8.P.8.3

  (F) 1.1 g/mL

  (G) 0.9 g/mL

  (H) 1.1 mL/g

  (I) 0.9 mL/g

7 Which physical property CANNOT be measured? SC.8.P.8.4

Ⓐ melting point

Ⓑ density

Ⓒ conductivity

Ⓓ color

8 A spoonful of sugar with a mass of 8.8 g is poured into a 10-mL graduated cylinder. The volume is 5.5 mL. What is the density of the sugar? SC.8.P.8.3

Ⓕ 1.6 g/mL

Ⓖ 0.6 g/mL

Ⓗ 1.6 mL/g

Ⓘ 0.6 mL/g

9 Which is a sign of a physical change? SC.8.P.9.2

Ⓐ Bread gets moldy with age.

Ⓑ Ice forms on a puddle in winter.

Ⓒ The metal on a car starts to rust.

Ⓓ Yeast causes bread dough to rise.

10 When a newspaper is left in direct sunlight for a few days, the paper begins to turn yellow. What is this change in color? SC.8.P.9.2

Ⓕ physical property

Ⓖ chemical property

Ⓗ physical change

Ⓘ chemical change

11 Suppose a candle is burned in a closed system where matter cannot enter or leave. Given this situation, what is equal to the mass of the original candle? SC.8.P.9.1

Ⓐ the mass of the burned candle

Ⓑ the mass of all gases in the closed system

Ⓒ the mass of the gases released while the candle is burned

Ⓓ the mass of the burned candle plus the mass of the gases released while the candle is burned

Use the table below to answer question 12.

Material	Density (g/cm³)
Aluminum	2.7
Iron	7.9
Gold	19.3

12 Densities of some metals are shown in the table. Find the density of an alloy that is 50% aluminum and 50% gold by weight. SC.8.P.8.3

Ⓕ 22 g/cm³

Ⓖ 13.6 g/cm³

Ⓗ 5.3 g/cm³

Ⓘ 11 g/cm³

NEED EXTRA HELP?

If You Missed Question...	1	2	3	4	5	6	7	8	9	10	11	12
Go to Lesson...	1	2	2	1	1,2	1	1	1	2	2	2	1

Benchmark Mini-Assessment   Chapter 5 • Lesson 1

mini BAT

Multiple Choice *Bubble the correct answer.*

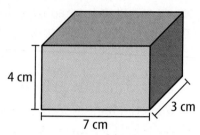

4 cm

3 cm

7 cm

1. The block shown above has a mass of 105 g. What is the density of the block? **SC.8.P.8.2**

   (A) 0.80 g/cm^3

   (B) 1.25 g/cm^3

   (C) 7.50 g/cm^3

   (D) 21.0 g/cm^3

2. A rock has which of these properties? **SC.8.P.8.3**

   (F) a changing shape and volume

   (G) a fixed shape and volume

   (H) a changing shape but a fixed volume

   (I) a changing volume but a fixed shape

3. Which physical property would NOT be useful for finding the difference between baking soda and cornstarch? **SC.8.P.8.2**

   (A) density

   (B) mass

   (C) melting point

   (D) solubility

4. Which formula below can be used to calculate the density of an object? **SC.8.P.8.3**

   (F) D = volume/mass

   (G) D = mass/volume

   (H) D = volume/weight

   (I) D = weight/mass

Multiple Choice *Bubble the correct answer.*

1. Which image shows a physical change?
SC.8.P.9.2

Ⓐ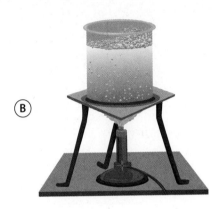

Ⓑ

Ⓒ

Ⓓ

2. You dissolve a white powder in a beaker of water. It is NOT likely that a chemical change took place if the water SC.8.P.9.2

Ⓕ gets hotter.

Ⓖ remains clear.

Ⓗ starts to bubble.

Ⓘ turns orange.

3. Which of these describes a chemical change? SC.8.P.9.2

Ⓐ baking cookies

Ⓑ melting ice

Ⓒ cutting apple slices

Ⓓ mowing the grass

Exp. No.	Description
1	Cardboard is cut into small pieces.
2	Bread is browned in a toaster.
3	Pressing a ball of clay into a square box.
4	Cheese is grated onto pasta.

4. In which of the experiments listed above can the change to the original substance be reversed? SC.8.P.9.2

Ⓕ Experiment 1

Ⓖ Experiment 2

Ⓗ Experiment 3

Ⓘ Experiment 4

Name _____ Date _____

Describing Atoms

Five friends were talking about the differences among solids, liquids, and gases. They each agreed that the differences have to do with the particles in each type of matter. However, they disagreed about which differences determine whether the matter is a solid, liquid, or gas. This is what they said:

Gwyneth: I think it has to do with the number of particles.

George: I think it has to do with the shape of the particles.

Hoda: I think it has to do with the size of the particles.

Natalie: I think it has to do with the movement of the particles.

William: I think it has to do with how hard or soft the particles are.

Whom do you agree with most? Explain why you agree with that friend.

Matter and ATOMS

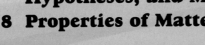

FLORIDA BIG IDEAS
1 **The Practice of Science**
3 **The Role of Theories, Laws, Hypotheses, and Models**
8 **Properties of Matter**

The Big Idea

Think About It!

How does the classification of matter depend on atoms?

From a distance, you might think this looks like a normal picture, but what happens when you look closely? Tiny photographs arranged in a specific way make a new image that looks very different from the individual pictures. The new image depends on the parts and the way they are arranged. Similarly, all the matter around you depends on its parts and the way they are arranged.

1 How do you think the image would be different if the individual pictures were arranged in another way?

_____

_____

_____

2 How do you think the image depends on the individual parts?

_____

_____

_____

Get Ready to Read

What do you think about matter and atoms?

Before you read, decide if you agree or disagree with each of these statements. As you read this chapter, see if you change your mind about any of the statements.

	AGREE	DISAGREE
1 Things that have no mass are not matter.	☐	☐
2 The arrangement of particles is the same throughout a mixture.	☐	☐
3 An atom that makes up gold is exactly the same as an atom that makes up aluminum.	☐	☐
4 An atom is mostly empty space.	☐	☐
5 If an atom gains electrons, the atom will have a positive charge.	☐	☐
6 Each electron is a cloud of charge that surrounds the center of an atom.	☐	☐

Substances and MIXTURES

 What is the relationship among atoms, elements, and compounds?

 How are some mixtures different from solutions?

 How do mixtures and compounds differ?

Vocabulary

matter p. 221

atom p. 221

substance p. 222

element p. 223

molecule p. 223

compound p. 224

mixture p. 226

heterogeneous mixture p. 227

homogeneous mixture p. 228

 Florida NGSSS

SC.8.P.8.5 Recognize that there are a finite number of elements and that their atoms combine in a multitude of ways to produce compounds that make up all of the living and nonliving things that we encounter.

SC.8.P.8.7 Explore the scientific theory of atoms (also known as atomic theory) by recognizing that atoms are the smallest unit of an element and are composed of sub-atomic particles (electrons surrounding a nucleus containing protons and neutrons).

SC.8.P.8.8 Identify basic examples of and compare and classify the properties of compounds, including acids, bases, and salts.

SC.8.P.8.9 Distinguish among mixtures (including solutions) and pure substances.

SC.8.N.1.1 Define a problem from the eighth grade curriculum using appropriate reference materials to support scientific understanding, plan and carry out scientific investigations of various types, such as systematic observations or experiments, identify variables, collect and organize data, interpret data in charts, tables, and graphics, analyze information, make predictions, and defend conclusions.

Also covers: LA.8.2.2.3, LA.8.4.2.2

 Launch Lab SC.8.P.8.9

10 minutes

Can you always see the parts of materials?

If you eat a pizza, you can see the cheese, the pepperoni, and the other parts it is made from. Can you always see the individual parts when you mix materials?

Procedure

1. Read and complete a lab safety form.
2. Observe the **materials** at the eight stations your teacher has set up.
3. Record the name and a short description of each material in the space below.

Data and Observations

Think About This

1. **Classify** Which materials have easily identifiable parts?

2. **Key Concept** Is it always easy to see the parts of materials that are mixed? Explain.

What is matter?

Think of how much fun it would be to go windsurfing! As the force of the wind pushes the sail, you lean back to balance the board. You feel the heat of the Sun and the spray of water against your face. Whether you are windsurfing on a lake or sitting at your desk in a classroom, everything around you is made of matter. **Matter** *is anything that has mass and takes up space.* Matter is everything you can see, such as water and trees. It is also some things you cannot see, such as air. You know that air is matter because you can feel its mass when it blows against your skin. You can see that it takes up space when it inflates a sail or a balloon.

Anything that does not have mass or volume is not matter. Types of energy, such as heat, sound, and electricity, are not matter. Forces, such as magnetism and gravity, also are not forms of matter.

What is matter made of?

The matter around you, including all solids, liquids, and gases, is made of atoms. *An* **atom** *is a small particle that is the building block of matter.* In this chapter, you will read that an atom is made of even smaller particles. There are many types of atoms. Each type of atom has a different number of smaller particles. You also will read that atoms can combine with each other in many ways. It is the many kinds of atoms and the ways they combine that form the different types of matter.

1. The worker is making a trophy by pouring hot, liquid metal iron into a mold. The molten metal is bronze, which is a mixture of several metals blended to make the trophy stronger. Why do you think a bronze metal would be stronger than a pure metal trophy?

_____

_____

_____

_____

WORD ORIGIN

atom
from Greek *atomos*, means "uncut"

Active Reading **2. Identify** Highlight two or three phrases in each paragraph that summarize the main ideas. After you have finished the lesson, review the highlighted text.

Active Reading **3. Explain** Why are there so many types of matter?

_____

_____

_____

Figure 1 You can classify matter as a substance or a mixture.

Matter
- Anything that has mass and takes up space
- Most matter is made up of atoms.

Substances
- Matter with a composition that is always the same

Mixtures
- Matter that can vary in composition

Active Reading 4. **Label** (Circle) the names of the two groups into which scientists place matter.

Classifying Matter

Because all the different types of matter around you are made of atoms, they must have common characteristics. But why do all types of matter look and feel different? How is the matter that makes up a pure gold ring similar to the matter that makes up your favorite soda or even the matter that makes up your body? How are these types of matter different?

As the chart in **Figure 1** shows, scientists place matter into one of two groups—substances or mixtures. Pure gold is in one group. Soda and your body are in the other. What determines whether a type of matter is a substance or a mixture? The difference is in the composition.

What is a substance?

What is the difference between a gold ring and a can of soda? What is the difference between table salt and trail mix? Pure gold is always made up of the same type of atom, but soda is not. Similarly, table salt, or sodium chloride, is always made up of the same types of atoms, but trail mix is not. This is because sodium chloride and gold are substances. *A **substance** is matter with a composition that is always the same.* A certain substance always contains the same kinds of atoms in the same combination. Soda and trail mix are another type of matter that you will read about later in this lesson.

Because gold is a substance, anything that is pure gold will have the same composition. Bars of gold are made of the same atoms as those in a pure gold ring, as shown in **Figure 2.** And since sodium chloride is a substance, if you are salting your food in Florida or in Ohio, the atoms that make up the salt will be the same in both places. If the composition of a given substance changes, you will have a new substance.

Active Reading 5. **Review** Why is gold classified as a substance?

_____

_____

Substances

Figure 2 A substance always contains the same kinds of atoms bonded in the same way.

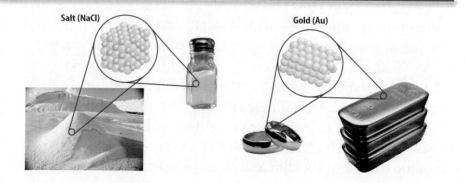

Salt (NaCl)   Gold (Au)

Elements

Some substances, such as gold, are made of only one kind of atom. Others, such as sodium chloride, are made of more than one kind. *An **element** is a substance made of only one kind of atom.* All atoms of an element are alike, but atoms of one element are different from atoms of other elements. For example, the element gold is made of only gold atoms, and all gold atoms are alike. But gold atoms are different from silver atoms, oxygen atoms, and atoms of every other element.

 6. NGSSS Check Explain How are atoms and elements related? SC.8.P.8.7

What is the smallest part of an element? If you could break down an element into its smallest part, that part would be one atom. Most elements, such as carbon and silver, consist of a large group of individual atoms. Some elements, such as hydrogen and bromine, consist of molecules. *A **molecule** (MAH lih kyewl) is two or more atoms that are held together by chemical bonds and act as a unit.* Examples of elements made of individual atoms and molecules are shown in **Figure 3.**

Elements on the Periodic Table You probably can name many elements, such as carbon, gold, and oxygen. Did you know that there are about 115 elements? As shown in **Figure 4,** each element has a symbol, such as C for carbon, Au for gold, and O for oxygen. The periodic table gives information about each element. You will learn more about elements in the next lesson.

7. Visual Check Identify What colors are used for the blocks of elements that have not yet been verified?

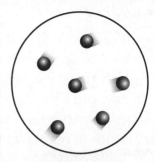

Individual atoms

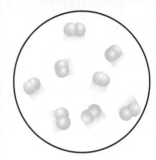

Molecules

Figure 3 The smallest part of all elements is an atom. In some elements, the atoms are grouped into molecules.

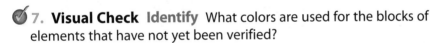

Figure 4 Element symbols have either one or two letters. Temporary symbols have three letters.

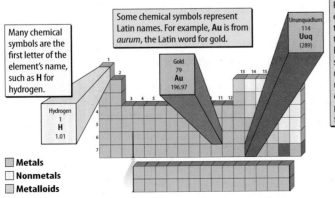

Many chemical symbols are the first letter of the element's name, such as **H** for hydrogen.

Some chemical symbols represent Latin names. For example, **Au** is from *aurum*, the Latin word for gold.

Gold
79
Au
196.97

Ununquadium
114
Uuq
(289)

Recently discovered elements have temporary three-letter symbols until they get official names. For example, **Uuq** is the symbol for element 114, ununquadium. The unusual names are based on a system of word parts. Un-un-quad-ium stands for 1-1-4.

Hydrogen
1
H
1.01

☐ Metals
☐ Nonmetals
☐ Metalloids

Compounds

Does it surprise you to learn that there are only about 115 different elements? After all, if you think about all the different things you see each day, you could probably name many more types of matter than this. Why are there so many kinds of matter when there are only about 115 elements? Most matter is made of atoms of different types of elements bonded together.

A **compound** *is a substance made of two or more elements that are chemically joined in a specific combination.* Because each compound is made of atoms in a specific combination, a compound is a substance. Pure water (H_2O) is a compound because every sample of pure water contains atoms of hydrogen and oxygen in the same combination—two hydrogen atoms to every oxygen atom. There are many types of matter because elements can join to form compounds.

Molecules Recall that a molecule is two or more atoms that are held together by chemical bonds and that act as a unit. Is a molecule the smallest part of a compound? For many compounds, this is true. Many compounds exist as molecules. An example is water. In water, two hydrogen atoms and one oxygen atom always exist together and act as a unit. Carbon dioxide (CO_2) and table sugar ($C_6H_{12}O_6$) are also examples of compounds that are made of molecules.

However, as shown in **Figure 5,** some compounds are not made of molecules. In some compounds, such as table salt or sodium chloride, no specific atoms travel together as a unit. However, table salt (NaCl) is still a substance because it always contains only sodium (Na) and chlorine (Cl) atoms.

8. NGSSS Check Summarize How do elements and compounds differ? SC.8.P.8.9

_____

_____

_____

Sugar

Salt

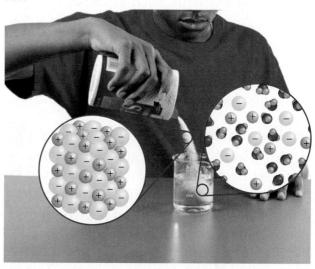

Figure 5 Sugar particles are molecules because they always travel together as a unit. Salt particles do not travel together as a unit.

9. Visual Check Analyze What happens to the salt particles when the boy mixes the salt in the water? What do you think would happen if the water evaporated?

_____

_____

_____

_____

Properties of Compounds The properties of a compound, such as table salt, are usually different from the properties of the elements from which it is made. Table salt is made of the elements sodium and chlorine. Sodium is a soft metal, and chlorine is a poisonous green gas. These properties are much different from the table salt you sprinkle on food!

Chemical Formulas Just as elements have chemical symbols, compounds have chemical formulas. A formula includes the symbols of each element in the compound. It also includes small numbers, called subscripts, that show the ratio of the elements in the compound. You can see the formulas for some compounds in **Table 1.**

Different Combinations of Atoms Sometimes the same elements combine to form different compounds. For example, nitrogen and oxygen can form six different compounds. The chemical formulas are N_2O, NO, N_2O_3, NO_2, N_2O_4, and N_2O_5. They contain the same elements, but because the combinations of atoms are different, each compound has different properties, as shown in **Table 1.**

Active Reading

FOLDABLES LA.8.2.2.3

Make a vertical two-tab book, and label it as shown. Use it to review properties of elements and compounds.

Properties of Elements

Properties of Compounds

Table 1	
Formula and Molecular Structure	**Properties/Uses**
Nitrous oxide N_2O	colorless gas used as an anesthetic
Nitrogen dioxide NO_2	brown gas, toxic, air pollutant
Dinitrogen trioxide N_2O_3	blue liquid

Table 1 Atoms can combine in different ways and form different compounds.

☑ 10. **Visual Check**
Describe How does changing the ratio of nitrogen atoms and oxygen atoms change the compounds?

_____

_____

_____

Figure 6 It's hard to tell which is in the glass— pure water (a substance) or lemon-lime soda (a mixture).

Active Reading

11. **Contrast** How do substances and mixtures differ?

_____

_____

_____

_____

What is a mixture?

By looking at the glass of clear liquid in **Figure 6,** can you tell whether it is lemon-lime soda or water? Lemon-lime soda is almost clear, and someone might confuse it with water, which is a substance. Recall that a substance is matter with a composition that is always the same. However, sodas are a combination of substances such as water, carbon dioxide, sugar, and other compounds. In fact, most solids, liquids, and gases you see each day are mixtures. _A_ **mixture** _is matter that can vary in composition._ It is made of two or more substances that are blended but are not chemically bonded.

What would happen if you added more sugar to a glass of soda? You would still have soda, but it would be sweeter. Changing the amount of one substance in a mixture does not change the identity of the mixture or its individual substances.

Air and tap water are also mixtures. Air is a mixture of nitrogen, oxygen, and other substances. However, the composition of air can vary. Air in a scuba tank usually contains more oxygen and less of the other substances. Tap water might look like pure water, but it is a mixture of pure water (H_2O) and small amounts of other substances. Since the substances that make up tap water are not bonded together, the composition of tap water can vary. This is true for all mixtures.

Inquiry SC.8.N.1.1, SC.8.P.8.9

LAB STATION **Try It!**

MiniLab _How do elements, compounds, and mixtures differ?_ at connectED.mcgraw-hill.com

Apply It! After you complete the lab, answer these questions.

1. Design a process to separate a mixture of sand and salt.

2. After separating the mixture of sand and salt, do the substances lose their identification?

Types of Mixtures

How do trail mix, soda, and air differ? One difference is that trail mix is a solid, soda is a liquid, and air is a gas. This tells you that a mixture can be any state of matter. Another difference is that you can see the *individual* parts that make up trail mix, but you cannot see the parts that make up soda or air. Trail mix is a different type of mixture than soda and air. There are two types of mixtures—heterogeneous (he tuh roh JEE nee us) and homogeneous (hoh muh JEE nee us). The prefix *hetero–* means "different," and the prefix *homo–* means "the same." Heterogeneous and homogeneous mixtures differ in how evenly the substances that compose them are mixed.

Heterogeneous Mixtures

Suppose you take a bag of trail mix and pour it into two identical bowls. What might you notice? At first glance, each bowl appears the same. However, if you look closely, you might notice that one bowl has more nuts and another bowl has more raisins. The contents of the bowls differ because trail mix is a heterogeneous mixture. *A* **heterogeneous mixture** *is a mixture in which the substances are not evenly mixed.* Therefore, if you take two samples from the same mixture, such as trail mix, the samples might have different amounts of the individual substances. The mixtures shown in **Figure 7** are examples of heterogeneous mixtures.

Active Reading 12. **Explain** Why is vegetable soup classified as a heterogeneous mixture?

_____

_____

ACADEMIC VOCABULARY

individual
(adjective) single; separate

Active Reading 13. **Explain** Does every bag of trail mix have the same composition?

_____

_____

_____

Figure 7 The different parts of a heterogeneous mixture are not evenly mixed.

Heterogeneous Mixtures

The numbers of peanuts, pretzels, raisins, and other types of food in trail mix could change, and it still would be trail mix.

You know that granite is a heterogeneous mixture because you can see the different minerals from which it is made.

With a microscope, you would be able to see that smoke is a heterogeneous mixture of gas and solid particles.

Figure 8 Salt is soluble in water. Pepper is insoluble in water. The pepper and water is a mixture, but not a solution.

Homogeneous Mixtures

If you pour soda into two glasses, the amounts of water, carbon dioxide, sugar, and other substances in the mixture would be the same in both glasses. Soda is an example of a **homogeneous mixture**—*a mixture in which two or more substances are evenly mixed, but not bonded together.*

Evenly Mixed Parts In a homogeneous mixture, the substances are so small and evenly mixed that you cannot see the boundaries between substances in the mixture. Brass, a mixture of copper and zinc, is a homogeneous mixture because the copper atoms and the zinc atoms are evenly mixed. You cannot see the boundaries between the different types of substances, even under most microscopes. Lemonade and air are also examples of homogeneous mixtures for the same reason.

Solution Another name for a homogeneous mixture is a solution. A solution is made of two parts—a solvent and one or more solutes. The solvent is the substance that is present in the largest amount. The solutes dissolve, or break apart, and mix evenly in the solvent. In **Figure 8,** water is the solvent, and salt is the solute. Salt is soluble in water. Notice also in the figure that pepper does not dissolve in water. No solution forms between pepper and water. Pepper is insoluble in water.

Other examples of solutions are described in **Figure 9.** Note that all three states of matter—solid, liquid, and gas—can be a solvent or a solute in a solution.

 14. NGSSS Check Conclude How are some mixtures different from solutions? SC.8.P.8.9

_____

_____

Figure 9 🔑 Solids, liquids, and gases can combine to make solutions.

Homogeneous Mixtures

A trumpet is made of brass, a solution of solid copper and solid zinc.

The natural gas used in a gas stove is a solution of methane, ethane, and other gases.

This ammonia cleaner is a solution of water and ammonia gas.

Compounds v. Mixtures

Think again about putting trail mix into two bowls. If you put more peanuts in one of the bowls, you still have trail mix in both bowls. Since the substances that make up a mixture are not bonded, adding more of one substance does not change the identity or the properties of the mixture. It also does not change the identity or the properties of each individual substance. In a heterogeneous mixture of peanuts, raisins, and pretzels, the properties of the individual parts don't change if you add more peanuts. The peanuts and the raisins don't bond together and become something new.

Similarly, in a solution such as soda or air, the substances do not bond together and form something new. Carbon dioxide, water, sugar, and other substances in soda are mixed together. Nitrogen, oxygen, and other substances in air also keep their separate properties because air is a mixture. If it were a compound, the parts would be bonded and would not keep their separate properties.

 15. NGSSS Check Explain How do mixtures and compounds differ? **SC.8.P.8.9**

Compounds and Solutions Differ

Compounds and solutions are alike in that they both look like pure substances. Look back at the lemon-lime soda and the water in **Figure 6.** The soda is a solution. A solution might look like a substance because the elements and the compounds that make up a solution are evenly mixed. However, compounds and solutions differ in one important way. The atoms that make up a given compound are bonded together. Therefore, the composition of a given compound is always the same. Changing the composition results in a new compound.

However, the substances that make up a solution, or any other mixture, are not bonded together. Therefore, adding more of one substance will not change the composition of the solution. It will just change the ratio of the substances in the solution. These differences are described in **Table 2.**

 16. Contrast In what way do compounds and solutions differ?

Table 2 Differences Between Solutions and Compounds 🔑

	Solutions	**Compound**
Composition	Made up of substances (elements and compounds) evenly mixed together; the composition can vary in a given mixture.	Made up of atoms bonded together; the combination of atoms is always the same in a given compound.
Changing the composition	The solution is still the same with similar properties. However, the relative amounts of substances might be different.	Changing the composition of a compound changes it into a new compound with new properties.
Properties of parts	The substances keep their own properties when they are mixed.	The properties of the compound are different from the properties of the atoms that make it up.

Separating Mixtures

Have you ever picked something you did not like off a slice of pizza? If you have, you have separated a mixture. Because the parts of a mixture are not combined chemically, you can use a physical process, such as removing them by hand, to separate the mixture. The identity of the parts does not change. Separating the parts of a compound is more difficult. The elements that make up a compound are combined chemically. Only a **chemical change** can separate them.

Separating Heterogeneous Mixtures Separating the parts of a pizza is easy because the pizza has large, solid parts. Two other ways to separate heterogeneous mixtures are shown in **Figure 10.** The strainer in the figure filters larger rocks from the mixture of rocks and dirt. The oil and vinegar is also a heterogeneous mixture because the oil floats on the vinegar. You can separate this mixture by carefully removing the floating oil.

Other properties also might be useful for separating the parts. For example, if one of the parts is magnetic, you could use a magnet to remove it. In a mixture of solid powders you might dissolve one part in water and then pour it out, leaving the other part behind. In each case, to separate a heterogeneous mixture you use differences in the physical properties of the parts.

REVIEW VOCABULARY

chemical change
a change in matter in which the substances that make up the matter change into other substances with different chemical and physical properties

Figure 10 You can separate heterogeneous and homogeneous mixtures.

✓ 17. **Visual Check**
Describe How could you separate the small rocks and dirt that passed through the strainer on the left?

_____

_____

_____

Active Reading 18. **Identify** Name three methods of separating heterogeneous mixtures.

_____

_____

_____

Separating Mixtures

A strainer removes large parts of the heterogeneous mixture of rocks and sediment. Only small rocks and dirt fall through.

In this heterogeneous mixture of oil and vinegar, the oil floats on the vinegar. You can separate them by lifting off the oil.

Making rock candy is a way of separating a solution. Solid sugar crystals form as a mixture of hot water and sugar cools.

Separating Homogeneous Mixtures Imagine trying to separate soda into water, carbon dioxide, sugar, and other substances it is made from. Because the parts are so small and evenly mixed, separating a homogeneous mixture such as soda can be difficult. However, you can separate some homogeneous mixtures by boiling or evaporation. For example, if you leave a bowl of sugar water outside on a hot day, the water will evaporate, leaving the sugar behind. An example of separating a homogeneous mixture by making rock candy is shown in **Figure 10.**

Visualizing Classification of Matter

Think about all the types of matter you have read about in this lesson. As shown in **Figure 11,** matter can be classified as either a substance or a mixture. Substances are either elements or compounds. The two kinds of mixtures are homogeneous mixtures and heterogeneous mixtures. Notice that all substances and mixtures are made of atoms. Matter is classified according to the types of atoms and the arrangement of atoms in matter. In the next lesson, you will study the structure of atoms.

Figure 11 You can classify matter based on its characteristics.

Active Reading 19. **Identify** (Circle) the type of matter that is not a mixture but contains two or more atoms.

Classifying Matter

Matter
- Anything that has mass and takes up space
- Most matter on Earth is made up of atoms.
- Two classifications of matter: substances and mixtures

Substances
- Matter with a composition that is always the same
- Two types of substances: elements and compounds

Element
- Consists of just one type of atom
- Organized on the periodic table
- Each element has a chemical symbol.

Compound
- Two or more types of atoms bonded together
- Properties are different from the properties of the elements that make it up
- Each compound has a chemical formula.

Substances physically combine to form mixtures.

Mixtures can be separated into substances by physical methods.

Mixtures
- Matter that can vary in composition
- Substances are not bonded together.
- Two types of mixtures: heterogeneous and homogeneous

Heterogeneous Mixture
- Two or more substances unevenly mixed
- Different substances are visible by an unaided eye or a microscope.

Homogeneous Mixture—Solution
- Two or more substances evenly mixed
- Different substances cannot be seen even by a microscope.

An element is a substance made of only one kind of atom.

The substances that make up a mixture are blended, but not chemically bonded.

Homogeneous mixtures have the same makeup of substances throughout a given sample.

Use Vocabulary

1 A small particle that is the building block of matter is a(n)

_____ .

2 **Use the term** *substance* in a sentence.

_____

_____

3 **Define** *molecule* in your own words.

_____

_____

Understand Key Concepts 🔑

4 **Describe** the relationship among atoms, elements, and compounds. SC.8.P.8.9

_____

_____

_____

5 Silver nitrate, $AgNO_3$, is which type of matter? SC.8.P.8.9

(A) element   (C) heterogeneous mixture

(B) compound   (D) homogeneous mixture

6 **Explain** how some mixtures are different from solutions. SC.8.P.8.9

_____

_____

Interpret Graphics

7 **Organize Information** Fill in the graphic organizer below with details about substances and mixtures. LA.8.2.2.3

Substances	Mixtures

Critical Thinking

8 **Decide** During a science investigation, a sample of matter breaks down into two kinds of atoms. Was the original sample an element or a compound? Explain. SC.8.P.8.9

_____

_____

Crude Oil

Separating Out Gasoline

Have you ever wondered where the gasoline used in automobiles comes from? Gasoline is part of a mixture of fuels called crude oil. How can workers separate gasoline from this mixture?

One way to separate a mixture is by boiling it. Crude oil is separated by a process called fractional distillation. First, the oil is boiled and allowed to cool. As the crude oil cools, each part changes from a gas to a liquid at a different temperature. Workers catch each fuel just as it changes back to a liquid. Eventually the crude oil is refined into all its useful parts.

❶ Crude oil often is taken from liquid deposits deep underground. It might also be taken from rocks or deposits mixed in sand. The crude oil is then sent to a furnace.

Crude oil

Gas 20°C

150°C → Gasoline

200°C → Kerosene

300°C → Diesel oil

370°C → Fuel oil

400°C

Distillation tower

Lubricating oil, paraffin wax, asphalt

❷ A furnace heats the oil inside a pipe until it begins to change from a liquid to a gas. The gas mixture then moves into the distillation tower.

Furnace

❸ The distillation tower is hot at the bottom and cooler higher up. As the gas mixture rises to fill the tower, it cools. It also passes over trays at different levels. Each fuel in the mixture changes to a liquid when it cools to a temperature that matches its boiling point. Gasoline changes to a liquid at the level in the tower at 150°C. A tray then catches the gasoline and moves it away.

It's Your Turn

CREATE A POSTER Blood is a mixture, too. Donated blood often is refined in laboratories to separate it into parts. What are those parts? What are they used for? How are they separated? Find the answers, and create a poster based on your findings.

LA.8.4.2.2

The Structure of ATOMS

 Where are protons, neutrons, and electrons located in an atom?

 How is the atomic number related to the number of protons in an atom?

 What effect does changing the number of particles in an atom have on the atom's identity?

Vocabulary

nucleus p. 236

proton p. 236

neutron p. 236

electron p. 236

electron cloud p. 237

atomic number p. 238

isotope p. 239

ion p. 239

 Florida NGSSS

LA.8.2.2.3 The student will organize information to show understanding or relationships among facts, ideas, and events (e.g., representing key points within text through charting, mapping, paraphrasing, summarizing, or comparing/contrasting);

SC.8.P.8.5 Recognize that there are a finite number of elements and that their atoms combine in a multitude of ways to produce compounds that make up all of the living and nonliving things that we encounter.

SC.8.P.8.7 Explore the scientific theory of atoms (also known as atomic theory) by recognizing that atoms are the smallest unit of an element and are composed of sub-atomic particles (electrons surrounding a nucleus containing protons and neutrons).

SC.8.N.1.1 Define a problem from the eighth grade curriculum using appropriate reference materials to support scientific understanding, plan and carry out scientific investigations of various types, such as systematic observations or experiments, identify variables, collect and organize data, interpret data in charts, tables, and graphics, analyze information, make predictions, and defend conclusions.

SC.8.N.3.1 Select models useful in relating the results of their own investigations.

Also covers: MA.6.A.3.6

 Inquiry Launch Lab SC.8.P.8.7

20 minutes

How can you make different things from the same parts?

Atoms are all made of the same parts. Atoms can be different from each other because they have different numbers of these parts. In this lab, you will investigate how you can make things that are different from each other even though you use the same parts to make them.

Procedure

1. Read and complete a lab safety form.

2. Think about how you can join **paper clips, toothpicks,** and **string** to make different types of objects. You must use at least one of each item, but not more than five of any kind.

3. Make the object. Use **tape** to connect the items.

4. Plan and make two more objects using the same three items, varying the numbers of each item.

5. Describe below how each of the objects you made are alike and different.

Data and Observations

Think About This

1. **Observe** What do the objects you made have in common? In what ways are they different?

2.  **Key Concept** What effect do you think increasing or decreasing the number of items you used would have on the objects you made?

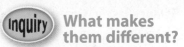

The Parts of an Atom

Now that you have read about ways to classify matter, you can probably recognize the different types you see each day. You might see pure elements, such as copper and iron, and you probably see many compounds, such as table salt. Table salt is a compound because it contains the atoms of two different elements—sodium and chlorine—in a specific combination. You also probably see many mixtures. The silver often used in jewelry is a homogeneous mixture of metals that are evenly mixed, but not bonded together.

As you read in Lesson 1, the many types of matter are possible because there are about 115 different elements. Each element is made up of a different type of atom. Atoms can combine in many different ways. They are the basic parts of matter.

What makes the atoms of each element different? Atoms are made of several types of tiny particles. The number of each of these particles in an atom is what makes atoms different from each other. It is what makes so many types of matter possible.

Active Reading

2. Explain What makes the atoms of different elements different from each other?

_____

_____

_____

Active Reading

3. Write As you read this lesson, use the table below to describe the parts of atoms including each part's location and charge.

Part	Description
Nucleus	
Proton	
Neutron	
Electron	

Inquiry What makes them different?

1. This ring is made of two of the most beautiful materials in the world—diamond and gold. Diamond is a clear, sparkling crystal made of only carbon atoms. Gold is a shiny, yellow metal made of only gold atoms. How can they be so different if each is made of just one type of atom? The structure of atoms makes significant differences in materials. Do you think a diamond could be used to make a ring band and gold be used to make a stone? Why or why not?

_____

_____

_____

Active Reading

FOLDABLES® LA.8.2.2.3

Make a vertical two-column chart book. Label it as shown. Use it to organize information about the particles in an atom.

Particles INSIDE the Nucleus	Particles OUTSIDE the Nucleus

The Nucleus—Protons and Neutrons

The basic structure of all atoms is the same. As shown in **Figure 12,** an atom has a center region with a positive **charge.** One or more negatively charged particles move around this center region. *The* **nucleus** *is the region at the center of an atom that contains most of the mass of the atom.* Two kinds of particles make up the nucleus. *A* proton *is a positively charged particle in the nucleus of an atom. A* neutron *is an uncharged particle in the nucleus of an atom.*

Active Reading **4. Review** <u>Underline</u> the particles that make up the nucleus.

Electrons

Atoms have no electric charge unless they change in some way. There must be a negative charge that balances the positive charge of the nucleus. *An* **electron** *is a negatively charged particle that occupies the space in an atom outside the nucleus.* Electrons are so small and move so quickly that scientists are unable to tell exactly where a given electron is located at any specific time. Therefore, scientists describe their positions around the nucleus as a cloud rather than specific points. A model of an atom and its parts is shown in **Figure 12.**

 5. NGSSS Check **Summarize** Where are protons, neutrons, and electrons located in an atom? SC.8.P.8.7

_____

_____

Figure 12 All atoms have a positively charged nucleus surrounded by one or more electrons.

Parts of an Atom 🔑

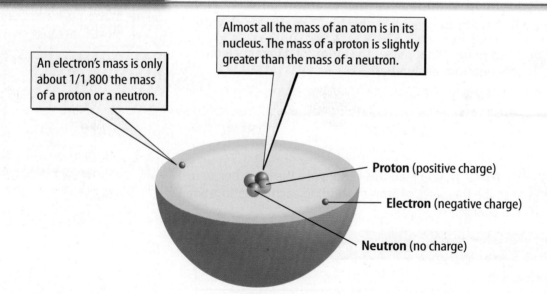

An electron's mass is only about 1/1,800 the mass of a proton or a neutron.

Almost all the mass of an atom is in its nucleus. The mass of a proton is slightly greater than the mass of a neutron.

Proton (positive charge)

Electron (negative charge)

Neutron (no charge)

6. Visual Check **Determine** How many protons and how many electrons does this atom have?

_____

An Electron Cloud Drawings of an atom, such as the one in **Figure 13,** often show electrons circling the nucleus like planets orbiting the Sun. Scientists have conducted experiments that show the movement of electrons is more complex than this. The modern idea of an atom is called the electron-cloud model. *An* **electron cloud** *is the region surrounding an atom's nucleus where one or more electrons are most likely to be found.* It is important to understand that an electron is not a cloud of charge. An electron is one tiny particle. An electron cloud is mostly empty space. At any moment in time, electrons are located at specific points within that area.

Electron Energy Electrons are constantly moving around the nucleus in a region called the electron cloud. However, some electrons are closer to the nucleus than others. Electrons occupy certain areas around the nucleus according to their energy, as shown in **Figure 13.** Electrons close to the nucleus are strongly attracted to it and have less energy. Electrons farther from the nucleus are less attracted to it and have more energy.

The Size of Atoms

As tiny as atoms are, electrons, protons, and neutrons are even smaller. The data in **Table 3** shows that protons and neutrons have about the same mass. Electrons have only about 1/2,000 the mass of a proton or a neutron. If you held a textbook and placed a paper clip on it, you wouldn't notice the added mass because the mass of a paper clip is small compared to the mass of the book. In a similar way, the masses of an atom's protons and neutrons are packed tightly into a tiny nucleus. Visualize the nucleus as the size of an ant. How large would the atom be? Amazingly, the atom would be the size of a football stadium.

Table 3 Properties of Atomic Particles		
Particle	**Charge**	**Mass (g)**
Proton	+1	1.007316
Neutron	0	1.008701
Electron	−1	0.000549

Active Reading

8. Infer Do you think all atoms are the exact same size? Explain.

_____

_____

_____

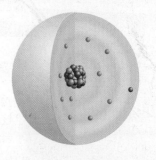

Figure 13 Electrons farther from the nucleus have more energy.

Math Skills   MA.6.A.3.6

Use Scientific Notation
Scientists write very large and very small numbers using scientific notation. A gram of carbon has about 50,000,000,000,000,000,000 atoms. Express this in scientific notation.

1. Move the decimal until one nonzero digit remains on the left: 5.0000000000000000000

2. Count the places you moved. Here it is 19 left.

3. Show that number as a power of 10. The exponent is negative if the decimal moves right and positive if it moves left. Answer: 5×10^{19}

4. Reverse the process to change scientific notation back to a whole number.

Practice

7. The diameter of a carbon atom is 2.2×10^{-8} cm. Write this as a whole number.

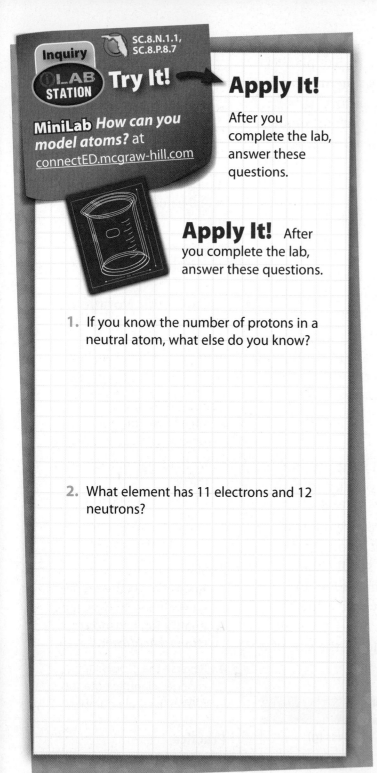

MiniLab *How can you model atoms?* at connectED.mcgraw-hill.com

Apply It!

After you complete the lab, answer these questions.

Apply It! After you complete the lab, answer these questions.

1. If you know the number of protons in a neutral atom, what else do you know?

2. What element has 11 electrons and 12 neutrons?

Differences in Atoms

In some ways atoms are alike. Each has a positively charged nucleus surrounded by a negatively charged electron cloud. But atoms can differ from each other in several ways. Atoms can have different numbers of protons, neutrons, or electrons.

Protons and Atomic Number

Look at the periodic table in the back of this book. In each block, the number under the element name shows how many protons each atom of the element has. For example, each oxygen atom has eight protons. *The* **atomic number** *is the number of protons in the nucleus of an atom of an element.* If there are 12 protons in the nucleus of an atom, that element's atomic number is 12. Examine **Figure 14.** Notice that the atomic number of magnesium is the whole number above its symbol. The atomic number of carbon is 6. This means that each carbon atom has 6 protons.

Every element in the periodic table has a different atomic number. You can identify an element if you know either its atomic number or the number of protons its atoms have. If an atom has a different number of protons, it is a different element.

Active Reading 9. **Describe** How is the atomic number related to the number of protons in an atom?

_____

_____

Active Reading 10. **Locate** (Circle) the atomic number in the cube that represents the carbon atom.

Figure 14 🔑 An atomic number is the number of protons in each atom of the element.

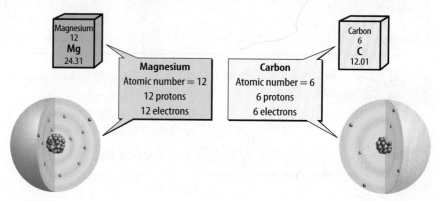

Neutrons and Isotopes

Each atom of an element contains the same number of protons, but the number of neutrons can vary. *An* **isotope** (I suh tohp) *is one of two or more atoms of an element having the same number of protons, but a different number of neutrons.* Boron-10 and boron-11 are isotopes of boron, as shown in **Figure 15.** Notice that boron-10 has 10 particles in its nucleus. Boron-11 has 11 particles in its nucleus.

Active Reading 11. **Differentiate** <u>Underline</u> how boron-10 and boron-11 are different.

Electrons and Ions

You read that atoms can differ by the number of protons or neutrons they have. **Figure 16** illustrates a third way atoms can differ—by the number of electrons. A neutral, or uncharged, atom has the same number of positively charged protons and negatively charged electrons. As atoms bond, their numbers of electrons can change. Because electrons are negatively charged, a neutral atom that has lost an electron has a positive charge. A neutral atom that has gained an electron has a negative charge. *An* **ion** (I ahn) *is an atom that has a charge because it has gained or lost electrons.* Because the number of protons is unchanged, an ion is the same element it was before.

In the previous lesson, you read that each particle of a compound is two or more atoms of different elements bonded together. One of the ways compounds form is when one or more electrons move from an atom of an element to an atom of a different element. This results in a positive ion for one element and a negative ion for the other element.

Isotopes

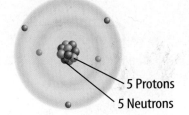

Boron-10

5 Protons
5 Neutrons

Boron-11

5 Protons
6 Neutrons

Figure 15 Boron-10 and boron-11 are isotopes. The number of protons is the same, but the number of neutrons is different.

Figure 16 A positive ion has fewer electrons than protons. A negative ion has more electrons than protons.

Ions

Neutral atom	Positive ion	Negative ion
4 Protons 4 Electrons	3 Protons 2 Electrons	7 Protons 10 Electrons
Beryllium	**Lithium**	**Nitrogen**
A neutral atom has the same number of electrons and protons. The atom has no charge.	If an atom loses an electron during chemical bonding, it has more protons than electrons. It is now positively charged.	If an atom gains an electron during chemical bonding, it has more electrons than protons. It is now negatively charged.

Active Reading 12. **Determine** Highlight the element that has more protons than electrons.

Table 4 Possible Changes in Atoms 🔑

Neutral Atom	Change	Results
Carbon **6** **C** **12.01** • 6 protons • 6 neutrons • 6 electrons	**Protons** add one proton	**New element—nitrogen** • ____ protons • 7 neutrons • 7 electrons
	Neutrons add one neutron	**Isotope** • 6 protons • ____ neutrons • 6 electrons
	Electrons add one electron	**Ion** • 6 protons • 6 neutrons • ____ electrons

13. **NGSSS Check** **Relate** Fill in the blanks in **Table 4** to show how changing the number of particles in an atom changes the atom's identity. SC.8.P.8.7

Atoms and Matter

You have now read that matter can be either a substance or a mixture. A substance has a composition that is always the same, but the composition of a mixture can vary. All types of matter are made of atoms. The atoms of a certain element always have the same number of protons, but the number of neutrons can vary. When elements combine to form compounds, the number of electrons in the atoms can change. The different ways in which atoms can change are summarized in **Table 4**.

Look back at the diamond and gold ring at the beginning of this lesson. Now can you answer the question of how they can be so different if each is made of just one type of atom? Each carbon atom in diamond has six protons. Each gold atom has 79 protons. The parts of an atom give an element its identity. The ways in which the atoms combine result in the many different kinds of matter.

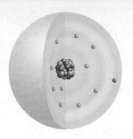

All matter is made of atoms. Atoms are made of protons, electrons, and neutrons.

An orange is about 100 million times wider than an atom.

Atoms of the same element can have different numbers of neutrons.

Inquiry

①LAB STATION **Try It!**

SC.8.N.1.1, SC.8.N.3.1, SC.8.P.8.5

MiniLab *Balloon Molecules* at connectED.mcgraw-hill.com

Use Vocabulary

1. **Distinguish** between a proton and a neutron.

_____

_____

2. An atom that has lost one or more electrons is a(n) _____ .

3. **Use the term** *isotope* in a complete sentence.

_____

_____

Understand Key Concepts 🔑

4. Which is located outside the nucleus of an atom? SC.8.P.8.7

   (A) electron   (C) neutron

   (B) ion   (D) proton

5. **Identify** the element that has nine protons. SC.8.P.8.7

_____

_____

6. **Explain** how atomic number relates to the number of particles in an atom's nucleus. SC.8.P.8.7

_____

_____

Interpret Graphics

7. **Organize** Summarize what you have learned about the parts, the sizes, and the differences of atoms. LA.8.2.2.3

Properties of Atoms	
Parts	
Size	
Differences	

Critical Thinking

8. **Decide** Can you tell which element an atom is if you know its charge and the number of electrons it has? Explain. SC.8.P.8.7

_____

_____

Think About It! There are a finite number of different types of atoms that combine in a multitude of ways. Matter is classified according to the combination and arrangement of atoms from which it is made.

🔑 Key Concepts Summary

Vocabulary

LESSON 1 **Substances and Mixtures**

- An **atom** is a building block of **matter**. An **element** is matter made of only one type of atom. A **compound** is a **substance** that contains two or more elements.
- A **heterogeneous mixture** is not a solution because the substances that make up a heterogeneous mixture are not evenly mixed. The substances that make up a solution, or a **homogeneous mixture**, are evenly mixed.
- **Mixtures** differ from compounds in their composition, whether their parts join, and the properties of their parts.

matter p. 221
atom p. 221
substance p. 222
element p. 223
molecule p. 223
compound p. 224
mixture p. 226
heterogeneous mixture
  p. 227
homogeneous mixture
  p. 228

LESSON 2 **The Structure of Atoms**

- The center of an atom is the **nucleus**. The nucleus contains **protons** and **neutrons**. **Electrons** occupy the space in an atom outside the nucleus.
- The identity of an atom is determined by its **atomic number**. The atomic number is the number of protons in the atom.
- The identity of an atom stays the same if the number of neutrons or electrons changes.

nucleus p. 236
proton p. 236
neutron p. 236
electron p. 236
electron cloud p. 237
atomic number p. 238
isotope p. 239
ion p. 239

FOLDABLES® **Chapter Project**

Assemble your lesson Foldables as shown to make a Chapter Project. Use the project to review what you have learned in this chapter.

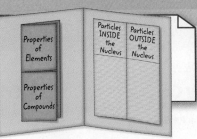

Use Vocabulary

1 A particle that consists of two or more atoms bonded together is a(n) _____ .

2 A salad is an example of a(n) _____ because it is a mixture in which you can easily remove the individual parts.

3 Matter is classified as a(n) _____ if it is made of two or more substances that are physically blended, not chemically bonded.

4 A positively charged particle in the nucleus of an atom is a(n) _____ .

5 Almost all of the mass of an atom is found in the _____ of an atom.

6 If a chlorine atom gains an electron, it becomes a(n) _____ of chlorine.

Link Vocabulary and Key Concepts

Use vocabulary terms from the previous page to complete the concept map.

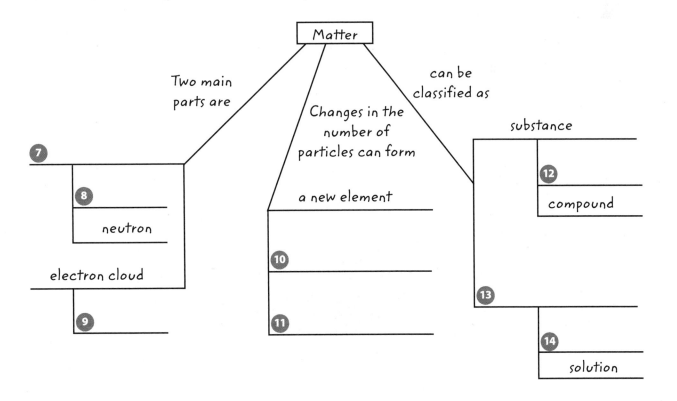

Fill in the correct answer choice.

🗝 Understand Key Concepts

1 Which is a substance? SC.8.P.8.9

(A) fruit salad

(B) granola cereal

(C) spaghetti

(D) table salt

2 Which is the best model for a homogeneous mixture? SC.8.P.8.9

(A)

(B)

(C)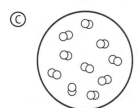

(D)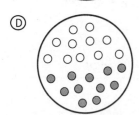

3 Which is a property of all atoms? SC.8.P.8.9

(A) more electrons than protons

(B) a nucleus with a positive charge

(C) a positively charged electron cloud

(D) same number of protons as neutrons

4 Which is another name for a solution? SC.8.P.8.9

(A) element

(B) compound

(C) heterogeneous mixture

(D) homogeneous mixture

Critical Thinking

5 **Classify** Look at the illustration below. Is this a model of a substance or a mixture? How do you know? SC.8.P.8.6

_____

_____

_____

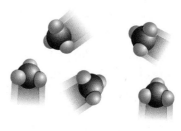

6 **Deduce** Each atom of protium has one proton, no neutrons, and one electron. Each atom of deuterium has one proton, two neutrons, and one electron. Are these the same or different elements? Why? SC.8.P.8.9

_____

_____

_____

7 **Decide** Suppose you mix several liquids in a jar. After a few minutes, the liquids form layers. Is this a homogeneous mixture or a heterogeneous mixture? Why? SC.8.P.8.9

_____

_____

_____

8 **Describe** a method for separating a mixture of salt water. SC.8.P.8.9

_____

_____

_____

_____

9 **Generalize** Consider the substances N_2O_5, H_2, CH_4, H_2O, KCl, and O_2. Is it possible to tell just from the symbols and the numbers which are elements and which are compounds? Explain. LA.8.2.2.3

_____

_____

_____

_____

10 **Suggest** how you can define an electron cloud differently from the definition in the chapter. SC.8.P.8.9

_____

_____

_____

11 **Analyze** A substance has an atomic number of 80. How many protons and electrons do atoms of the substance have? What is the substance? SC.8.P.8.7

_____

_____

_____

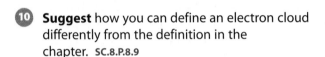
Writing in Science

12 **Write** a paragraph on a separate sheet of paper in which you explain the modern atomic model to an adult who has never heard of it before. Include two questions he or she might ask, and write answers to the questions. SC.8.P.8.7

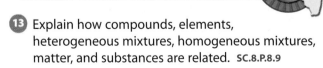
Big Idea Review

13 Explain how compounds, elements, heterogeneous mixtures, homogeneous mixtures, matter, and substances are related. SC.8.P.8.9

_____

_____

_____

14 The photograph below depends on its parts. This is similar to the relationship of matter and atoms. How does the classification of matter depend on atoms? SC.8.P.8.5

_____

_____

_____

Math Skills MA.6.A.3.6

Use Scientific Notation

15 The mass of one carbon atom is 0.00000000000000000000001994 g. Express this number in scientific notation.

16 In 1 L of hydrogen gas, there are about 54,000,000,000,000,000,000,000 hydrogen atoms. Express the number of atoms using scientific notation.

Fill in the correct answer choice.

Multiple Choice

1 Gloria is making a model of an atom. She uses three different colors to represent the three basic particles that make up the atom. Which particles should she display in the nucleus of the atom? SC.8.P.8.7

Ⓐ neutrons only

Ⓑ electrons only

Ⓒ protons and neutrons

Ⓓ electrons and protons

Use the illustration below to answer question 2.

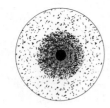

2 The illustration shows the electron configuration of a hydrogen atom. Why is this diagram a good representation of the locations of electrons within an atom? SC.8.P.8.7

Ⓕ The outer energy level is the electric field produced by a given electron.

Ⓖ The outer energy level represents the number of electrons around the nucleus.

Ⓗ The outer energy level is the region where electrons are least likely to be found.

Ⓘ The outer energy level is the region where electrons are most likely to be found.

3 Which class of matter is the least evenly mixed? SC.8.P.8.9

Ⓐ compounds

Ⓑ heterogeneous mixtures

Ⓒ homogeneous mixtures

Ⓓ solutions

4 Which correctly describes a compound but not a mixture? SC.8.P.8.9

Ⓕ All the atoms are of the same element.

Ⓖ All the molecules have at least two atoms.

Ⓗ The combination of substances never changes.

Ⓘ The substances can be separated without breaking bonds.

5 A girl pours a spoonful of sugar into a glass of warm water. She stirs the water until the sugar disappears. When she tastes the water, she notices that it is now sweet. Which describes the kind of matter in the glass? SC.8.P.8.9

Ⓐ a compound

Ⓑ an element

Ⓒ a solution

Ⓓ a substance

Use the diagram below to answer questions 6 and 7.

6 In the diagram of an atom, what is located in the cloud region? SC.8.P.8.7

Ⓕ electrons

Ⓖ neutrons

Ⓗ nucleus

Ⓘ protons

7 In the atom shown above, which particle or particles make up the nucleus? SC.8.P.8.7

Ⓐ protons only

Ⓑ electrons only

Ⓒ protons and neutrons

Ⓓ neutrons and electrons

8 Why are atoms electrically neutral? SC.8.P.8.7

   Ⓕ They contain more neutrons than protons.

   Ⓖ They contain more electrons than protons.

   Ⓗ They contain an equal number of protons and neutrons.

   Ⓘ They contain an equal number of protons and electrons.

Use the table below to answer questions 9 and 10.

	Number of Protons	Number of Neutrons	Number of Electrons
A	8	8	8
B	8	8	10
C	8	10	8
D	8	10	9

9 The table shows the numbers of protons, neutrons, and electrons for four atoms. Which atom has a negative charge? SC.8.P.8.7

   Ⓐ A

   Ⓑ B

   Ⓒ C

   Ⓓ D

10 Which atom is a different element than the others? SC.8.P.8.7

   Ⓕ A

   Ⓖ B

   Ⓗ C

   Ⓘ D

11 Which statement best describes the structure of an atom? SC.8.P.8.7

   Ⓐ An atom consists of a neutral center surrounded by a cloud of positive particles.

   Ⓑ An atom consists of a positive center surrounded by a cloud of negative particles.

   Ⓒ An atom consists of a negative center surrounded by a cloud of negative particles.

   Ⓓ An atom consists of a positive center surrounded by a cloud of neutral particles.

Use the figures below to answer question 12.

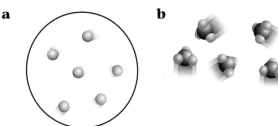
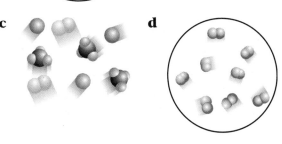

12 Which two models are elements? SC.8.P.8.9

   Ⓕ Models a and d

   Ⓖ Models b and c

   Ⓗ Models a and c

   Ⓘ Models b and d

NEED EXTRA HELP?

If You Missed Question...	1	2	3	4	5	6	7	8	9	10	11	12
Go to Lesson...	2	2	1	1	1	2	2	2	2	2	2	1

Benchmark Mini-Assessment **Chapter 6 • Lesson 1**

Multiple Choice *Bubble the correct answer.*

W	$C_6H_{12}O_6$
X	CO_2
Y	H_2O
Z	NaCl

1. In the table above, which compound is not made of molecules? **SC.8.P.8.9**

   (A) W

   (B) X

   (C) Y

   (D) Z

2. Which is a homogeneous mixture? **SC.8.P.8.9**

   (F) fruit salad

   (G) potting soil

   (H) macaroni and cheese

   (I) salt dissolved in water

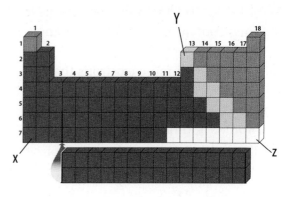

3. The figure above shows the periodic table. What does X represent? **SC.8.P.8.6**

   (A) atoms

   (B) metalloids

   (C) metals

   (D) nonmetals

4. NaCl is an example of a(n) **SC.8.P.8.9**

   (F) atom.

   (G) compound.

   (H) element.

   (I) mixture.

Benchmark Mini-Assessment   Chapter 6 • Lesson 2

Multiple Choice *Bubble the correct answer.*

Use the image below to answer questions 1 and 2.

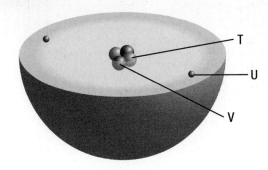

1. Which correctly identifies a part of the atom shown in the figure above? **SC.8.P.8.7**

(A) T: electron

(B) T: nucleus

(C) U: electron

(D) U: proton

2. The loss of the structure labeled U in the figure above results in the formation of a(n) **SC.8.P.8.7**

(F) atom.

(G) element.

(H) ion.

(I) isotope.

Boron-10

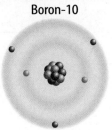

Boron-11

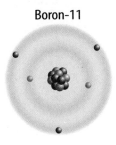

3. If the two atoms shown above are the same element, which statement is true? **SC.8.P.8.7**

(A) They have a different number of electron clouds.

(B) They have a different number of electrons.

(C) They have a different number of neutrons.

(D) They have a different number of protons.

4. Which correctly describes a part of an atom? **SC.8.P.8.7**

(F) The electrons are positively charged.

(G) The neutrons are negatively charged.

(H) The nucleus does not have a charge.

(I) The protons are positively charged.

Name _____ Date _____

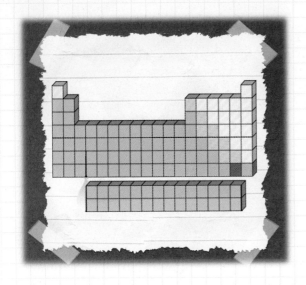

How is it arranged?

Five students were looking at a large poster of the periodic table. They noticed the elements were arranged in rows and columns. They also noticed the elements were arranged in order from 1–118. They had different ideas about the arrangement of the elements on the periodic table. This is what they said:

Damon: I think the elements are arranged by increasing mass.

Flo: I think the elements are arranged according to their properties.

Sienna: I think the elements are arranged by when they were discovered.

Kyle: I think the elements are arranged according to how common they are.

Glenda: I don't agree with any of you. I think there must be a different reason for the arrangement.

Which student do you agree with? Explain why you agree with that student.

The Periodic TABLE

FLORIDA BIG IDEAS

1 The Practice of Science

8 Properties of Matter

Think About It!

How is the periodic table used to classify and provide information about all known elements?

Things are made out of specific materials for a reason. A weather balloon can rise high in the atmosphere and gather weather information. The plastic that forms this weather balloon and the helium gas that fills it were chosen after scientists researched and studied the properties of these materials.

1 What property of helium do you think makes the balloon rise through the air?

_____

_____

2 Why do you think the periodic table is a useful tool when determining properties of different materials?

_____

_____

_____

Get Ready to Read

What do you think about the periodic table?

Before you read, decide if you agree or disagree with each of these statements. As you read this chapter, see if you change your mind about any of the statements.

	AGREE	DISAGREE
1 The elements on the periodic table are arranged in rows in the order they were discovered.	☐	☐
2 The properties of an element are related to the element's location on the periodic table.	☐	☐
3 Fewer than half of the elements are metals.	☐	☐
4 Metals are usually good conductors of electricity.	☐	☐
5 Most of the elements in living things are nonmetals.	☐	☐
6 Even though they look very different, oxygen and sulfur share some similar properties.	☐	☐

There's More Online!

Video • Audio • Review • ⓘLab Station • WebQuest • Assessment • Concepts in Motion • Multilingual eGlossary

Using the PERIODIC TABLE

ESSENTIAL QUESTIONS

How are elements arranged on the periodic table?

What can you learn about elements from the periodic table?

Vocabulary

periodic table p. 255

group p. 260

period p. 260

Florida NGSSS

LA.8.2.2.3 The student will organize information to show understanding or relationships among facts, ideas, and events (e.g., representing key points within text through charting, mapping, paraphrasing, summarizing, or comparing/contrasting);

MA.6.A.3.6 Construct and analyze tables, graphs, and equations to describe linear functions and other simple relations using both common language and algebraic notation.

SC.8.P.8.6 Recognize that elements are grouped in the periodic table according to similarities of their properties.

SC.8.N.1.1 Define a problem from the eighth grade curriculum using appropriate reference materials to support scientific understanding, plan and carry out scientific investigations of various types, such as systematic observations or experiments, identify variables, collect and organize data, interpret data in charts, tables, and graphics, analyze information, make predictions, and defend conclusions.

SC.8.N.1.6 Understand that scientific investigations involve the collection of relevant empirical evidence, the use of logical reasoning, and the application of imagination in devising hypotheses, predictions, explanations and models to make sense of the collected evidence.

 Launch Lab SC.8.N.1.6, SC.8.P.8.6

15 minutes

How can objects be organized?

What would it be like to shop at a grocery store where all the products are mixed up on the shelves? Cereal might be next to the dish soap and bread might be next to the canned tomatoes. It would take a long time to find the groceries that you need. How does organizing objects help you find and use what you need?

Procedure

1. Read and complete a lab safety form.

2. Empty the **interlocking plastic bricks** from the **plastic bag** onto your desk and observe their properties. Think about ways you might group and sequence the bricks so that they are organized.

3. Organize the bricks according to your plan.

4. Compare your pattern of organization with those used by several other students.

Think About This

1. Describe the way you grouped your bricks. Why did you choose that way of grouping?

2. Describe how you sequenced your bricks.

3. **Key Concept** How does organizing things help you to use them more easily?

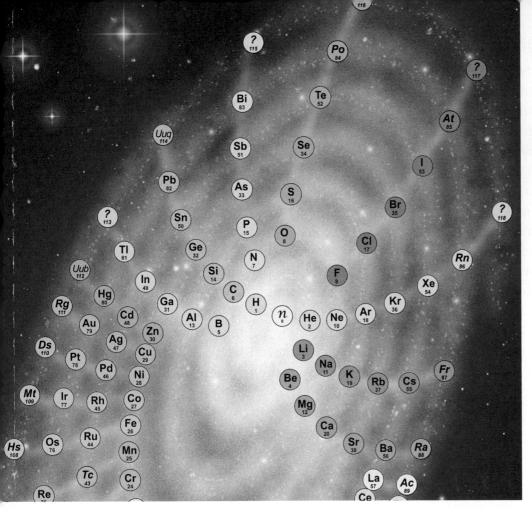

Inquiry **Same Information?**

1. You probably have seen a copy of a table that is used to organize the elements. Does it look like this chart? Describe the symbols in this image. How are they arranged?

_____

_____

_____

_____

_____

_____

What is the periodic table?

The "junk drawer" in **Figure 1** is full of pens, notepads, rubber bands, and other supplies. It would be difficult to find a particular item in this messy drawer. How might you organize it? First, you might dump the contents onto the counter. Then you could sort everything into piles. Pens and pencils might go into one pile. Notepads and paper go into another. Organizing the contents of the drawer makes it easier to find the things you need, also shown in **Figure 1.**

Just as sorting helps to organize the objects in the junk drawer, sorting can help scientists organize information about the elements. Recall that there are more than 100 elements, each with a unique set of physical and chemical properties.

Scientists use a table called the periodic (pihr ee AH dihk) table to organize elements. _The_ **periodic table** _is a chart of the elements arranged into rows and columns according to their physical and chemical properties._ It can be used to determine the relationships among the elements.

In this chapter, you will read about how the periodic table was developed. You will also read about how you can use the periodic table to learn about the elements.

Figure 1 Sorting objects by their similarities makes it easier to find what you need.

Developing a Periodic Table

In 1869, a Russian chemist and teacher named Dimitri Mendeleev (duh MEE tree · men duh LAY uf) was working on a way to classify elements. At that time, more than 60 elements had been discovered. He studied the physical properties such as density, color, melting point, and atomic mass of each element. Mendeleev also noted chemical properties such as how each element reacted with other elements. Mendeleev arranged the elements in a list using their atomic masses. He noticed that the properties of the elements seemed to repeat in a pattern.

When Mendeleev placed his list of elements into a table, he arranged them in rows of increasing atomic mass. Elements with similar properties were grouped in the same column. The columns in his table are like the piles of sorted objects in your junk drawer. Both contain groups of things with similar properties.

 Active Reading

2. List What physical property did Mendeleev use to place the elements in rows on the periodic table?

Patterns in Properties

The term *periodic* means "repeating pattern." For example, seasons and months are periodic because they follow a repeating pattern every year. The days of the week are periodic since they repeat every seven days.

What were some of the repeating patterns Mendeleev noticed in his table? Melting point is one property that shows a repeating pattern. Recall that melting point is the temperature at which a solid changes to a liquid. The blue line in **Figure 2** represents the melting points of the elements in row 2 of the periodic table. Notice that the melting point of carbon is higher than the melting point of lithium. However, the melting point of fluorine, at the far right of the row, is lower than that of carbon. How do these melting points show a pattern? Look at the red line in **Figure 2.** This line represents the melting points of the elements in row 3 of the periodic table. The melting points follow the same increasing and then decreasing pattern as the blue line, or row 2. Boiling point and reactivity also follow a periodic pattern.

A Periodic Property 🔑

Figure 2 Melting points increase, then decrease, across a period on the periodic table.

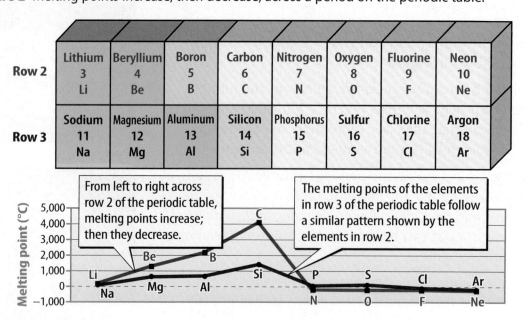

Predicting Properties of Undiscovered Elements

When Mendeleev arranged all known elements by increasing atomic mass, there were large gaps between some elements. He predicted that scientists would discover elements that would fit into these spaces. Mendeleev also predicted that the properties of these elements would be similar to the known elements in the same columns. Both of his predictions turned out to be true.

Changes to Mendeleev's Table

Mendeleev's periodic table enabled scientists to relate the properties of the known elements to their positions on the table. However, the table had a problem—some elements seemed out of place. Mendeleev believed that the atomic masses of certain elements must be invalid because the elements appeared in the wrong place on the periodic table. For example, Mendeleev placed tellurium before iodine despite the fact that tellurium has a greater atomic mass than iodine. He did so because iodine's properties more closely resemble those of fluorine and chlorine, just as copper's properties are closer to those of silver and gold, as shown in **Figure 3.**

Active Reading

FOLDABLES® LA.8.2.2.3

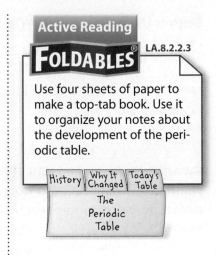

Use four sheets of paper to make a top-tab book. Use it to organize your notes about the development of the periodic table.

History | Why It Changed | Today's Table

The Periodic Table

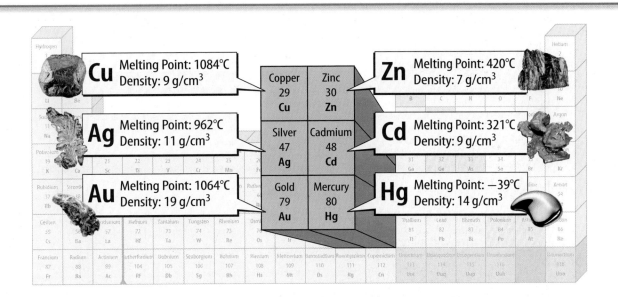

	Melting Point: 1084°C Density: 9 g/cm³ (Cu)
	Melting Point: 962°C Density: 11 g/cm³ (Ag)
	Melting Point: 1064°C Density: 19 g/cm³ (Au)
	Melting Point: 420°C Density: 7 g/cm³ (Zn)
	Melting Point: 321°C Density: 9 g/cm³ (Cd)
	Melting Point: −39°C Density: 14 g/cm³ (Hg)

Copper 29 Cu — Zinc 30 Zn
Silver 47 Ag — Cadmium 48 Cd
Gold 79 Au — Mercury 80 Hg

The Importance of Atomic Number

In the early 1900s, scientist Henry Moseley solved the problem with Mendeleev's table. Moseley found that if elements were listed according to increasing atomic number instead of listing atomic mass, columns would contain elements with similar properties. Recall that the atomic number of an element is the number of protons in the nucleus of each of that element's atoms.

Active Reading

3. Recall <u>Underline</u> what determines where an element is located on the periodic table.

Figure 3 On today's periodic table, copper is in the same column as silver and gold. Zinc is in the same column as cadmium and mercury.

Today's Periodic Table

You can identify many of the properties of an element from its placement on the **periodic** table. The table, as shown in **Figure 4,** is organized into columns, rows, and blocks, which are based on certain patterns of properties. In the next two lessons, you will learn how an element's position on the periodic table can help you interpret the element's physical and chemical properties.

Figure 4 🔑 The periodic table is used to organize elements according to increasing atomic number and properties.

PERIODIC TABLE OF THE ELEMENTS

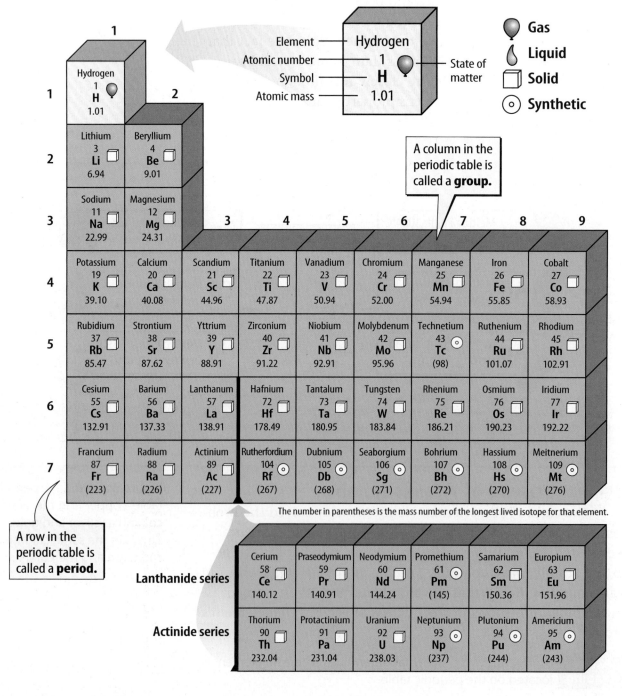

The number in parentheses is the mass number of the longest lived isotope for that element.

A column in the periodic table is called a **group.**

A row in the periodic table is called a **period.**

	Gas
	Liquid
	Solid
	Synthetic

What is on an element key?

The element key shows an element's chemical symbol, atomic number, and atomic mass. The key also contains a symbol that shows the state of matter at room temperature. Look at the element key for helium in **Figure 5.** Helium is a gas at room temperature. Some versions of the periodic table give additional information, such as density, conductivity, or melting point.

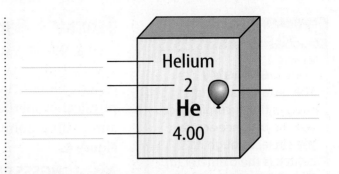

Figure 5 An element key shows important information about each element.

Active Reading
4. Label What does this key tell you about helium?

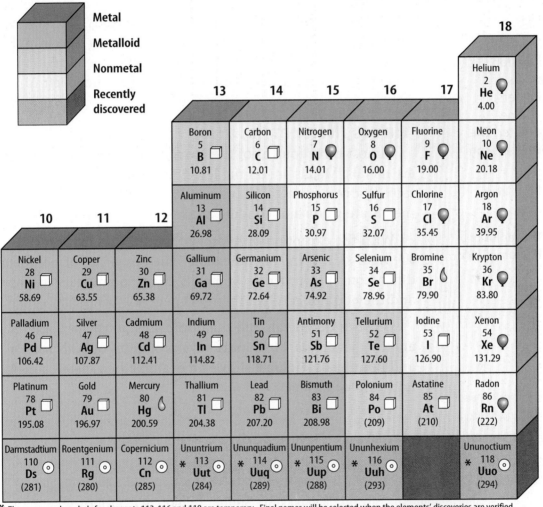

* The names and symbols for elements 113-116 and 118 are temporary. Final names will be selected when the elements' discoveries are verified.

Use Geometry

The distance around a circle is the circumference (*C*). The distance across the circle, through its center, is the diameter (*d*). The radius (*r*) is half of the diameter. The circumference divided by the diameter for any circle is equal to π (pi), or 3.14. The formula for determining the circumference is:

$$C = \pi d \quad \text{or} \quad C = 2\pi r$$

For example, an iron (Fe) atom has a radius of **126 pm** (picometers; 1 picometer = one-trillionth of a meter) The circumference of an iron atom is:

$$C = 2 \times 3.14 \times 126 \text{ pm}$$

$$C = 791 \text{ pm}$$

5. **Practice** The radius of a uranium (U) atom is 156 pm. What is its circumference?

Groups

A **group** *is a column on the periodic table.* Elements in the same group have similar chemical properties and react with other elements in similar ways. There are patterns in the physical properties of a group such as density, melting point, and boiling point. The groups are numbered 1–18, as shown in **Figure 4.**

 6. **NGSSS Check** **Classify** What can you infer about the properties of two elements in the same group? SC.8.P.8.6

_____

_____

Periods

The rows on the periodic table are called **periods**. The atomic number of each element increases by one as you read from left to right across each period. The physical and chemical properties of the elements also change as you move left to right across a period.

Metals, Nonmetals, and Metalloids

Almost three-fourths of the elements on the periodic table are metals. Metals are on the left side and in the middle of the table. Individual metals have some properties that differ, but all metals are shiny and conduct thermal energy and electricity.

With the exception of hydrogen, nonmetals are located on the right side of the periodic table. The properties of nonmetals differ from the properties of metals. Many nonmetals are gases, and they do not conduct thermal energy or electricity.

Between the metals and the nonmetals on the periodic table are the metalloids. Metalloids have properties of both metals and nonmetals. **Figure 6** shows an example of a metal, a metalloid, and a nonmetal.

Figure 6 In period 3, magnesium is a metal, silicon is a metalloid, and sulfur is a nonmetal.

Sodium 11 Na	Magnesium 12 Mg	Aluminum 13 Al	Silicon 14 Si	Phosphorus 15 P	Sulfur 16 S	Chlorine 17 Cl	Argon 18 Ar

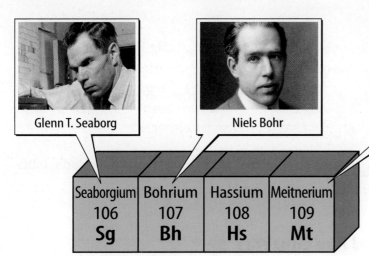

Glenn T. Seaborg   Niels Bohr   Lise Meitner

Seaborgium	Bohrium	Hassium	Meitnerium
106	107	108	109
Sg	**Bh**	**Hs**	**Mt**

Figure 7 Three of these synthetic elements are named to honor important scientists.

How Scientists Use the Periodic Table

Even today, new elements are created in laboratories, named, and added to the present-day periodic table. Four of these elements are shown in **Figure 7.** These elements are all synthetic, or made by people, and do not occur naturally on Earth. Sometimes scientists can create only a few atoms of a new element. Yet scientists can use the periodic table to predict the properties of new elements they create. Look back at the periodic table in **Figure 4.** What group would you predict to contain element 117? You would probably expect element 117 to be in group 17 and to have similar properties to other elements in the group. Scientists hope to one day synthesize element 117.

The periodic table contains more than 100 elements. Each element has unique properties that differ from the properties of other elements. But each element also shares similar properties with nearby elements. The periodic table shows how elements relate to each other and fit together into one organized chart. Scientists use the periodic table to understand and predict elements' properties. You can, too.

Active Reading 7. **Explain** How is the periodic table used to predict the properties of an element?

Inquiry **LAB STATION** SC.8.N.1.1, SC.8.P.8.6 **Try It!**

MiniLab *How does atom size change across a period?* at connectED.mcgraw-hill.com

Apply It!

After you complete the lab, answer these questions.

1. **Describe** the relationship between your observations of the atomic radii and the atomic number (the number of protons and neutrons) in period 2.

2. **Predict** the pattern of atomic radii of the elements in group 2.

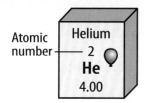

Atomic number

Helium
2
He
4.00

On the periodic table, elements are arranged according to increasing atomic number and similar properties.

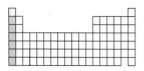

A column of the periodic table is called a group. Elements in the same group have similar properties.

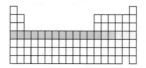

A row of the periodic table is called a period. Properties of elements repeat in the same pattern from left to right across each period.

SC.8.N.1.1, SC.8.N.1.6, SC.8.P.8.6

Inquiry

LAB STATION **Try It!**

Skill Lab *How is the periodic table arranged?* at connectED.mcgraw-hill.com

Use Vocabulary

① **Identify** the scientific term used for rows on the periodic table.

_____

② **Name** the scientific term used for columns on the periodic table.

_____

Understand Key Concepts 🔑

③ The _____ increases by one for each element as you move left to right across a period. SC.8.P.8.6

④ What does the decimal number in an element key represent?
- (A) atomic mass
- (C) chemical symbol
- (B) atomic number
- (D) state of matter

Interpret Graphics

⑤ **Identify** Fill in the graphic organizer below to identify the color-coded regions of the periodic table. LA.8.2.2.3

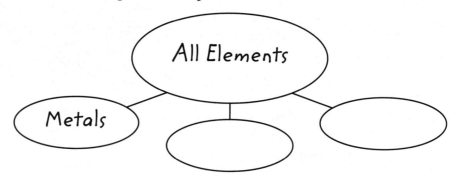

All Elements

Metals

Critical Thinking

⑥ **Predict** Look at the perioidic table and predict three elements that have lower melting points than calcium (Ca). SC.8.P.8.6

_____

_____

Math Skills MA.6.A.3.6

⑦ Carbon (C) and silicon (Si) are in group 4 of the periodic table. The atomic radius of carbon is 77 pm and sulfur is 117 pm. What is the circumference of each atom?

🔑 **Organize** Show how the periodic table is arranged by completing the concept map.

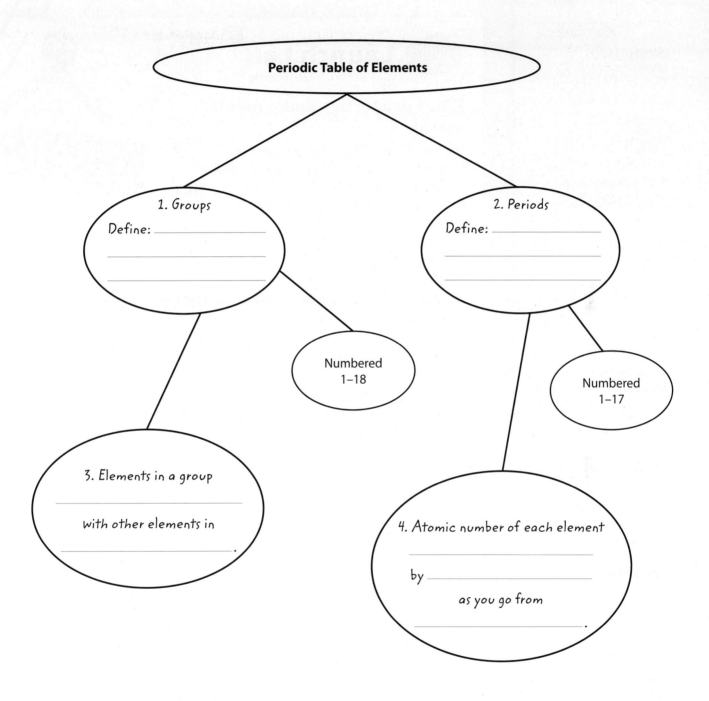

METALS

 What elements are metals?

 What are the properties of metals?

Vocabulary

metal p. 265

luster p. 265

ductility p. 266

malleability p. 266

alkali metal p. 267

alkaline earth metal p. 267

transition element p. 268

 Florida NGSSS

LA.8.2.2.3 The student will organize information to show understanding or relationships among facts, ideas, and events (e.g., representing key points within text through charting, mapping, paraphrasing, summarizing, or comparing/contrasting);

LA.8.4.2.2 The student will record information (e.g., observations, notes, lists, charts, legends) related to a topic, including visual aids to organize and record information and include a list of sources used;

SC.8.P.8.4 Classify and compare substances on the basis of characteristic physical properties that can be demonstrated or measured; for example, density, thermal or electrical conductivity, solubility, magnetic properties, melting and boiling points, and know that these properties are independent of the amount of the sample.

SC.8.P.8.6 Recognize that elements are grouped in the periodic table according to similarities of their properties.

SC.8.N.1.1 Define a problem from the eighth grade curriculum using appropriate reference materials to support scientific understanding, plan and carry out scientific investigations of various types, such as systematic observations or experiments, identify variables, collect and organize data, interpret data in charts, tables, and graphics, analyze information, make predictions, and defend conclusions.

 Launch Lab SC.8.N.1.1, SC.8.P.8.4

20 minutes

What properties make metals useful?

The properties of metals determine their uses. Copper conducts thermal energy, which makes it useful for cookware. Aluminum has low density, so it is used in aircraft bodies. What other properties make metals useful?

Procedure

1. Read and complete a lab safety form.

2. With your group, observe the **metal objects** in your **container.** For each object, discuss what properties allow the metal to be used in that way.

3. Observe the **photographs of gold and silver jewelry.** What properties make these two metals useful in jewelry?

4. Examine **other objects around the room** that you think are made of metal. Do they share the same properties as the objects in your container? Do they have other properties that make them useful?

Data and Observations

Think About This

1. What properties do all the metals share? What properties are different?

2. **Key Concept** List at least four properties of metals that determine their uses.

1. Lightning strikes the top of the Empire State Building approximately 100 times per year. Why do you think lightning hits the top of this building instead of the city streets or buildings below?

_____

_____

_____

_____

What is a metal?

What do stainless steel knives and forks, copper wire, aluminum foil, and gold jewelry have in common? They are all made from metals.

As you read in Lesson 1, most of the elements on the periodic table are metals. In fact, of all the known elements, more than three-quarters are metals. With the exception of hydrogen, all of the elements in groups 1–12 on the periodic table are metals. In addition, some of the elements in groups 13–15 are metals. To be a metal, an element must have certain properties.

 2. NGSSS Check Identify Highlight where metals are found on the periodic table. SC.8.P.8.6

Physical Properties of Metals

Recall that physical properties are characteristics used to describe or identify something without changing its makeup. All metals share certain physical properties.

A **metal** _is an element that is generally shiny. It is easily pulled into wires or hammered into thin sheets. A metal is a good conductor of electricity and thermal energy._ Gold exhibits the common properties of metals.

Luster and Conductivity People use gold for jewelry because of its beautiful color and metallic luster. **Luster** _describes the ability of a metal to reflect light._ Gold is also a good conductor of thermal energy and electricity. However, gold is too expensive to use in normal electrical wires or metal cookware. Copper is often used instead.

Figure 8 Gold has many uses based on its properties.

Gold

Active Reading 3. **Connect** Match the properties of gold with the corresponding images below.

Ductility   Malleability   Luster   Conductivity   Unreactive

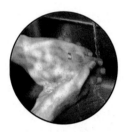    

WORD ORIGIN
ductility
from Latin *ductilis*, means "may be led or drawn"

REVIEW VOCABULARY
density
the mass per unit volume of a substance

Ductility and Malleability Gold is the most ductile metal. **Ductility** (duk TIH luh tee) *is the ability to be pulled into thin wires.* A piece of gold with the mass of a paper clip can be pulled into a wire that is more than 3 km long.

Malleability (ma lee uh BIH luh tee) *is the ability of a substance to be hammered or rolled into sheets.* Gold is so malleable that it can be hammered into thin sheets. A pile of a million thin sheets would be only as high as a coffee mug.

Other Physical Properties of Metals In general the **density**, strength, boiling point, and melting point of a metal are greater than those of other elements. Except for mercury, all metals are solid at room temperature. Many uses of a metal are determined by the metal's physical properties, as shown in **Figure 8**.

 4. **NGSSS Check** Find (Circle) some physical properties of metals. SC.8.P.8.4

Chemical Properties of Metals

Recall that a chemical property is the ability or inability of a substance to change into one or more new substances. The chemical properties of metals can differ greatly. However, metals in the same group usually have similar chemical properties. For example, gold and other elements in group 11 do not easily react with other substances.

Active Reading

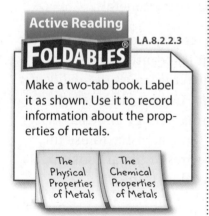

FOLDABLES LA.8.2.2.3

Make a two-tab book. Label it as shown. Use it to record information about the properties of metals.

| The Physical Properties of Metals | The Chemical Properties of Metals |

Group 1: Alkali Metals

The elements in group 1 are called **alkali** *(AL kuh li)* **metals**. The alkali metals include lithium, sodium, potassium, rubidium, cesium, and francium.

Because they are in the same group, alkali metals have similar chemical properties. Alkali metals react quickly with other elements, such as oxygen. Therefore, in nature, they occur only in compounds. Pure alkali metals must be stored so that they do not come in contact with oxygen and water vapor in the air. **Figure 9** shows potassium and sodium reacting with water.

Alkali metals also have similar physical properties. Pure alkali metals have a silvery appearance. As shown in **Figure 9,** they are soft enough to cut with a knife. The alkali metals also have the lowest densities of all metals. A block of pure sodium metal could float on water because of its very low density.

Active Reading

5. Characterize Assess information about alkali metals. (Circle) the correct choice in each set of parentheses.

Characteristics of Alkali Metals
• React (quickly, slowly) with other elements
• Found in nature (as elements, in compounds)
• Have a (dull, shiny) appearance
• (Soft, hard)
• Have the (highest, lowest) densities of all metals

Figure 9 Alkali metals react violently with water. They are also soft enough to be cut with a knife.

Potassium

Sodium

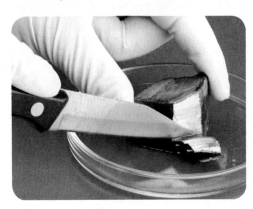

Lithium

Group 2: Alkaline Earth Metals

The elements in group 2 on the periodic table are called **alkaline** *(AL kuh lun)* **earth metals**. These metals are beryllium, magnesium, calcium, strontium, barium, and radium.

Alkaline earth metals also react quickly with other elements. However, they do not react as quickly as the alkali metals do. Like the alkali metals, pure alkaline earth metals do not occur naturally. Instead, they combine with other elements and form compounds. The physical properties of the alkaline earth metals are also similar to those of the alkali metals. Alkaline earth metals are soft and silvery. They also have low density, but they have greater density than alkali metals.

Active Reading

6. Predict Which element reacts faster with oxygen—barium or potassium?

_____

_____

Figure 10 Transition elements are in blocks at the center of the periodic table. Many colorful materials contain small amounts of transition elements.

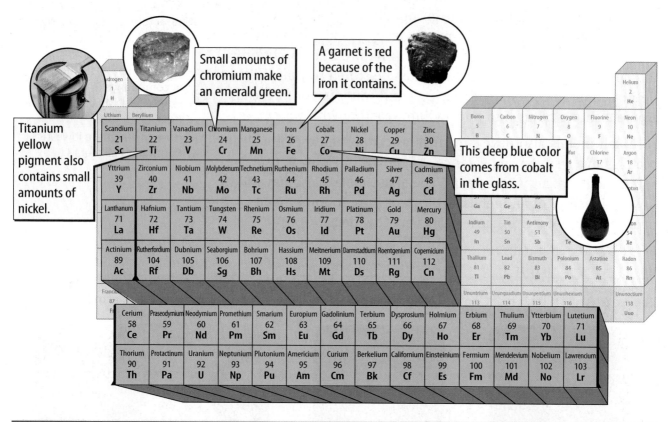

Titanium yellow pigment also contains small amounts of nickel.

Small amounts of chromium make an emerald green.

A garnet is red because of the iron it contains.

This deep blue color comes from cobalt in the glass.

Groups 3-12: Transition Elements

The elements in groups 3–12 are called **transition elements**. The transition elements are in two blocks on the periodic table. The main block is in the center of the periodic table. The other block includes the two rows at the bottom of the periodic table, as shown in **Figure 10.**

Properties of Transition Elements

All transition elements are metals. They have higher melting points, greater strength, and higher densities than the alkali metals and the alkaline earth metals. Transition elements also react less quickly with oxygen. Some transition elements can exist in nature as free elements. An element is a free element when it occurs in pure form, not in a compound.

Uses of Transition Elements

Transition elements in the main block of the periodic table have many important uses. Because of their high densities, strength, and resistance to corrosion, transition elements such as iron make good building materials. Copper, silver, nickel, and gold are used to make coins. These metals are also used for jewelry, electrical wires, and many industrial applications.

Main-block transition elements can react with other elements and form many compounds. Many of these compounds are colorful. Artists use transition-element compounds in paints and pigments. The colors of many gems, such as garnets and emeralds, come from the presence of small amounts of transition elements, as illustrated in **Figure 10.**

Lanthanide and Actinide Series

Two rows of transition elements are at the bottom of the periodic table, as shown in **Figure 10.** These elements were removed from the main part of the table so that periods 6 and 7 were not longer than the other periods. If these elements were included in the main part of the table, the first row, called the lanthanide series, would stretch between lanthanum and halfnium. The second row, called the actinide series, would stretch between actinium and rutherfordium.

Some lanthanide and actinide series elements have valuable properties. For example, lanthanide series elements are used to make strong magnets. Plutonium, one of the actinide series elements, is used as a fuel in some nuclear reactors.

Patterns in Properties of Metals

Recall that the properties of elements follow repeating patterns across the periods of the periodic table. In general, elements increase in metallic properties such as luster, malleability, and electrical conductivity from right to left across a period, as shown in **Figure 11.** The elements on the far right of a period have no metallic properties at all. Potassium (K), the element on the far left in period 4, has the highest luster, is the most malleable, and conducts electricity better than all the elements in this period.

There are also patterns within groups. Metallic properties tend to increase as you move down a group, also shown in **Figure 11.** You could predict that the malleability of gold is greater than the malleability of either silver or copper because it is below these two elements in group 11.

Active Reading
7. Indicate Draw a line around the area on the periodic table in **Figure 11** where you would expect to find elements with few or no metallic properties.

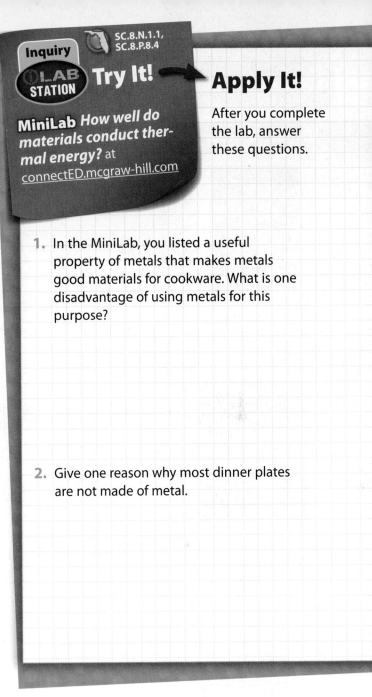

Inquiry **LAB STATION** SC.8.N.1.1, SC.8.P.8.4
Try It!

MiniLab *How well do materials conduct thermal energy?* at connectED.mcgraw-hill.com

Apply It!
After you complete the lab, answer these questions.

1. In the MiniLab, you listed a useful property of metals that makes metals good materials for cookware. What is one disadvantage of using metals for this purpose?

2. Give one reason why most dinner plates are not made of metal.

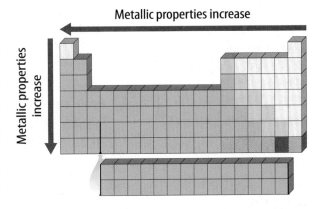

Figure 11 Metallic properties of elements increase as you move to the left and down on the periodic table.

Metallic properties increase

Metallic properties increase

Visual Summary

Properties of metals include conductivity, luster, malleability, and ductility.

Alkali metals and alkaline earth metals react easily with other elements. These metals make up groups 1 and 2 on the periodic table.

Transition elements make up groups 3–12 and the lanthanide and actinide series on the periodic table.

Use Vocabulary

1 **Use the term** *luster* in a sentence.

_____

_____

2 **Identify** the property that makes copper metal ideal for wiring.

_____

3 Elements that have the lowest densities of all the metals are called

_____.

Understand Key Concepts 🔑

4 **List** the physical properties that most metals have in common. SC.8.P.8.6

_____

_____

5 Which is a chemical property of transition elements?
- (A) brightly colored
- (B) great ductility
- (C) denser than alkali metals
- (D) reacts little with oxygen

6 **Organize** the following metals from least metallic to most metallic: barium, zinc, iron, and strontium. SC.8.P.8.6

Interpret Graphics

7 **Examine** this section of the periodic table. What metal will have properties most similar to those of chromium (Cr)? Why? SC.8.P.8.6

_____

_____

_____

_____

Vanadium	Chromium	Manganese
23	24	25
V	**Cr**	**Mn**
Niobium	Molybdenum	Technetium
41	42	43
Nb	**Mo**	**Tc**

Critical Thinking

8 **Investigate** your classroom and locate five examples of materials made from metal.

_____

_____

9 **Evaluate** the physical properties of potassium, magnesium, and copper. Select the best choice to use for a building project. Explain why this metal is the best building material to use. SC.8.P.8.4

_____

_____

_____

Fireworks

Metals add variety to color.

About 1,000 years ago, the Chinese discovered the chemical formula for gunpowder. Using this formula, they invented the first fireworks. One of the primary ingredients in gunpowder is saltpeter, or potassium nitrate. Find potassium on the periodic table. Notice that potassium is a metal. How does the chemical behavior of a metal contribute to a colorful fireworks show?

Purple: mix of strontium and copper compounds

Blue: copper compounds

Yellow: sodium compounds

Gold: iron burned with carbon

White-hot: barium-oxygen compounds or aluminum or magnesium burn

Orange: calcium compounds

Metal compounds contribute to the variety of colors you see at a fireworks show. Recall that metals have special chemical and physical properties. Compounds that contain metals also have special properties. For example, each metal turns a characteristic color when burned. Lithium, an alkali metal, forms compounds that burn red. Copper compounds burn blue. Aluminum and magnesium burn white.

Green: barium compounds

Red: strontium and lithium compounds

It's Your Turn

FORM AN OPINION Fireworks contain metal compounds. Are they bad for the environment or for your health? Research the effects of metals on human health and on the environment. Decide if fireworks are safe to use at holiday celebrations.

LA.8.4.2.2

Nonmetals and METALLOIDS

 Where are nonmetals and metalloids on the periodic table?

 What are the properties of nonmetals and metalloids?

Vocabulary

nonmetal p. 273

halogen p. 275

noble gas p. 276

metalloid p. 277

semiconductor p. 277

 Florida NGSSS

LA.8.2.2.3 The student will organize information to show understanding or relationships among facts, ideas, and events (e.g., representing key points within text through charting, mapping, paraphrasing, summarizing, or comparing/contrasting);

SC.8.P.8.4 Classify and compare substances on the basis of characteristic physical properties that can be demonstrated or measured; for example, density, thermal or electrical conductivity, solubility, magnetic properties, melting and boiling points, and know that these properties are independent of the amount of the sample.

SC.8.P.8.6 Recognize that elements are grouped in the periodic table according to similarities of their properties.

SC.8.P.8.8 Identify basic examples of and compare and classify the properties of compounds, including acids, bases, and salts.

SC.8.N.1.1 Define a problem from the eighth grade curriculum using appropriate reference materials to support scientific understanding, plan and carry out scientific investigations of various types, such as systematic observations or experiments, identify variables, collect and organize data, interpret data in charts, tables, and graphics, analyze information, make predictions, and defend conclusions.

SC.8.N.1.6 Understand that scientific investigations involve the collection of relevant empirical evidence, the use of logical reasoning, and the application of imagination in devising hypotheses, predictions, explanations and models to make sense of the collected evidence.

 Inquiry **Launch Lab**

SC.8.N.1.1, SC.8.P.8.6

20 minutes

What are some properties of nonmetals?

You now know what the properties of metals are. What are the properties of nonmetals?

Procedure

1. Read and complete a lab safety form.

2. Examine pieces of **copper, carbon, aluminum,** and **sulfur**. Describe the appearance of these elements.

3. Use a **conductivity tester** to check how well these elements conduct electricity. Record your observations.

4. Wrap each element sample in a **paper towel**. Carefully hit the sample with a **hammer**. Unwrap the towel and observe the sample. Record your observations.

Data and Observations

Think About This

1. Locate these elements on the periodic table. From their locations, which elements are metals? Which elements are nonmetals?

2. 🔑 **Key Concept** Using your results, compare the properties of metals and nonmetals.

3. 🔑 **Key Concept** What property of a nonmetal makes it useful to insulate electrical wires?

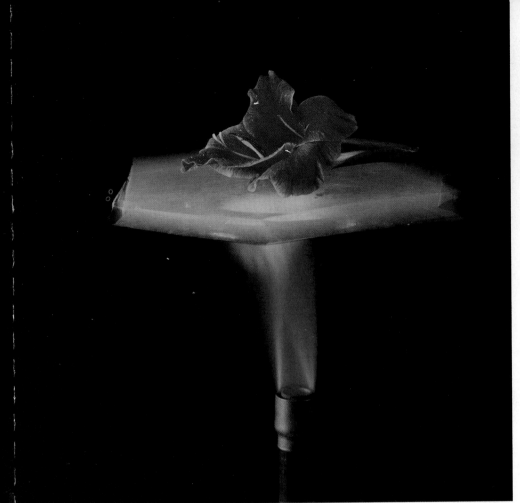

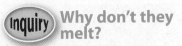
1. What do you expect to happen to something when a flame is placed against it? As you can see, the material this flower sits on protects the flower from the flame. Do you think the material is made of metal? Why or why not?

_____

_____

_____

_____

_____

The Elements of Life

Would it surprise you to learn that more than 96 percent of the mass of your body comes from just four elements? As shown in **Figure 12,** all four of these elements—oxygen, carbon, hydrogen, and nitrogen—are nonmetals. **Nonmetals** _are elements that have no metallic properties._

Of the remaining elements in your body, the two most common elements also are nonmetals—phosphorus and sulfur. These six elements form the compounds in proteins, fats, nucleic acids, and other large molecules in your body and in all other living things.

 2. Recall List the six most common elements in the human body.

_____

_____

_____

Figure 12 Like other living things, this woman's mass comes mostly from nonmetals.

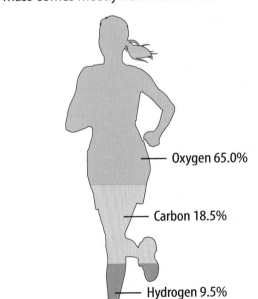

Oxygen 65.0%

Carbon 18.5%

Hydrogen 9.5%

Nitrogen 3.3%

Other elements 3.7%

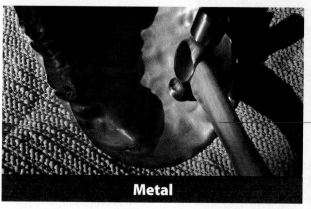
Metal

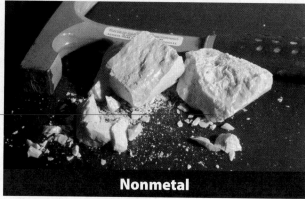
Nonmetal

Figure 13 Solid metals, such as copper, are malleable. Solid nonmetals, such as sulfur, are brittle.

Figure 14 Nonmetals have properties that are different from those of metals. Phosphorus and carbon are dull, brittle solids that do not conduct thermal energy or electricity.

How are nonmetals different from metals?

Recall that metals have luster. They are ductile, malleable, and good conductors of electricity and thermal energy. All metals except mercury are solids at room temperature.

The properties of nonmetals are different from those of metals. Many nonmetals are gases at room temperature. Those that are solid at room temperature have a dull surface, which means they have no luster. Because nonmetals are poor conductors of electricity and thermal energy, they are good insulators. For example, nose cones on space shuttles are insulated from the intense thermal energy of reentry by a material made from carbon, a nonmetal. **Figure 13** and **Figure 14** show several properties of nonmetals.

 3. NGSSS Check List Highlight the properties of nonmetals. SC.8.P.8.4

Properties of Nonmetals 🔑

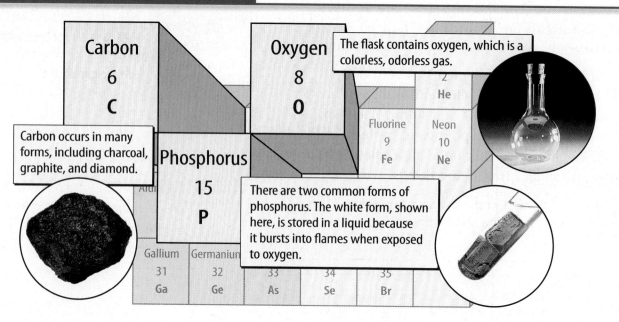

Carbon occurs in many forms, including charcoal, graphite, and diamond.

The flask contains oxygen, which is a colorless, odorless gas.

There are two common forms of phosphorus. The white form, shown here, is stored in a liquid because it bursts into flames when exposed to oxygen.

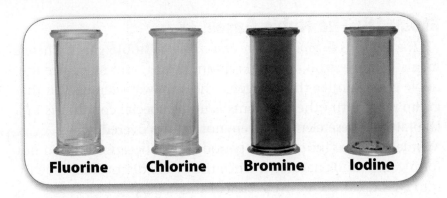

Fluorine   Chlorine   Bromine   Iodine

Figure 15 These glass containers each hold a halogen gas. Although they are different colors in their gaseous state, they react similarly with other elements.

Nonmetals in Groups 14–16

Look back at the periodic table in **Figure 4**. Notice that groups 14–16 contain metals, nonmetals, and metalloids. The chemical properties of the elements in each group are similar. However, the physical properties of the elements can be quite different.

Carbon is the only nonmetal in group 14. It is a solid that has different forms. Carbon is in most of the compounds that make up living things. Nitrogen, a gas, and phosphorus, a solid, are the only nonmetals in group 15. These two elements form many different compounds with other elements, such as oxygen. Group 16 contains three nonmetals. Oxygen is a gas that is essential for many organisms. Sulfur and selenium are solids that have the physical properties of other solid nonmetals.

Group 17: The Halogens

An element in group 17 of the periodic table is called a **halogen** (HA luh jun). **Figure 15** shows the halogens fluorine, chlorine, bromine, and iodine. The term *halogen* refers to an element that can react with a metal and form a salt. For example, chlorine gas reacts with solid sodium and forms sodium chloride, or table salt. Calcium chloride is another salt often used on icy roads.

Halogens react readily with other elements and form compounds. They react so readily that halogens only can occur naturally in compounds. They do not exist as free elements. They even form compounds with other nonmetals, such as carbon. In general, the halogens are less reactive as you move down the group.

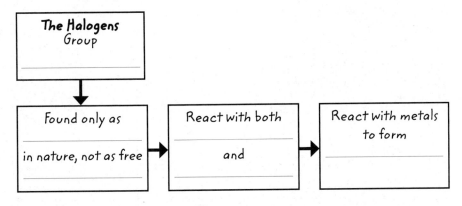

The Halogens
Group
_____

↓

Found only as
_____
in nature, not as free
_____

→

React with both
_____
and
_____

→

React with metals
to form
_____

Active Reading

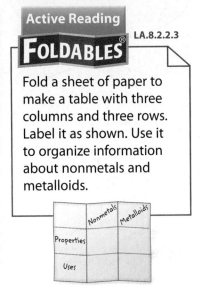

FOLDABLES®   LA.8.2.2.3

Fold a sheet of paper to make a table with three columns and three rows. Label it as shown. Use it to organize information about nonmetals and metalloids.

WORD ORIGIN

halogen
from Greek **hals**, means "salt"; and **–gen**, means "to produce"

Active Reading 4. **Explain** Will bromine react with sodium? Why or why not?

_____

_____

Active Reading 5. **Describe** Using the organizer on the left, show the properties of halogens.

6. Organize
What properties do the noble gases have?

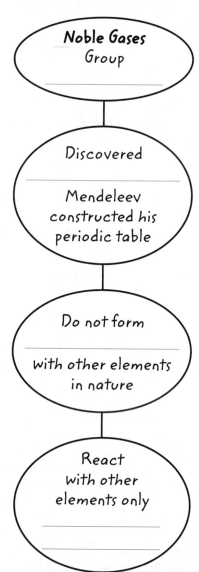

Noble Gases
Group

Discovered
_____
Mendeleev
constructed his
periodic table

Do not form
_____
with other elements
in nature

React
with other
elements only
_____
_____

Figure 16 More than 90 percent of all the atoms in the universe are hydrogen atoms. Hydrogen is the main fuel for the nuclear reactions that occur in stars.

Group 18: The Noble Gases

The elements in group 18 are known as the **noble gases**. The elements helium, neon, argon, krypton, xenon, and radon are the noble gases. Unlike the halogens, the only way elements in this group react with other elements is under special conditions in a laboratory. These elements were not yet discovered when Mendeleev **constructed** his periodic table because they do not form compounds naturally. Once they were discovered, they fit into a group at the far-right side of the table.

Hydrogen

Figure 16 shows the element key for hydrogen. Of all the elements, hydrogen has the smallest atomic mass. It is also the most common element in the universe.

Is hydrogen a metal or a nonmetal? Hydrogen is most often classified as a nonmetal because it has many properties like those of nonmetals. For example, like some nonmetals, hydrogen is a gas at room temperature. However, hydrogen also has some properties similar to those of the group 1 alkali metals. In its liquid form, hydrogen conducts electricity just like a metal does. In some chemical reactions, hydrogen reacts as if it were an alkali metal. However, under conditions on Earth, hydrogen usually behaves like a nonmetal.

7. Explain Why is hydrogen usually classified as a nonmetal?

_____

_____

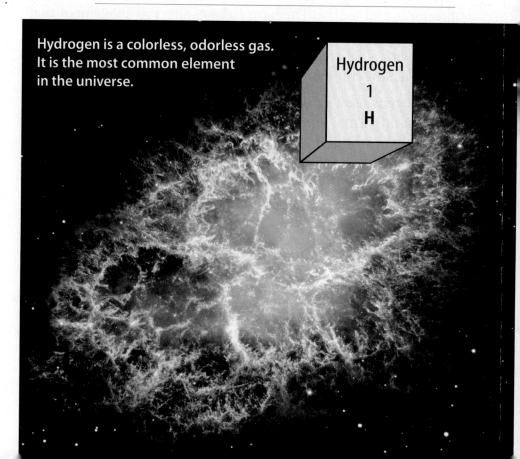

Hydrogen is a colorless, odorless gas. It is the most common element in the universe.

Hydrogen
1
H

Metalloids

Between the metals and the nonmetals on the periodic table are elements known as metalloids. *A* **metalloid** *(MEH tul oyd) is an element that has physical and chemical properties of both metals and nonmetals.* The elements boron, silicon, germanium, arsenic, antimony, tellurium, polonium, and astatine are metalloids. Silicon is the most abundant metalloid in the universe. Most sand is made of a compound containing silicon. Silicon is also used in many different products, some of which are shown in **Figure 17.**

 8. NGSSS Check Identify Where are metalloids found on the periodic table? SC.8.P.8.6

Semiconductors

Recall that metals are good conductors of thermal energy and electricity. Nonmetals are poor conductors of thermal energy and electricity but are good insulators. A property of metalloids is the ability to act as a semiconductor. *A* **semiconductor** *conducts electricity at high temperatures, but not at low temperatures.* At high temperatures, metalloids act like metals and conduct electricity. But at lower temperatures, metalloids act like nonmetals and stop electricity from flowing. This property is useful in electronic devices such as computers, televisions, and solar cells.

Active Reading

9. Classify Show how the characteristics of metalloids are like metals and like nonmetals.

Metalloids
Like Metals
Conduct electricity at
temperatures
Like Nonmetals
Stop electricity from flowing at
temperatures

WORD ORIGIN

semiconductor
from Latin *semi–*, means "half"; and *conducere*, means "to bring together"

Uses of Silicon **Figure 17** The properties of silicon make it useful for many different products.

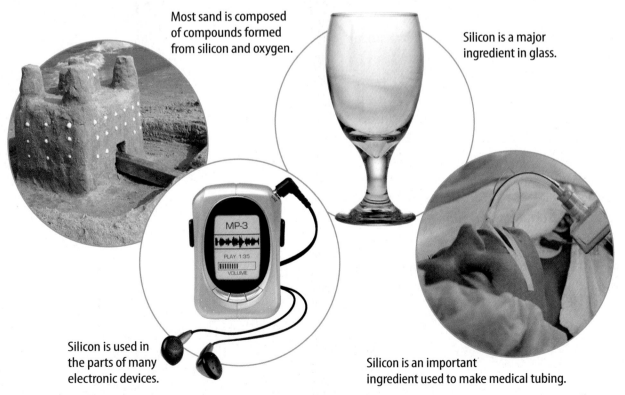

Most sand is composed of compounds formed from silicon and oxygen.

Silicon is a major ingredient in glass.

Silicon is used in the parts of many electronic devices.

Silicon is an important ingredient used to make medical tubing.

Properties and Uses of Metalloids

Pure silicon is used in making semiconductor devices for computers and other electronic products. Germanium is also used as a semiconductor. However, metalloids have other uses, as shown in **Figure 18.** Pure silicon and Germanium are used in semiconductors. Boron is used in water softeners and laundry products. Boron also glows bright green in fireworks. Silicon is one of the most abundant elements on Earth. Sand, clay, and many rocks and minerals are made of silicon compounds.

Figure 18 This microchip conducts electricity at high temperatures using a semiconductor.

Metals, Nonmetals, and Metalloids

You have read that all metallic elements have common characteristics, such as malleability, conductivity, and ductility. However, each metal has unique properties that make it different from other metals. The same is true for nonmetals and metalloids. How can knowing the properties of an element help you evaluate its uses?

Look again at the periodic table. An element's position on the periodic table tells you a lot about the element. By knowing that sulfur is a nonmetal, for example, you know that it breaks easily and does not conduct electricity. You would not choose sulfur to make a wire. You would not try to use oxygen as a semiconductor or sodium as a building material. You know that transition elements are strong, malleable, and do not react easily with oxygen or water. These metals make good building materials because they are strong, malleable, and less reactive than other elements. Understanding the properties of elements can help you decide which element to use in a given situation.

Active Reading 10. **Explain** Why would you not use an element on the right side of the periodic table as a building material?

_____

_____

_____

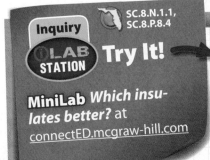

Inquiry **LAB STATION** **Try It!**   SC.8.N.1.1, SC.8.P.8.4

MiniLab *Which insulates better?* at connectED.mcgraw-hill.com

Apply It!

After you complete the lab, answer the question below.

1. Insulation is placed inside the outside walls of most homes to hold a layer of air in the walls. Based on what you learned in the MiniLab, why is this beneficial?

Visual Summary

A nonmetal is an element that has no metallic properties. Solid nonmetals are dull, brittle, and do not conduct thermal energy or electricity.

Halogens and noble gases are nonmetals. These elements are found in group 17 and group 18 of the periodic table.

Metalloids have some metallic properties and some nonmetallic properties. The most important use of metalloids is as semiconductors.

SC.8.N.1.1, SC.8.N.1.6, SC.8.P.8.6

Inquiry **Try It!**

LAB STATION

Inquiry Lab *Alien Insect Periodic Table?* at connectED.mcgraw-hill.com

Use Vocabulary

1 **Distinguish** between a nonmetal and a metalloid.

_____

_____

2 An element in group 17 of the periodic table is called a(n) _____

_____ .

3 An element in group 18 of the periodic table is called a(n)

_____

Understand Key Concepts

4 The ability of a halogen to react with a metal to form a salt is an example of a _____ property.

(A) chemical   (C) periodic

(B) noble gas   (D) physical

5 **Classify** each of the following elements as a metal, a nonmetal, or a metalloid: boron, carbon, aluminum, and silicon. SC.8.P.8.6

_____

_____

6 **Infer** which group you would expect to contain element 117. Use the periodic table to help you answer this question. SC.8.P.8.6

_____

Interpret Graphics

7 **Sequence** nonmetals, metals, and metalloids in order from left to right across the periodic table by completing the graphic organizer below. LA.8.2.2.3

Critical Thinking

8 **Hypothesize** how your classroom would be different if there were no metalloids.

_____

_____

9 **Analyze** why hydrogen is sometimes classified as a metal. SC.8.P.8.6

_____

_____

Chapter 7 Study Guide

 Think About It! Elements are grouped and organized on the periodic table according to increasing atomic number and similarities of their properties.

🔑 Key Concepts Summary

LESSON 1 Using the Periodic Table

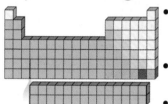

- Elements are organized on the **periodic table** by increasing atomic number and similar properties.
- Elements in the same **group**, or column, of the periodic table have similar properties.
- Elements' properties change across a **period**, which is a row of the periodic table.
- Each element key on the periodic table provides the name, symbol, atomic number, and atomic mass for an element.

LESSON 2 Metals

- **Metals** are located on the left and middle side of the periodic table.
- Metals are elements that have **ductility**, **malleability**, **luster**, and conductivity.
- The **alkali metals** are in group 1 of the periodic table, and the **alkaline earth metals** are in group 2.
- **Transition elements** are metals in groups 3–12 of the periodic table, as well as the lanthanide and actinide series.

LESSON 3 Nonmetals and Metalloids

- **Nonmetals** are on the right side of the periodic table, and **metalloids** are located between metals and nonmetals.
- Nonmetals are elements that have no metallic properties. Solid nonmetals are dull in appearance, brittle, and do not conduct electricity. Metalloids are elements that have properties of both metals and nonmetals.
- Some metalloids are **semiconductors**.
- Elements in group 17 are called **halogens**, and elements in group 18 are **noble gases**.

Active Reading

FOLDABLES® **Chapter Project**

Assemble your lesson Foldables as shown to make a Chapter Project. Use the project to review what you have learned in this chapter.

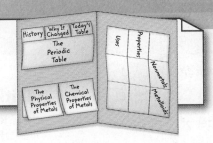

Use Vocabulary

1 The element magnesium (Mg) is in _____ 3 of the periodic table.

2 An element that is shiny, is easily pulled into wires or hammered into thin sheets, and is a good conductor of electricity and heat is a(n) _____.

3 Copper is used to make wire because it has the property of _____.

4 An element that is sometimes a good conductor of electricity and sometimes a good insulator is a(n) _____.

5 An element that is a poor conductor of heat and electricity but is a good insulator is a(n) _____.

Link Vocabulary and Key Concepts

Use vocabulary terms from the previous page to complete the concept map.

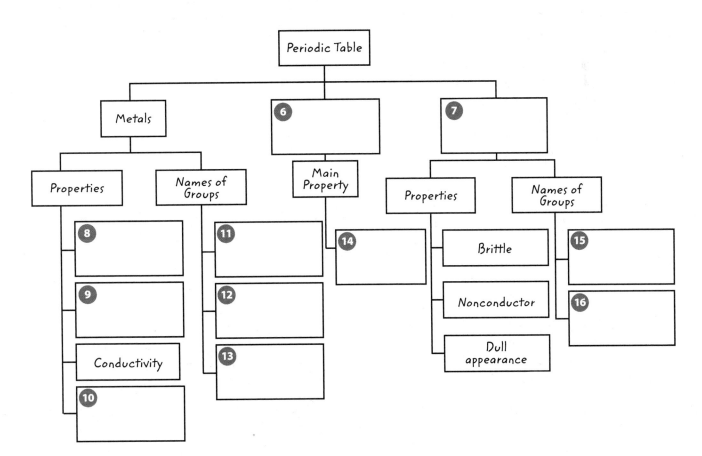

Fill in the correct answer choice.

🔑 Understand Key Concepts

1 What determines the order of elements on today's periodic table? SC.8.P.8.6
- Ⓐ increasing atomic mass
- Ⓑ decreasing atomic mass
- Ⓒ increasing atomic number
- Ⓓ decreasing atomic number

2 The element key for nitrogen is shown below.

Nitrogen
7
N
14.01

From this key, determine the atomic mass of nitrogen. SC.8.P.8.6
- Ⓐ 7
- Ⓑ 7.01
- Ⓒ 14.01
- Ⓓ 21.01

3 Look at the periodic table in Lesson 1. Which of the following lists of elements forms a group on the periodic table? SC.8.P.8.6
- Ⓐ Li, Be, B, C, N, O, F, and Ne
- Ⓑ He, Ne, Ar, Kr, Xe, and Rn
- Ⓒ B, Si, As, Te, and At
- Ⓓ Sc, Ti, V, Cr, Mn, Fe, Co, Cu, Ni, and Zn

4 Which is NOT a property of metals? SC.8.P.8.4
- Ⓐ brittleness
- Ⓑ conductivity
- Ⓒ ductility
- Ⓓ luster

5 What are two properties that make a metal a good choice to use as wire in electronics? SC.8.P.8.4
- Ⓐ conductivity, malleability
- Ⓑ ductility, conductivity
- Ⓒ luster, malleability
- Ⓓ malleability, high density

Critical Thinking

6 **Recommend** an element to use to fill bottles that contain ancient paper. The element should be a gas at room temperature, should be denser than helium, and should not easily react with other elements. SC.8.P.8.4

_____

_____

7 **Apply** Why is mercury the only metal to have been used in thermometers? SC.8.P.8.4

_____

_____

_____

8 **Evaluate** the following types of metals as a choice to make a sun reflector: alkali metals, alkaline earth metals, or transition metals. The metal cannot react with water or oxygen and must be shiny and strong. SC.8.P.8.4

_____

_____

_____

9 The figure below shows a pattern of densities.

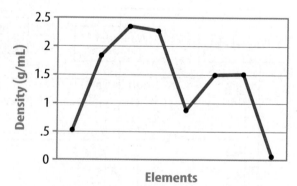

Infer whether you are looking at a graph of elements within a group or across a period. Explain your answer. SC.8.P.8.6

_____

10 **Contrast** aluminum and nitrogen. Show why aluminum is a metal and nitrogen is not. SC.8.P.8.6

_____

_____

_____

_____

11 **Classify** A student sorted six elements. He placed iron, silver, and sodium in group A. He placed neon, oxygen, and nitrogen in group B. Name one other element that fits in group A and another element that belongs in group B. Explain your answer. SC.8.P.8.6

_____

_____

_____

Writing in Science

12 **Write** a plan on a separate sheet of paper that shows how a metal, a nonmetal, and a metalloid could be used when constructing a building. LA.8.2.2.3

Big Idea Review

13 Explain how atomic number and properties are used to determine where element 115 is placed on the periodic table. SC.8.P.8.6

_____

_____

_____

14 The photo on page 252 shows how the properties of materials determine their uses. How can the periodic table be used to help you find elements with properties similar to that of helium? SC.8.P.8.6

_____

_____

_____

Math Skills   MA.6.A.3.6

Use Geometry

15 The table below shows the atomic radii of three elements in group 1 on the periodic table.

Element	Atomic radius
Li	152 pm
Na	186 pm
K	227 pm

a. What is the circumference of each atom?

b. Rubidium (Rb) is the next element in Group 1. What would you predict about the radius and circumference of a rubidium atom?

Florida NGSSS

Fill in the correct answer choice.

Multiple Choice

1 Where are most nonmetals located on the periodic table? SC.8.P.8.6

Ⓐ in the bottom row

Ⓑ on the left side and in the middle

Ⓒ on the right side

Ⓓ in the top row

2 The illustration below shows elements in four groups in the periodic table.

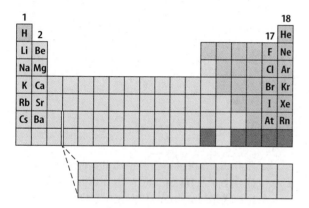

Which group of elements most readily combines with Group 17 elements? SC.8.P.8.6

Ⓕ Group 1

Ⓖ Group 2

Ⓗ Group 17

Ⓘ Group 18

3 Which element is most likely to react with potassium? SC.8.P.8.4

Ⓐ bromine

Ⓑ calcium

Ⓒ nickel

Ⓓ sodium

Use the table below about group 13 elements to answer question 4.

Element Symbol	Atomic Number	Density (g/cm³)	Atomic Mass
B	5	2.34	10.81
Al	13	2.70	26.98
Ga	31	5.90	69.72
In	49	7.30	114.82

4 How do density and atomic mass change as atomic number increases? SC.8.P.8.6

Ⓕ Density and atomic mass decrease.

Ⓖ Density and atomic mass increase.

Ⓗ Density decreases and atomic mass increases.

Ⓘ Density increases and atomic mass decreases.

5 Which elements have high densities, strength, and resistance to corrosion? SC.8.P.8.4

Ⓐ alkali metals

Ⓑ alkaline earth metals

Ⓒ metalloids

Ⓓ transition elements

6 Many elements that are essential for life, including nitrogen, oxygen, and carbon, are part of what classification? SC.8.P.8.6

Ⓕ metalloids

Ⓖ metals

Ⓗ noble gases

Ⓘ nonmetals

Use the figure below to answer questions 7 and 8.

17

7 The figure shows a group in the periodic table. What is the name of this group of elements? SC.8.P.8.6

Ⓐ halogens

Ⓑ metalloids

Ⓒ metals

Ⓓ noble gases

8 Which is a property of these elements? SC.8.P.8.4

Ⓕ They are conductors.

Ⓖ They are semiconductors.

Ⓗ They are nonreactive with other elements.

Ⓘ They react easily with other elements.

9 What is one similarity among elements in a group? SC.8.P.8.6

Ⓐ atomic mass

Ⓑ atomic weight

Ⓒ chemical properties

Ⓓ practical uses

Use the figure below to answer question 10.

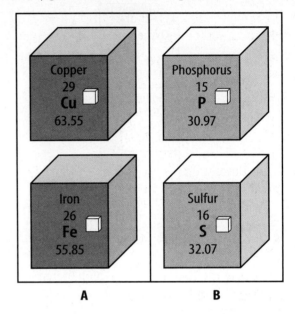

A   **B**

10 Groups A and B each contain two elements. Identify each group as metals, nonmetals, or metalloids. SC.8.P.8.6

Ⓕ Group A contains metals and Group B contains nonmetals.

Ⓖ Group A contains nonmetals and Group B contains metalloids.

Ⓗ Group A contains metalloids and Group B contains metals.

Ⓘ Group A contains metals and Group B contains metals.

NEED EXTRA HELP?

If You Missed Question...	1	2	3	4	5	6	7	8	9	10
Go to Lesson...	1	3	3	1	2	3	3	3	1	2,3

Benchmark Mini-Assessment   Chapter 7 • Lesson 1

Multiple Choice *Bubble the correct answer.*

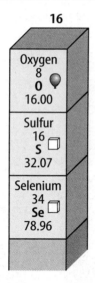

16

Oxygen
8
O
16.00

Sulfur
16
S
32.07

Selenium
34
Se
78.96

1. Using the image above, which statement is TRUE about oxygen, sulfur, and selenium? **SC.8.P.8.6**

Ⓐ They belong to the same group.

Ⓑ They belong to the same period.

Ⓒ The mass number doubles between each element.

Ⓓ The atomic number increases by eight between each element.

2. If the atomic number for oxygen is 8, how many protons are in the nucleus of each of its atoms? **SC.8.P.8.6**

Ⓕ 2

Ⓖ 4

Ⓗ 8

Ⓘ 16

3. Elements on the periodic table are listed according to **SC.8.P.8.6**

Ⓐ atomic mass.

Ⓑ atomic number.

Ⓒ chemical symbol.

Ⓓ state of matter.

| Thorium 90 Th 232.04 | Protactinium 91 Pa 231.04 | Uranium 92 U 238.03 | Neptunium 93 Np (237) |

4. Examine the section of the periodic table in the image above. Which element does NOT occur naturally on Earth? **SC.8.P.8.6**

Ⓕ neptunium

Ⓖ protactinium

Ⓗ thorium

Ⓘ uranium

Multiple Choice *Bubble the correct answer.*

Use the image below to answer questions 1 and 2.

| Potassium 19 **K** 39.10 | Calcium 20 **Ca** 40.08 | Scandium 21 **Sc** 44.96 | Titanium 22 **Ti** 47.87 |

1. Identify which element is the MOST metallic. **SC.8.P.8.4**

   (A) calcium

   (B) potassium

   (C) scandium

   (D) titanium

2. Identify which element is the LEAST malleable. **SC.8.P.8.4**

   (F) calcium

   (G) potassium

   (H) scandium

   (I) titanium

3. On the periodic table, metallic properties increase from **SC.8.P.8.6**

   (A) left to right and bottom to top.

   (B) left to right and top to bottom.

   (C) right to left and bottom to top.

   (D) right to left and top to bottom.

4. Angelo wants to develop a new series of pots and pans for use in cooking. Which property of metals should he be MOST concerned about? **SC.8.P.8.4**

   (F) conductivity

   (G) ductility

   (H) luster

   (I) malleability

Benchmark Mini-Assessment   **Chapter 7 • Lesson 3**   mini BAT

Multiple Choice *Bubble the correct answer.*

Use the image below to answer questions 1 and 2.

					18
					Helium 2 **He** 4.00
13	**14**	**15**	**16**	**17**	
Boron 5 **B** 10.81	Carbon 6 **C** 12.01	Nitrogen 7 **N** 14.01	Oxygen 8 **O** 16.00	Fluorine 9 **F** 19.00	Neon 10 **Ne** 20.18
Aluminum 13 **Al** 26.98	Silicon 14 **Si** 28.09	Phosphorus 15 **P** 30.97	Sulfur 16 **S** 32.07	Chlorine 17 **Cl** 35.45	Argon 18 **Ar** 39.95
Gallium 31 **Ga** 69.72	Germanium 32 **Ge** 72.64	Arsenic 33 **As** 74.92	Selenium 34 **Se** 78.96	Bromine 35 **Br** 79.90	Krypton 36 **Kr** 83.80
Indium 49 **In** 114.82	Tin 50 **Sn** 118.71	Antimony 51 **Sb** 121.76	Tellurium 52 **Te** 127.60	Iodine 53 **I** 126.90	Xenon 54 **Xe** 131.29
Thallium 81 **Tl** 204.38	Lead 82 **Pb** 207.20	Bismuth 83 **Bi** 208.98	Polonium 84 **Po** (209)	Astatine 85 **At** (210)	Radon 86 **Rn** (222)
Ununtrium 113 **Uut** (284)	Ununquadium 114 **Uuq** (289)	Ununpentium 115 **Cn** (288)	Ununhexium 116 **Uuh** (293)		Ununoctium 118 **Uuo** (294)

1. Use the periodic table to identify to which group the noble gases belong. **SC.8.P.8.6**

(A) 13

(B) 14

(C) 17

(D) 18

2. Identify which element is a metalloid. **SC.8.P.8.6**

(F) aluminum

(G) arsenic

(H) carbon

(I) helium

3. Nonmetals often are used in insulating material because they do NOT **SC.8.P.8.4**

(A) break apart easily.

(B) conduct electricity and heat.

(C) form compounds with other elements.

(D) have luster.

4. How do halogens differ from noble gases? **SC.8.P.8.4**

(F) Halogens are metalloids, whereas noble gases are nonmetals.

(G) Halogens are semiconductors, whereas noble gases are conductors.

(H) Halogens are solids at room temperature, whereas noble gases are gases.

(I) Halogens form compounds, whereas noble gases generally do not.

Notes

Unit 3
CHANGES IN MATTER

WITH A GIANT LEMON HEADING FOR EARTH'S OCEANS...

...THE WORLD IS IN A PANIC.

THE PLANET'S TOP SCIENTISTS CALL AN EMERGENCY MEETING.

"WE NEED A BASE TO NEUTRALIZE THE ACID."

500 B.C. 1600 1700

1000 B.C.
Chemistry is considered more of an art than a science. Chemical arts include the smelting of metals and the making of drugs, dyes, iron, and bronze.

1661
A clear distinction is made between chemistry and alchemy when *The Sceptical Chymist* is published by Robert Boyle. Modern chemistry begins to emerge.

1789
Antoine Lavoisier, the "father of modern chemistry," clearly outlines the law of conservation of mass.

1803
John Dalton publishes his atomic theory, which states that all matter is composed of atoms, which are small and indivisible and can join together to form chemical compounds. Dalton is considered the originator of modern atomic theory.

Inquiry

Visit ConnectED for this unit's STEM activity.

1869
The first periodic table is published by Dmitri Mendeleev. The table arranges elements into vertical columns and horizontal rows and is arranged by atomic number.

1953
James Watson and Francis Crick develop the double-helix model of DNA. This discovery leads to a spike in research into the biochemistry of life.

1983
Kary Mullis devises the polymerase chain reaction (PCR), a technique for copying a small portion of DNA in a lab environment. PCR can be used to synthesize specific pieces of DNA and makes the sequencing of DNA of organisms possible.

SC.8.N.1.6,
SC.8.N.4.1,
SC.8.N.4.2

Health and Science

Have an upset stomach? Chew on some charcoal. Have a headache? Rub a little peppermint oil on your temples. As shown in **Figure 1,** people have used chemicals to fix physical ailments for thousands of years. Many cures were discovered by accident. People did not understand why the cures worked, only that they did work.

Asking Questions About Health

Over time, people asked questions about which cures worked and which cures did not work. They made observations, recorded their findings, and had discussions. This process was the start of the scientific investigation of health. **Health** is the overall condition of an organism or group of organisms at any given time. Early studies focused on treating the physical parts of the body. The study of how chemicals interact in organisms and affect health opened a whole new field of study known as biochemistry. The time line in **Figure 2** shows some of the discoveries people made that led to the development of medicines.

Figure 1 Thousands of years ago, people believed that evil spirits were responsible for illness. They often treated the physical symptoms with herbs or other natural materials.

Figure 2 The time line shows several significant discoveries and developments in the history of medicine.

> **Active Reading**
> **1. Discuss** How has the process of scientific investigation of health improved people's health?

4,200 years ago Clay tablets describe using sesame oil on wounds to treat infection.

3,500 years ago An ancient papyrus described how Egyptians applied moldy bread to wounds to prevent infection.

More than 3,300 years later, scientists found that a chemical in mold broke down the cell membranes of bacteria, killing them. Similar discoveries led to the development of antibiotics.

Year 900 The first pharmacy opened in Persia, which is now Iraq.

2,500 years ago Hippocrates, the "Father of Medicine," is the first physician known to separate medical knowledge from myth and superstition.

1740s A doctor found that the disease called scurvy was caused by a lack of Vitamin C.

Early explorers on long sea voyages often lost their teeth or developed deadly sores. Ships could not carry many fruits and vegetables, which contain Vitamin C, because they spoil quickly. Scientists suspect that many early explorers might have died because their diets did not include the proper vitamins.

Benefits and Risks of Medicines

Scientists might recognize that a person's body is missing a necessary chemical, but they cannot always fix the problem. For example, people used to get necessary vitamins and minerals by eating natural, whole foods. Today, food processing destroys many nutrients. Foods might last longer, but they might not provide nutrients the body needs.

Researchers still do not understand the role of many chemicals in the body. Taking a medicine to fix one problem sometimes causes others, called side effects. For example, antibiotics kill some disease-causing bacteria. However, widespread use of antibiotics has resulted in "super bugs"—bacteria that are resistant to treatment.

Histamines are chemicals that have many functions in the body, including regulating sleep and decreasing sensitivity to allergens. However, low levels of histamines have been linked to some serious illnesses. Many medicines have long-term effects on health. Before you take a medicine, recognize that you are adding a chemical to your body. Be as informed as possible about any side effects.

Active Reading **2. Recognize** What are some possible advantages and disadvantages of taking medicines?

MA.6.S.6.2, LA.8.4.2.2

Inquiry **LAB STATION** **Try It!**

MiniLab *Is everyone's chemistry the same?* at connectED.mcgraw-hill.com

Apply It! After you complete the lab, answer this question.

1. **Analyze** Discuss why doctors should not always prescribe the same medicine for every patient. Include information about personal biochemistry and pH, as well as prevention of side effects and medication-resistant "super bugs."

Scientists studying digestion in dogs noticed that ants were attracted to the urine of a dog whose pancreas had been removed. They determined the dog's urine contained sugar, which attracted ants. Eventually, scientists discovered that diabetes resulted from a lack of insulin, a chemical produced in the pancreas that regulates blood sugar. Today, some people with diabetes wear an insulin pump that monitors their blood sugar and delivers insulin to their bodies.

1770s The first vaccination is developed and administered.

1800s Nitrous oxide is first used as an anesthetic by dentists.

1920s Insulin is identified as the missing hormone in people with diabetes.

1920s Penicillin is discovered, but not developed for treatment of disease until the mid-1940s.

2000s First vaccine to target a cause of cancer

Name _____ Date _____

How do the atoms form bonds?

The picture above shows a spoon of sugar granules. The sugar granules are made up of countless molecules of sugar that contain the elements carbon, hydrogen, and oxygen. Chemical bonds hold atoms of these elements together and form sugar molecules. Which best describes how atoms form bonds?

A: When two atoms join, their nuclei form a bond.

B: An attractive force between atoms holds them together but they do not touch.

C: Each atom has a structure that enables it to join with one or more other atoms.

D: A sticky substance holds atoms together when they form a molecule.

Explain your thinking. Describe your ideas about how atoms form chemical bonds.

Elements and Chemical BONDS

FLORIDA BIG IDEAS

1 The Practice of Science

3 The Role of Theories, Laws, Hypotheses, and Models

8 Properties of Matter

Think About It!

How do elements join together to form chemical compounds?

How many different words could you type using just the letters on a keyboard? The English alphabet has only 26 letters, but a dictionary lists hundreds of thousands of words using these letters! Similarly only about 115 different elements make all kinds of matter.

1 How do you think so few elements form so many different kinds of matter?

_____

_____

_____

2 Why do you think different types of matter have different properties?

_____

_____

_____

3 How do you think atoms are held together to produce different types of matter?

_____

_____

_____

Get Ready to Read

What do you think about chemical bonding?

Before you read, decide if you agree or disagree with each of these statements. As you read this chapter, see if you change your mind about any of the statements.

	AGREE	DISAGREE
1 Elements rarely exist in pure form. Instead, combinations of elements make up most of the matter around you.	☐	☐
2 Chemical bonds that form between atoms involve electrons.	☐	☐
3 The atoms in a water molecule are more chemically stable than they would be as individual atoms.	☐	☐
4 Many substances dissolve easily in water because opposite ends of a water molecule have opposite charges.	☐	☐
5 Losing electrons can make some atoms more chemically stable.	☐	☐
6 Metals are good electric conductors because they tend to hold onto their valence electrons very tightly.	☐	☐

There's More Online!

Video • Audio • Review • ⓘLab Station • WebQuest • Assessment • Concepts in Motion • Multilingual eGlossary

Electrons and Energy LEVELS

 How is an electron's energy related to its distance from the nucleus?

 Why do atoms gain, lose, or share electrons?

Vocabulary

chemical bond p. 300

valence electron p. 302

electron dot diagram p. 303

 Florida NGSSS

LA.8.2.2.3 The student will organize information to show understanding or relationships among facts, ideas, and events (e.g., representing key points within text through charting, mapping, paraphrasing, summarizing, or comparing/contrasting);

LA.8.4.2.2 The student will record information (e.g., observations, notes, lists, charts, legends) related to a topic, including visual aids to organize and record information, as appropriate, and attribute sources of information;

SC.8.P.8.5 Recognize that there are a finite number of elements and that their atoms combine in a multitude of ways to produce compounds that make up all of the living and nonliving things that we encounter.

SC.8.P.8.6 Recognize that elements are grouped in the periodic table according to similarities of their properties.

SC.8.P.8.7 Explore the scientific theory of atoms (also known as atomic theory) by recognizing that atoms are the smallest unit of an element and are composed of sub-atomic particles (electrons surrounding a nucleus containing protons and neutrons).

SC.8.N.3.1 Select models useful in relating the results of their own investigations.

SC.8.P.8.6

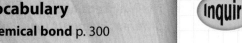 **Launch Lab**

20 minutes

How is the periodic table organized?

How do you begin to put together a puzzle of a thousand pieces? You first sort similar pieces into groups. All edge pieces might go into one pile. All blue pieces might go into another pile. Similarly, scientists place the elements into groups based on their properties. The periodic table organizes information about all the elements.

Procedure

1. Obtain six **index cards** from your teacher. Using one card for each element name, write the names *beryllium, sodium, iron, zinc, aluminum,* and *oxygen* at the top of a card.

2. Using the periodic table, locate the element key for each element written on your cards.

3. For each element, find the following information and write it on the index card: symbol, atomic number, atomic mass, state of matter, and element type.

Think About This

1. What do the elements in the blue blocks have in common? In the green blocks? In the yellow blocks?

2. **Key Concept** Each element in a column on the periodic table has similar chemical properties and forms bonds in similar ways. Based on this, for the element you wrote on each card, name another element on the periodic table that has similar chemical properties.

The Periodic Table

Imagine trying to find a book in a library if all the books were unorganized. Books are organized in a library to help you easily find the information you need. The periodic table is like a library of information about all chemical elements.

The periodic table has more than 100 blocks—one for each known element. Each block on the periodic table includes basic properties of each element such as the element's state of matter at room temperature and its atomic number. The atomic number is the number of protons in each atom of the element. Each block also lists an element's atomic mass, or the average mass of all the different isotopes of that element.

Periods and Groups

You can learn about some properties of an element from its position on the periodic table. Elements are organized in periods (rows) and groups (columns). The periodic table lists elements in order of atomic number. The atomic number increases from left to right as you move across a period. Elements in each group have similar chemical properties and react with other elements in similar ways. In this lesson, you will read more about how an element's position on the periodic table can be used to predict its properties.

Inquiry Are pairs more stable?

1. Rowing can be hard work, especially if you are part of a racing team. The job is made easier because the rowers each pull on the water with a pair of oars. How do pairs make the boat more stable?

_____

_____

_____

_____

_____

_____

_____

Active Reading **2. Describe** As you read this page and the next, complete the sentences below about the periodic table.

The atomic number of an element is the

_____

_____ .

On the periodic table, groups are arranged _____ , and periods are arranged

_____ .

One difference between metals and nonmetals is

_____

_____

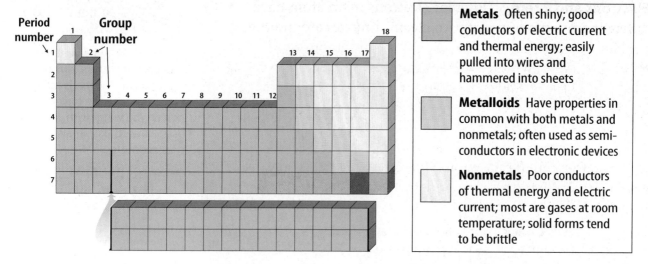

| Metals | Often shiny; good conductors of electric current and thermal energy; easily pulled into wires and hammered into sheets |

Metalloids Have properties in common with both metals and nonmetals; often used as semi-conductors in electronic devices

Nonmetals Poor conductors of thermal energy and electric current; most are gases at room temperature; solid forms tend to be brittle

Figure 1 Elements on the periodic table are classified as metals, nonmetals, or metalloids.

Figure 2 Protons and neutrons are in an atom's nucleus. Electrons move around the nucleus.

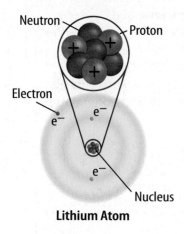

Lithium Atom

Metals, Nonmetals, and Metalloids

The three main regions of elements on the periodic table are shown in **Figure 1.** Except for hydrogen, elements on the left side of the table are metals. Nonmetals are on the right side of the table. Metalloids form the narrow stair-step region between metals and nonmetals.

Active Reading **3. Label** Place an *X* on all the metalloid blocks in the periodic table above.

Atoms Bond

In nature, pure elements are rare. Instead, atoms of different elements chemically combine and form **compounds**. Compounds make up most of the matter around you, including living and nonliving things. There are only about 115 elements, but these elements combine and form millions of compounds. Chemical bonds hold them together. *A **chemical bond** is a force that holds two or more atoms together.*

Electron Number and Arrangement

Recall that atoms contain protons, neutrons, and electrons, as shown in **Figure 2.** Each proton has a positive charge; each neutron has no charge; and each electron has a negative charge. The atomic number of an element is the number of protons in each atom of that element. In a neutral (uncharged) atom, the number of protons equals the number of electrons.

The exact position of electrons in an atom cannot be determined. This is because electrons are in constant motion around the nucleus. However, each electron is usually in a certain area of space around the nucleus. Some are in areas close to the nucleus, and some are in areas farther away.

Electrons and Energy Different electrons in an atom have different amounts of energy. An electron moves around the nucleus at a distance that corresponds to its amount of energy. Areas of space in which electrons move around the nucleus are called energy levels. Electrons closest to the nucleus have the least amount of energy. They are in the lowest energy level. Electrons farthest from the nucleus have the greatest amount of energy. They are in the highest energy level. The energy levels of an atom are shown in **Figure 3.** Notice that only two electrons can be in the lowest energy level. The second energy level can hold up to eight.

 4. NGSSS Check Describe How is an electron's energy related to the electron's position in an atom? SC.8.P.8.7

_____

_____

_____

Electrons and Bonding Imagine two magnets. The closer they are to each other, the stronger the attraction of their opposite ends. Negatively charged electrons have a similar attraction to the positively charged nucleus of an atom. The electrons in energy levels closest to the nucleus of the same atom have a strong attraction to that nucleus. However, electrons farther from that nucleus are weakly attracted to it. These outermost electrons can be attracted to the nucleus of other atoms. This attraction between the positive nucleus of one atom and the negative electrons of another is what causes a chemical bond.

Active Reading

FOLDABLES® LA.8.2.2.3

Make two quarter-sheet note cards from a sheet of paper. Use them to organize your notes on valence electrons and electron dot diagrams.

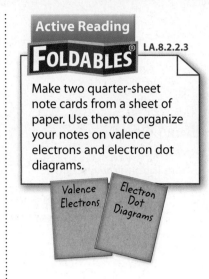

Valence Electrons

Electron Dot Diagrams

Active Reading **5. Identify** Correctly fill in the blanks in the diagram below with words from the following list.

- farthest
- positively
- closest
- negatively

Figure 3 Electrons are in certain energy levels within an atom.

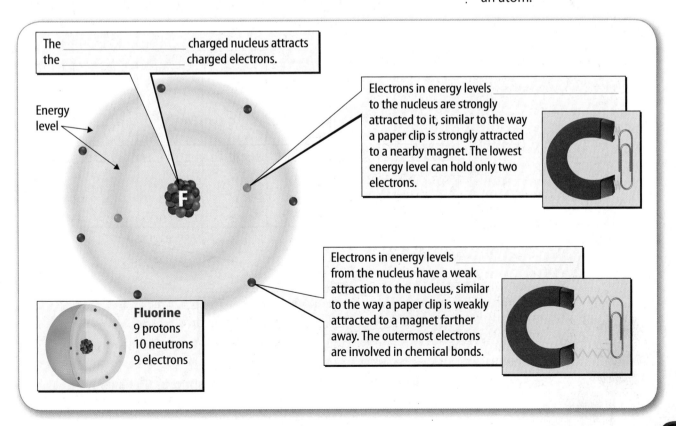

The _____ charged nucleus attracts the _____ charged electrons.

Energy level

F

Electrons in energy levels _____ to the nucleus are strongly attracted to it, similar to the way a paper clip is strongly attracted to a nearby magnet. The lowest energy level can hold only two electrons.

Electrons in energy levels _____ from the nucleus have a weak attraction to the nucleus, similar to the way a paper clip is weakly attracted to a magnet farther away. The outermost electrons are involved in chemical bonds.

Fluorine
9 protons
10 neutrons
9 electrons

Valence Electrons

You have read that electrons farthest from their nucleus are easily attracted to the nuclei of nearby atoms. These outermost electrons are the only electrons involved in chemical bonding. Even atoms that have only a few electrons, such as hydrogen or lithium, can form chemical bonds. This is because these electrons are still the outermost electrons and are exposed to the nuclei of other atoms. A **valence electron** *is an outermost electron of an atom that participates in chemical bonding.* Valence electrons have the most energy of all electrons in an atom.

The number of valence electrons in each atom of an element can help determine the type and the number of bonds it can form. How do you know how many valence electrons an atom has? The periodic table can tell you. Except for helium, elements in certain groups have the same number of valence electrons. **Figure 4** illustrates how to use the periodic table to determine the number of valence electrons in the atoms of groups 1, 2, and 13–18. Determining the number of valence electrons for elements in groups 3–12 is more complicated. You will learn about these groups in later chemistry courses.

WORD ORIGIN

valence

from Latin *valentia*, means "strength, capacity"

Figure 4 🔑 You can use the group numbers at the top of the columns to determine the number of valence electrons in atoms of groups 1, 2, and 13–18.

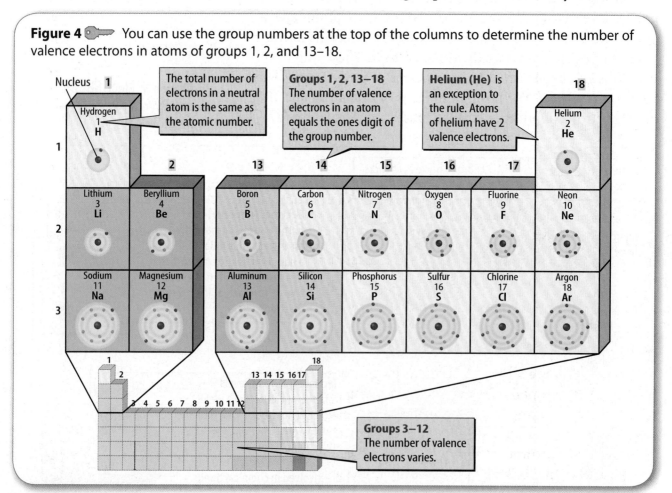

Active Reading

6. **Decide** How many valence electrons does an atom of phosphorus (P) have?

_____

Writing and Using Electron Dot Diagrams

Steps for writing a dot diagram	Beryllium	Carbon	Nitrogen	Argon
1 Identify the element's group number on the periodic table.		14	15	
2 Identify the number of valence electrons. • This equals the ones digit of the group number.	2	4		
3 Draw the electron dot diagram. • Place one dot at a time on each side of the symbol (top, right, bottom, left). Repeat until all dots are used.	Be·		·N̈·	:Är:
4 Determine if the atom is chemically stable. • An atom is chemically stable if all dots on the electron dot diagram are paired.	Chemically Unstable	Chemically Unstable	Chemically Unstable	Chemically Stable
5 Determine how many bonds this atom can form. • Count the dots that are unpaired.	2		3	

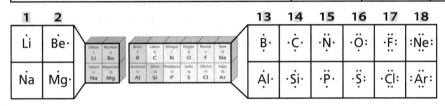

Figure 5 Electron dot diagrams show the number of valence electrons in an atom.

Electron Dot Diagrams

In 1916 an American chemist named Gilbert Lewis developed a method to show an element's valence electrons. He developed the **electron dot diagram**, *a model that represents valence electrons in an atom as dots around the element's chemical symbol.*

Electron dot diagrams can help you predict how an atom will bond with other atoms. Dots, representing valence electrons, are placed one-by-one on each side of an element's chemical symbol until all the dots are used. Some dots will be paired up; others will not. The number of unpaired dots is often the number of bonds an atom can form. The steps for writing dot diagrams are shown in **Figure 5.**

Recall that each element in a group has the same number of valence electrons. As a result, every element in a group has the same number of dots in its electron dot diagram.

Notice in **Figure 5** that an argon atom, Ar, has eight valence electrons, or four pairs of dots, in the diagram. There are no unpaired dots. Atoms with eight valence electrons do not easily react with other atoms. They are chemically stable. Atoms that have between one and seven valence electrons are reactive, or chemically unstable. These atoms easily bond with other atoms and form chemically stable compounds.

Atoms of hydrogen and helium have only one energy level. These atoms are chemically stable with two valence electrons.

Active Reading **7. Summarize** Complete the table above with the correct numbers and symbol.

Active Reading **8. Explain** Why are electron dot diagrams useful?

_____

_____

_____

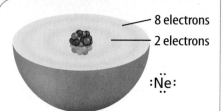

8 electrons
2 electrons

:N̈e:

Neon has 10 electrons: 2 inner electrons and 8 valence electrons. A neon atom is chemically stable because it has 8 valence electrons.

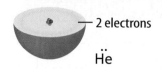

2 electrons

Ḧe

Helium has 2 electrons. Because an atom's lowest energy level can hold only 2 electrons, the 2 dots in the dot diagram are paired. Helium is stable.

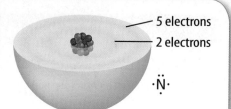

5 electrons
2 electrons

·N̈·

Nitrogen has 7 electrons: 2 inner electrons and 5 valence electrons. Its dot diagram has 3 unpaired dots. Nitrogen atoms become more stable by forming chemical bonds.

Figure 6 Atoms gain, lose, or share electrons to become chemically stable.

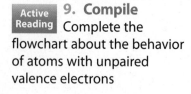

Inquiry   SC.8.N.3.1

LAB STATION Try It!

MiniLab *How does the electron's energy relate to its position in an atom?* at connectED.mcgraw-hill.com

Apply It!

After you complete the lab, answer the question below.

1. Look at the periodic table. Notice that krypton (Kr) is in group 18. Why do atoms of krypton not easily bond with other atoms?

Active Reading 9. **Compile** Complete the flowchart about the behavior of atoms with unpaired valence electrons

Noble Gases

The elements in Group 18 are called noble gases. With the exception of helium, noble gases have eight valence electrons and are chemically stable. Chemically stable atoms do not easily react, or form bonds, with other atoms. The electron structures of two noble gases—neon and helium—are shown in **Figure 6.** Notice that all dots are paired in the dot diagrams of these atoms.

Stable and Unstable Atoms

Atoms with unpaired dots in their electron dot diagrams are reactive, or chemically unstable. For example, nitrogen, shown in **Figure 6,** has three unpaired dots in its electron dot diagram, and it is reactive. Nitrogen, like many other atoms, becomes more stable by forming chemical bonds with other atoms.

When an atom forms a bond, it gains, loses, or shares valence electrons with other atoms. By forming bonds, atoms become more chemically stable. Recall that atoms are most stable with eight valence electrons. Therefore, atoms with fewer than eight valence electrons form chemical bonds and become stable. In Lessons 2 and 3, you will read which atoms gain, lose, or share electrons when forming stable compounds.

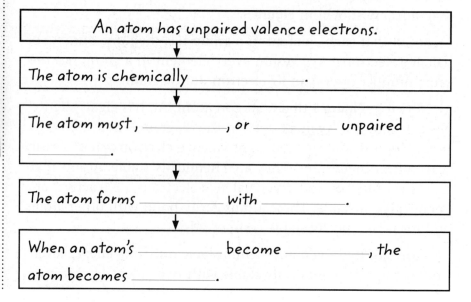

An atom has unpaired valence electrons.

↓

The atom is chemically _____.

↓

The atom must, _____, or _____ unpaired _____.

↓

The atom forms _____ with _____.

↓

When an atom's _____ become _____, the atom becomes _____.

Visual Summary

Electrons are less strongly attracted to a nucleus the farther they are from it, similar to the way a magnet attracts a paper clip.

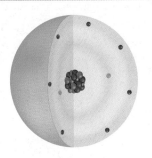

Electrons in atoms are in energy levels around the nucleus. Valence electrons are involved in chemical bonding.

All noble gases, except He, have four pairs of dots in their electron dot diagrams. Noble gases are chemically stable.

Use Vocabulary

1 **Use the term** *chemical bond* in a complete sentence.

_____

_____

2 The electrons of an atom that participate in chemical bonding are called _____.

Understand Key Concepts 🗝

3 **Identify** the number of valence electrons in each atom:

calcium _____ carbon _____ and sulfur _____ . SC.8.P.8.7

4 Which part of the atom forms a chemical bond? SC.8.P.8.7

(A) electron   (C) nucleus

(B) neutron   (D) proton

5 **Draw** electron dot diagrams for oxygen, iodine, and beryllium. SC.8.P.8.6

Interpret Graphics

6 **Determine** the number of valence electrons in each diagram shown below.

Magnesium
12
Mg

_____

Chlorine
17
Cl

_____

7 **Organize Information** Fill in the graphic organizer below to describe one or more details for each concept: electron energy, valence electrons, stable atoms. LA.8.2.2.3

Concept	Description

Critical Thinking

8 **Compare** krypton and bromine in terms of chemical stability. SC.8.P.8.6

_____

_____

_____

New Green Airships

The Difference of One Valence Electron

Faster than ocean liners and safer than airplanes, airships used to be the best way to travel. The largest, the *Hindenburg,* was nearly the size of the *Titanic.* To this day, no larger aircraft has ever flown. So, what happened to the giant airship? The answer lies in a valence electron.

The builders of the *Hindenburg* filled it with a lighter-than-air gas, hydrogen, so that it would float. Their plan was to use helium, a noble gas. However, helium was scarce. They knew hydrogen was explosive, but it was easier to get. For nine years, hydrogen airships floated safely back and forth across the Atlantic. But in 1937, disaster struck. Just before it landed, the *Hindenburg* exploded in flames. The age of the airship was over.

Since the *Hindenburg,* airplanes have become the main type of air transportation. A big airplane uses hundreds of gallons of fuel to take off and fly. As a result, it releases large amounts of pollutants into the atmosphere. Some people are looking for other types of air transportation that will be less harmful to the environment. Airships may be the answer. An airship floats and needs very little fuel to take off and stay airborne. Airships also produce far less pollution than other aircraft.

Today, however, airships use helium—not hydrogen. With two valence electrons instead of one, as hydrogen has, helium is unreactive. Thanks to helium's chemical stability, someday you might be a passenger on a new, luxurious, but not explosive, version of the *Hindenburg.*

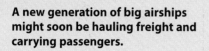

A new generation of big airships might soon be hauling freight and carrying passengers.

It's Your Turn

RESEARCH Precious documents deteriorate with age as their surfaces react with air. Parchment turns brown and crumbles. Find out how our founding documents have been saved from this fate by noble gases. LA.8.4.2.2

Compounds, Chemical Formulas, and Covalent BONDS

 How do elements differ from the compounds they form?

 What are some common properties of a covalent compound?

 Why is water a polar compound?

Vocabulary

covalent bond p. 309

molecule p. 310

polar molecule p. 311

chemical formula p. 312

 Florida NGSSS

LA.8.2.2.3 The student will organize information to show understanding or relationships among facts, ideas, and events (e.g., representing key points within text through charting, mapping, paraphrasing, summarizing, or comparing/contrasting);

SC.8.P.8.5 Recognize that there are a finite number of elements and that their atoms combine in a multitude of ways to produce compounds that make up all of the living and nonliving things that we encounter.

SC.8.P.8.8 Identify basic examples of and compare and classify the properties of compounds, including acids, bases, and salts.

SC.8.N.1.1 Define a problem from the eighth grade curriculum using appropriate reference materials to support scientific understanding, plan and carry out scientific investigations of various types, such as systematic observations or experiments, identify variables, collect and organize data, interpret data in charts, tables, and graphics, analyze information, make predictions, and defend conclusions.

SC.8.N.3.1 Select models useful in relating the results of their own investigations.

 Inquiry Launch Lab SC.8.P.8.5

20 minutes

How is a compound different from its elements?

The sugar you use to sweeten foods at home is probably sucrose. Sucrose contains the elements carbon, hydrogen, and oxygen. How does table sugar differ from the elements that it contains?

Procedure

1. Read and complete a lab safety form.

2. Air is a mixture of several gases, including oxygen and hydrogen. Charcoal is a form of carbon. Write some properties of oxygen, hydrogen, and carbon.

3. Obtain from your teacher a piece of **charcoal** and a **beaker** with **table sugar** in it.

4. Observe the charcoal. Describe the way it looks and feels.

5. Observe the table sugar in the beaker. What does it look and feel like? Record your observations.

Data and Observations

Think About This

1. **Compare and contrast** the properties of charcoal, hydrogen, and oxygen.

2. **Key Concept** **Explain** How do you think the physical properties of carbon, hydrogen, and oxygen change when they combine to form sugar?

From Elements to Compounds

Have you ever baked cupcakes? First, combine flour, baking soda, and a pinch of salt. Then, add sugar, eggs, vanilla, milk, and butter. Each ingredient has unique physical and chemical properties. When you mix the ingredients together and bake them, a new product results—cupcakes. The cupcakes have properties that are different from the ingredients.

In some ways, compounds are like cupcakes. Recall that a compound is a substance made up of two or more different elements. Just as cupcakes are different from their ingredients, compounds are different from their elements. An element is made of one type of atom, but compounds are chemical combinations of different types of atoms. Compounds and the elements that make them up often have different properties.

Chemical **bonds** join atoms together. Recall that a chemical bond is a force that holds atoms together in a compound. In this lesson, you will learn that one way that atoms can form bonds is by sharing valence electrons. You will also learn how to write and read a chemical formula.

Inquiry How do they combine?

1. A jigsaw puzzle has pieces that connect in a certain way. The pieces fit together by sharing tabs with other pieces. All of the pieces combine and form a complete puzzle. Like pieces of a puzzle, atoms can join together and form a compound by sharing electrons. What do you think the world would be like if atoms could not join together?

_____

_____

_____

_____

SCIENCE USE V. COMMON USE

bond

Science Use a force that holds atoms together in a compound

Common Use a close personal relationship between two people

Active Reading **2. Recall** Read each statement in the table below. If it is true, write *T* in the center column. If it is false, write *F* in the center column. In the third column, replace the underlined words to make the statement true.

Statement	T or F	Correction
Compounds are chemical combinations of elements.		
Compounds usually have the same properties as the bonds they are made from.		
Atoms form bonds by sharing physical properties.		

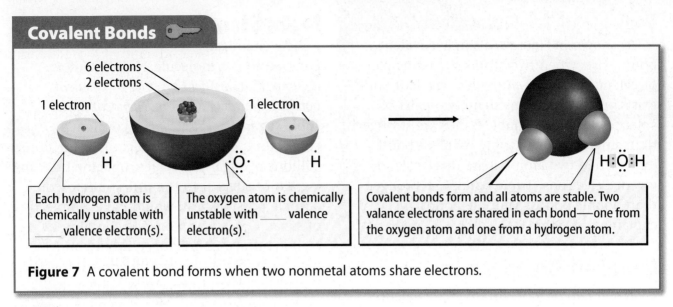

Covalent Bonds 🔑

6 electrons

2 electrons

1 electron

Ḣ

·Ö·

1 electron

Ḣ

Each hydrogen atom is chemically unstable with ____ valence electron(s).

The oxygen atom is chemically unstable with ____ valence electron(s).

H:Ö:H

Covalent bonds form and all atoms are stable. Two valance electrons are shared in each bond—one from the oxygen atom and one from a hydrogen atom.

Figure 7 A covalent bond forms when two nonmetal atoms share electrons.

Active Reading

3. Identify Fill in the blanks in the two left boxes in **Figure 7** with number of valence electrons for each atom.

Covalent Bonds—Electron Sharing

As you read in Lesson 1, one way that atoms can become more chemically stable is by sharing valence electrons. When unstable, nonmetal atoms bond together, they bond by sharing valence electrons. *A **covalent bond** is a chemical bond formed when two atoms share one or more pairs of valence electrons.* The atoms then form a stable covalent compound.

A Noble Gas Electron Arrangement

Look at the reaction between hydrogen and oxygen in **Figure 7.** Before the reaction, each hydrogen atom has one valence electron. The oxygen atom has six valence electrons. Recall that most atoms are chemically stable with eight valence electrons—the same electron arrangement as a noble gas. An atom with less than eight valence electrons becomes stable by forming chemical bonds until it has eight valence electrons. Therefore, an oxygen atom forms two bonds to become stable. A hydrogen atom is stable with two valence electrons. It forms one bond to become stable.

Shared Electrons

If the oxygen atom and each hydrogen atom share their unpaired valence electrons, they can form two covalent bonds and become a stable covalent compound. Each covalent bond contains two valence electrons—one from the hydrogen atom and one from the oxygen atom. Since these electrons are shared, they count as valence electrons for both atoms in the bond. Each hydrogen atom now has two valence electrons. The oxygen atom now has eight valence electrons because it bonds to two hydrogen atoms. All three atoms have the electron arrangement of a noble gas, and the compound is stable.

Active Reading

4. Find How does an atom with less than eight valence electrons become stable? Highlight your answer in the text.

Active Reading

FOLDABLES® LA.8.2.2.3

Make three quarter-sheet note cards from a sheet of paper to organize information about single, double, and triple covalent bonds.

Triple Covalent Bonds
Double Covalent Bonds
Single Covalent Bonds

Double and Triple Covalent Bonds

As shown in **Figure 8,** a single covalent bond exists when two atoms share one pair of valence electrons. A double covalent bond exists when two atoms share two pairs of valence electrons. Double bonds are stronger than single bonds. A triple covalent bond exists when two atoms share three pairs of valence electrons. Triple bonds are stronger than double bonds. Multiple bonds are explained in **Figure 8.**

Covalent Compounds

When two or more atoms share valence electrons, they form a stable covalent compound. The covalent compounds carbon dioxide, water, and sugar are very different, but they also share similar properties. Covalent compounds usually have low melting points and low boiling points. They are usually gases or liquids at room temperature, but they can also be solids. Covalent compounds are poor conductors of thermal energy and electric current.

Molecules

The chemically stable unit of a covalent compound is a molecule. A **molecule** is *a group of atoms held together by covalent bonding that acts as an independent unit.* Table sugar ($C_{12}H_{22}O_{11}$) is a covalent compound. One grain of sugar is made up of trillions of sugar molecules. Imagine breaking a grain of sugar into the tiniest microscopic particle possible. You would have a molecule of sugar. One sugar molecule contains 12 carbon atoms, 22 hydrogen atoms, and 11 oxygen atoms all covalently bonded together. The only way to further break down the molecule would be to chemically separate the carbon, hydrogen, and oxygen atoms. These atoms alone have very different properties from the compound sugar.

  5. NGSSS Check Explain What are some common properties of covalent compounds? SC.8.P.8.5

_____

_____

_____

Figure 8 The more valence electrons that two atoms share, the stronger the covalent bond is between the atoms.

When two hydrogen atoms bond, they form a single covalent bond.	**One Single Covalent Bond** $\dot{H} + \dot{H} \longrightarrow H\!:\!H$	In a single covalent bond, 1 pair of electrons is shared between two atoms. Each H atom shares 1 valence electron with the other.
When one carbon atom bonds with two oxygen atoms, two double covalent bonds form.	**Two Double Covalent Bonds** $\cdot\ddot{O}\!: + \cdot\dot{C}\cdot + \cdot\ddot{O}\!: \longrightarrow :\!\ddot{O}\!::\!C\!::\!\ddot{O}\!:$	In a double covalent bond, 2 pairs of electrons are shared between two atoms. One O atom and the C atom each share 2 valence electrons with the other.
When two nitrogen atoms bond, they form a triple covalent bond.	**One Triple Covalent Bond** $\cdot\ddot{N}\cdot + \cdot\ddot{N}\cdot \longrightarrow :\!N\!::\!N\!:$	In a triple covalent bond, 3 pairs of electrons are shared between two atoms. Each N atom shares 3 valence electrons with the other.

Active Reading **6. Compare** Is the bond stronger between atoms in hydrogen gas (H_2) or nitrogen gas (N_2)? Why?

_____

_____

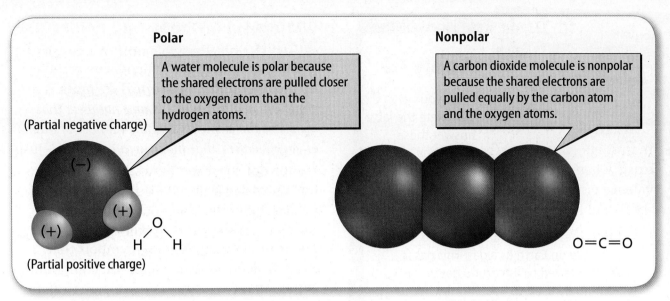

Polar

A water molecule is polar because the shared electrons are pulled closer to the oxygen atom than the hydrogen atoms.

(Partial negative charge)

(−)

(+)

(+)

(Partial positive charge)

$$H \quad \overset{O}{\underset{}{}} \quad H$$

Nonpolar

A carbon dioxide molecule is nonpolar because the shared electrons are pulled equally by the carbon atom and the oxygen atoms.

$O=C=O$

Figure 9 Atoms of a polar molecule share their valence electrons unequally. Atoms of a nonpolar molecule share their valence electrons equally.

Water and Other Polar Molecules

In a covalent bond, one atom can attract the shared electrons more strongly than the other atom can. Think about the valence electrons shared between oxygen and hydrogen atoms in a water molecule. The oxygen atom attracts the shared electrons more strongly than each hydrogen atom does. As a result, the shared electrons are pulled closer to the oxygen atom, as shown in **Figure 9.** Since electrons have a negative charge, the oxygen atom has a partial negative charge. The hydrogen atoms have a partial positive charge. *A molecule that has a partial positive end and a partial negative end because of unequal sharing of electrons is a* **polar molecule**.

The charges on a polar molecule affect its properties. Sugar, for example, dissolves easily in water because both sugar and water are polar. The negative end of a water molecule pulls on the positive end of a sugar molecule. Also, the positive end of a water molecule pulls on the negative end of a sugar molecule. This causes the sugar molecules to separate from one another and mix with the water molecules.

Nonpolar Molecules

A hydrogen molecule, H_2, is a nonpolar molecule. Because the two hydrogen atoms are identical, their attraction for the shared electrons is equal. The carbon dioxide molecule, CO_2, in **Figure 9** is also nonpolar. A nonpolar compound will not easily dissolve in a polar compound, but it will dissolve in other nonpolar compounds. Oil is an example of a nonpolar compound. It will not dissolve in water. Have you ever heard someone say "like dissolves like"? This means that polar compounds can dissolve in other polar compounds. Similarly, nonpolar compounds can dissolve in other nonpolar compounds.

WORD ORIGIN

polar
from Latin *polus*, means "pole"

Active Reading 7. **Summarize** Fill in the flowchart below to describe the structure of polar molecules.

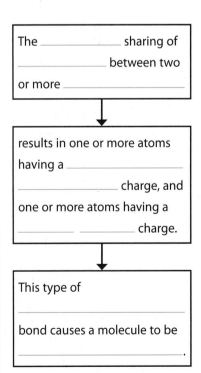

The _____ sharing of _____ between two or more _____

↓

results in one or more atoms having a _____ _____ charge, and one or more atoms having a _____ _____ charge.

↓

This type of _____ bond causes a molecule to be _____ .

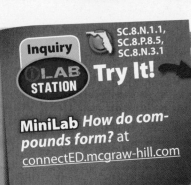

Inquiry
SC.8.N.1.1,
SC.8.P.8.5,
SC.8.N.3.1
LAB STATION
Try It!

MiniLab *How do compounds form?* at connectED.mcgraw-hill.com

Apply It!

After you complete the lab, answer these questions.

1. CH_4 is methane. Methane is found naturally on Earth as a gas. This gas is often referred to as natural gas and is commonly used to heat homes. In the table below, draw the dot diagram and structural formula for this molecule.

Dot Diagram	Structural Formula

2. Study the two models in **Figure 10.** What does each dot and each line represent?

Chemical Formulas and Molecular Models

How do you know which elements make up a compound? *A **chemical formula** is a group of chemical symbols and numbers that represent the elements and the number of atoms of each element that make up a compound.* Just as a recipe lists ingredients, a chemical formula lists the elements in a compound. For example, the chemical formula for carbon dioxide shown in **Figure 10** is CO_2. The formula uses chemical symbols that show which elements are in the compound. Notice that CO_2 is made up of carbon (C) and oxygen (O). A subscript, or small number after a chemical symbol, shows the number of atoms of each element in the compound. Carbon dioxide (CO_2) contains two atoms of oxygen bonded to one atom of carbon.

A chemical formula describes the types of atoms in a compound or a molecule, but it does not explain the shape or appearance of the molecule. There are many ways to model a molecule. Each one can show the molecule in a different way. Common types of models for CO_2 are shown in **Figure 10.**

Active Reading 8. **State** What information is given in a chemical formula?

Figure 10 Chemical formulas and molecular models provide information about molecules.

Chemical Formula

A carbon dioxide molecule is made up of carbon (C) and oxygen (O) atoms.

CO₂

A symbol without a subscript indicates one atom. Each molecule of carbon dioxide has one carbon atom.

The subscript 2 indicates two atoms of oxygen. Each molecule of carbon dioxide has two oxygen atoms.

Dot Diagram
- Shows atoms and valence electrons

$\ddot{O}::C::\ddot{O}$

Structural Formula
- Shows atoms and lines; each line represents one shared pair of electrons

$O=C=O$

Ball-and-Stick Model
- Balls represent atoms and sticks represent bonds; used to show bond angles

Space-Filling Model
- Spheres represent atoms; used to show three-dimensional arrangement of atoms

CO_2

A chemical formula is one way to show the elements that make up a compound.

O=C=O

A covalent bond forms when atoms share valence electrons. The smallest particle of a covalent compound is a molecule.

(−)

(+)

(+)

Water is a polar molecule because the oxygen and hydrogen atoms unequally share electrons.

Inquiry SC.8.N.1.1, SC.8.P.8.5, SC.8.P.8.8, SC.8.N.3.1

LAB STATION Try It!

Skill Lab *How can you model compounds?* at connectED.mcgraw-hill.com

Use Vocabulary

1 **Define** *covalent bond* in your own words.

_____

2 The group of symbols and numbers that shows the types and numbers of atoms that make up a compound is a _____.

Understand Key Concepts 🔑

3 **Contrast** Name at least one way water (H_2O) is different from the elements that make up water. SC.8.P.8.5

_____

4 **Explain** why water is a polar molecule.

_____

5 A sulfur dioxide molecule has one sulfur atom and two oxygen atoms. Which is its correct chemical formula? SC.8.P.8.5

Ⓐ SO_2 Ⓒ S_2O_2

Ⓑ $(SO)_2$ Ⓓ S_2O

Interpret Graphics

6 **Examine** the electron dot diagram for chlorine below.

$\cdot\ddot{\underset{..}{Cl}}:$ In chlorine gas, two chlorine atoms join to form a Cl_2 molecule. How many pair(s) of valence electrons do the atoms share? SC.8.P.8.5

7 **Compare and Contrast** Fill in the graphic organizer below to identify at least one way polar and nonpolar molecules are similar and one way they are different. LA.8.2.2.3

Polar and Nonpolar Molecules	
Similarities	
Differences	

Critical Thinking

8 **Develop** an analogy to explain the unequal sharing of valence electrons in a water molecule.

_____

Complete the Lesson 1 and Lesson 2 graphic organizers now. As you read the following lessons come back to this page and complete the graphic organizers for Lesson 3.

Lesson 1

1. **Analyze** details about valence electrons.

farthest from _____		weakest attraction to _____
most energy	**Valence Electrons**	involved in _____ _____

same number for all elements in _____ (with the exception of _____)

Sequence the steps in constructing and interpreting an electron dot diagram.

2 Identify the element's _____ .

3 Identify the number of _____ , which is the same as the _____ of the _____ .

4 Place _____ dot at a time on each _____ of the _____ . Pair up the dots until all are used.

5 Identify an atom as _____ if all _____ are _____ .

6 Count the _____ to determine how many _____ an unstable atom can form.

Lesson 2

7. **Summarize** the relationship between an electron's energy level and its location in an atom. (Circle) the word that makes each statement true.

The closer to the nucleus, the lower / higher an electron's energy level.	The farther from the nucleus, the lower / higher an electron's energy level.

8. **Describe** types of covalent bonds.

Covalent Bond	Description of Valence Electron Sharing	Comment on the Strength of the Bond
Single		weakest type of covalent bond
Double		
Triple		

9. **Identify** four common properties of covalent compounds.

_____

_____

_____

_____

Lesson 3

10. **Describe** three properties of metallic compounds.

Properties of Metallic Compounds		

11. **Contrast** three ways atoms can bond and become stable.

Process	Electron Pooling	Electron Transfer	Electron Sharing
Type of chemical bond	metallic	ionic	covalent
Description			

Ionic and Metallic BONDS

ESSENTIAL QUESTIONS

 What is an ionic compound?

 How do metallic bonds differ from covalent and ionic bonds?

Vocabulary

ion p. 316

ionic bond p. 318

metallic bond p. 319

 Florida NGSSS

LA.8.2.2.3 The student will organize information to show understanding or relationships among facts, ideas, and events (e.g., representing key points within text through charting, mapping, paraphrasing, summarizing, or comparing/contrasting);

MA.6.A.3.6 Construct and analyze tables, graphs, and equations to describe linear functions and other simple relations using both common language and algebraic notation.

SC.8.P.8.5 Recognize that there are a finite number of elements and that their atoms combine in a multitude of ways to produce compounds that make up all of the living and nonliving things that we encounter.

SC.8.P.8.8 Identify basic examples of and compare and classify the properties of compounds, including acids, bases, and salts.

SC.8.N.1.1 Define a problem from the eighth grade curriculum using appropriate reference materials to support scientific understanding, plan and carry out scientific investigations of various types, such as systematic observations or experiments, identify variables, collect and organize data, interpret data in charts, tables, and graphics, analyze information, make predictions, and defend conclusions.

SC.8.N.1.6 Understand that scientific investigations involve the collection of relevant empirical evidence, the use of logical reasoning, and the application of imagination in devising hypotheses, predictions, explanations and models to make sense of the collected evidence.

SC.8.N.3.1 Select models useful in relating the results of their own investigations.

 **Inquiry Launch Lab**

SC.8.P.8.5, SC.8.N.3.1

15 minutes

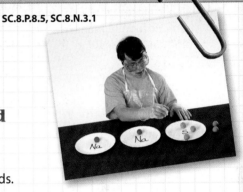

How can atoms form compounds by gaining and losing electrons?

Metals on the periodic table often lose electrons when forming stable compounds. Nonmetals often gain electrons.

Procedure

1. Read and complete a lab safety form.

2. Make two model atoms of sodium and one model atom each of calcium, chlorine, and sulfur. To do this, write each element's chemical symbol with a **marker** on a **paper plate.** Surround the symbol with small balls of **clay** to represent valence electrons. Use one color of clay for the metals (groups 1 and 2 elements) and another color of clay for nonmetals (groups 16 and 17 elements).

3. To model sodium sulfide (Na_2S), place the two sodium atoms next to the sulfur atom. To form a stable compound, move each sodium atom's valence electron to the sulfur atom.

4. Form as many other compound models as you can by removing valence electrons from the groups 1 and 2 plates and placing them on the groups 16 and 17 plates.

Think About This

1. **Name** What other compounds were you able to form?

2. **Key Concept** **Differentiate** How do you think your models are different from covalent compounds?

1. This scene might look like snow along a shoreline, but it is actually thick deposits of salt on a lake. Over time, salt dissolved in river water flowed into this lake and built up as water evaporated. Salt is a compound that forms when elements form bonds by gaining or losing valence electrons, not sharing them. How do humans use this common substance?

Active Reading

2. Explain Why do atoms that gain electrons become an ion with a negative charge?

_____

_____

_____

_____

Active Reading

FOLDABLES LA.8.2.2.3

Make two quarter-sheet note cards as shown. Use the cards to summarize information about ionic and metallic compounds.

Metallic Compounds   Ionic Compounds

Understanding Ions

As you read in Lesson 2, the atoms of two or more nonmetals form compounds by sharing valence electrons. However, when a metal and a nonmetal bond, they do not share electrons. Instead, one or more valence electrons transfers from the metal atom to the nonmetal atom. After electrons transfer, the atoms bond and form a chemically stable compound. Transferring valence electrons results in atoms with the same number of valence electrons as a noble gas.

When an atom loses or gains a valence electron, it becomes an ion. *An **ion** is an atom that is no longer electrically neutral because it has lost or gained valence electrons.* Because electrons have a negative charge, losing or gaining an electron changes the overall charge of an atom. An atom that loses valence electrons becomes an ion with a positive charge. This is because the number of electrons is now less than the number of protons in the atom. An atom that gains valence electrons becomes an ion with a negative charge. This is because the number of protons is now less than the number of electrons.

Losing Valence Electrons

Look at the periodic table on the inside back cover of this book. What information about sodium (Na) can you infer from the periodic table? Sodium is a metal. Its atomic number is 11. This means each sodium atom has 11 protons and 11 electrons. Sodium is in group 1 on the periodic table. Therefore, sodium atoms have one valence electron, and they are chemically unstable.

Metal atoms, such as sodium, become more stable when they lose valence electrons and form a chemical bond with a nonmetal. If a sodium atom loses its one valence electron, it will have a total of ten electrons. Which element on the periodic table has atoms with ten electrons? Neon (Ne) atoms have a total of ten electrons. Eight of these are valence electrons. When a sodium atom loses one valence electron, the electrons in the next-lower energy level are now the new valence electrons. The sodium atom then has eight valence electrons, the same as the noble gas neon, and is chemically stable.

Gaining Valence Electrons

In Lesson 2, you read that nonmetal atoms can share valence electrons with other nonmetal atoms. Nonmetal atoms can also gain valence electrons from metal atoms. Either way, they achieve the electron arrangement of a noble gas. Find the nonmetal chlorine (Cl) on the periodic table. Its atomic number is 17. Atoms of chlorine have seven valence electrons. If a chlorine atom gains one valence electron, it will have eight valence electrons. It will also have the same electron arrangement as the noble gas argon (Ar).

When a sodium atom loses a valence electron, it becomes a positively charged ion. This is shown by a plus (+) sign. When a chlorine atom gains a valence electron, it becomes a negatively charged ion. This is shown by a negative (−) sign. **Figure 11** illustrates this process.

 3. Choose Are atoms of group 16 elements more likely to gain or lose valence electrons?

Losing and Gaining Electrons

Figure 11 Sodium atoms have a tendency to lose a valence electron. Chlorine atoms have a tendency to gain a valence electron.

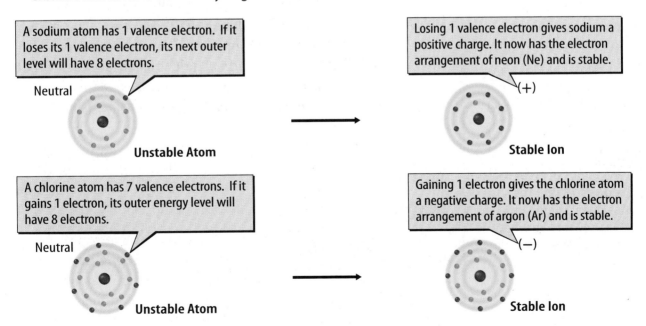

A sodium atom has 1 valence electron. If it loses its 1 valence electron, its next outer level will have 8 electrons.

Neutral — Unstable Atom

Losing 1 valence electron gives sodium a positive charge. It now has the electron arrangement of neon (Ne) and is stable.

(+) Stable Ion

A chlorine atom has 7 valence electrons. If it gains 1 electron, its outer energy level will have 8 electrons.

Neutral — Unstable Atom

Gaining 1 electron gives the chlorine atom a negative charge. It now has the electron arrangement of argon (Ar) and is stable.

(−) Stable Ion

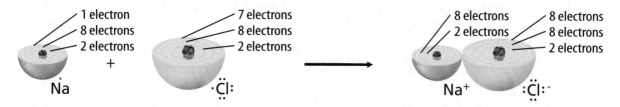

1 electron
8 electrons
2 electrons

Na

7 electrons
8 electrons
2 electrons

:Cl̈:

8 electrons
2 electrons

8 electrons
8 electrons
2 electrons

Na$^+$:C̈l:$^-$

+

A sodium atom loses one valence electron and becomes stable. A chlorine atom gains one valence electron and becomes stable.

The positively charged sodium ion and the negatively charged chlorine ion attract each other. Together they form a strong ionic bond.

Figure 12 An ionic bond forms between Na and Cl when an electron transfers from Na to Cl.

Active Reading

Math Skills   MA.6.A.3.6

Use Percentage

A picometer (pm) is 1 trillion times smaller than a meter. When an atom becomes an ion, its radius increases or decreases. For example, a Na atom has a radius of **186 pm**. A Na$^+$ ion has a radius of **102 pm**. By what percentage does the radius change?

Subtract the atom's radius from the ion's radius.

102 pm − 186 pm = _____

Divide the difference above by the atom's radius.

−84 pm ÷ 186 pm = _____

Multiply the answer above by 100 and add a % sign.

−0.45 × 100 = _____

A negative value is a decrease in size. A positive value is an increase.

Practice

5. The radius of an oxygen (O) atom is 73 pm. The radius of an oxygen ion (O^{2-}) is 140 pm. By what percentage does the radius change?

Determining an Ion's Charge

Atoms are electrically neutral because they have the same number of protons and electrons. Once an atom gains or loses electrons, it becomes a charged ion. For example, the atomic number for nitrogen (N) is 7. Each N atom has 7 protons and 7 electrons and is electrically neutral. However, an N atom often gains 3 electrons when forming an ion. The N ion then has 10 electrons. To determine the charge, subtract the number of electrons in the ion from the number of protons.

7 protons − 10 electrons = −3 charge

A nitrogen ion has a −3 charge. This is written as N^{3-}.

Ionic Bonds—Electron Transferring

Recall that metal atoms typically lose valence electrons and nonmetal atoms typically gain valence electrons. When forming a chemical bond, the nonmetal atoms gain the electrons lost by the metal atoms. Take a look at **Figure 12.** In NaCl, or table salt, an atom of the metal sodium loses a valence electron. The electron is transferred to a nonmetal chlorine atom. The sodium atom becomes a positively charged ion. The chlorine atom becomes a negatively charged ion. These ions attract each other and form a stable ionic compound. *The attraction between positively and negatively charged ions in an ionic compound is an* **ionic bond**.

Active Reading   4. Underline Mark the sentences in the text that explain how metal atoms and nonmetal atoms chemically bond.

Ionic Compounds

Ionic compounds are usually solid and brittle at room temperature. They also have relatively high melting and boiling points. Many ionic compounds dissolve in water. Water that contains dissolved ionic compounds is a good conductor of electricity. This is because an electrical charge can pass from ion to ion in the solution.

Active Reading

6. Analyze What happens to sodium and chlorine atoms in the formation of the compound sodium chloride?

	Na (sodium)	**Cl** (chlorine)
Type of element		
Atomic number		
Number of valence electrons		
Stable or unstable?		
Loses or gains electrons		
Type of ion		

Comparing Ionic and Covalent Compounds

Recall that in a covalent bond, two or more nonmetal atoms share electrons and form a unit, or molecule. Covalent compounds, such as water, are made up of many molecules. However, when nonmetal ions bond to metal ions in an ionic compound, there are no molecules. Instead, there is a large collection of oppositely charged ions. All of the ions attract each other and are held together by ionic bonds.

Metallic Bonds— Electron Pooling

Recall that metal atoms typically lose valence electrons when forming compounds. What happens when metal atoms bond to other metal atoms? Metal atoms form compounds with one another by combining, or pooling, their valence electrons. A **metallic bond** *is a bond formed when many metal atoms share their pooled valence electrons.*

The pooling of valence electrons in aluminum is shown in **Figure 13.** The aluminum atoms lose their valence electrons and become positive ions, indicated by the plus (+) signs. The negative (−) signs indicate the valence electrons, which move from ion to ion. Valence electrons in metals are not bonded to one atom. Instead, a "sea of electrons" surrounds the positive ions.

Inquiry **LAB STATION** **Try It!** SC.8.N.1.1, SC.8.N.8.5

MiniLab *How many ionic compounds can you make?* at connectED.mcgraw-hill.com

Apply It!

After you complete the lab, answer these questions.

1. Why are electric wires not made of ionic substances?

2. Explain why two sodium ions (Na^+) do not bond to form an ionic compound.

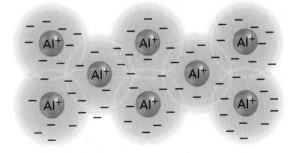

Figure 13 Valence electrons move among all the aluminum (Al) ions.

 7. NGSSS Check Explain How do metal atoms bond with one another? SC.8.P.8.5

Properties of Metallic Compounds

Metals are good conductors of thermal energy and electricity. Because the valence electrons can move from ion to ion, they can easily **conduct** an electric charge. When a metal is hammered into a sheet or drawn into a wire, it does not break. The metal ions can slide past one another in the electron sea and move to new positions. Metals are shiny because the valence electrons at the surface of a metal interact with light. **Table 1** compares the covalent, ionic, and metallic bonds that you studied in this chapter.

Active Reading **8. Paraphrase** How does valence electron pooling explain why metals can be hammered into a sheet?

_____

_____

Active Reading **9. Choose** Use the correct words to fill in the blanks to complete **Table 1.**

Table 1 Covalent, Ionic, and Metallic Bonds

Type of Bond	What is bonding?	Properties of Compounds
Water	nonmetal atoms; nonmetal atoms	• gas, liquid, or solid • low melting and boiling points • often not able to dissolve in water • poor conductors of thermal energy and electric current • dull appearance
Salt	_____ ions; _____ ions	• solid crystals • high melting and boiling points • dissolves in water • solids are poor conductors of thermal energy and electric current • ionic compounds in water solutions conduct electric current
Aluminum	_____ ions; _____ ions	• usually solid at room temperature • high melting and boiling points • do not dissolve in water • good conductors of thermal energy and electric current • shiny surface • can be hammered into sheets and pulled into wires

Visual Summary

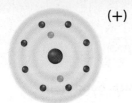

(+)

Metal atoms lose electrons, and nonmetal atoms gain electrons and form stable compounds. An atom that has gained or lost an electron is an ion.

(+)   (−)

Na⁺   :Cl:⁻

An ionic bond forms between positively and negatively charged ions.

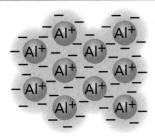

A metallic bond forms when many metal atoms share their pooled valence electrons.

Inquiry

LA.8.2.2.3,
SC.8.N.1.1,
SC.8.N.1.6,
SC.8.N.8.5

LAB STATION Try It!

Inquiry Lab *Ions in Solution* at
connectED.mcgraw-hill.com

Use Vocabulary

1. **Define** *ionic bond* in your own words.

2. An atom that changes so that it has an electric charge is a(n) _____.

3. **Use the term** *metallic bond* in a sentence.

Understand Key Concepts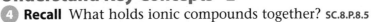

4. **Recall** What holds ionic compounds together? SC.8.P.8.5

5. Which element would most likely bond with lithium and form an ionic compound? SC.8.P.8.5
   - (A) beryllium
   - (B) calcium
   - (C) fluorine
   - (D) sodium

6. **Contrast** Why are metals good conductors of electricity but covalent compounds are poor conductors? SC.8.P.8.6

Interpret Graphics

7. **Organize** Copy and fill in the graphic organizer below. In each oval, list a common property of an ionic compound. LA.8.2.2.3

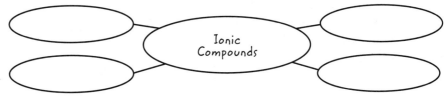

Ionic Compounds

Critical Thinking

8. **Evaluate** What type of bonding does a material most likely have if it has a high melting point, is solid at room temperature, and easily dissolves in water?

Math Skills   MA.6.A.3.6

9. The radius of the aluminum (Al) atom is 143 pm. The radius of the aluminum ion (Al^{3+}) is 54 pm. By what percentage did the radius change as the ion formed?

Chapter 8 — Study Guide

 Think About It! There are a finite number of elements, and their atoms join together through sharing, transferring, or pooling their electrons to make chemical compounds.

Key Concepts Summary

Vocabulary

LESSON 1 Electrons and Energy Levels

- Electrons with more energy are farther from the atom's nucleus and are in a higher energy level.

- Atoms with fewer than eight **valence electrons** gain, lose, or share valence electrons and form stable compounds. Atoms in stable compounds have the same electron arrangement as a noble gas.

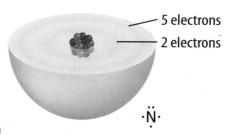

5 electrons

2 electrons

·N̈·

chemical bond p. 300

valence electron p. 302

electron dot diagram p. 303

LESSON 2 Compounds, Chemical Formulas, and Covalent Bonds

- A compound and the elements it is made from have different chemical and physical properties.

- A **covalent bond** forms when two nonmetal atoms share valence electrons. Common properties of covalent compounds include low melting points and low boiling points. They are usually gas or liquid at room temperature and poor conductors of electricity.

- Water is a polar compound because the oxygen atom pulls more strongly on the shared valence electrons than the hydrogen atoms do.

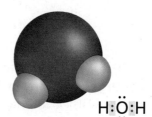

H:Ö:H

covalent bond p. 309

molecule p. 310

polar molecule p. 311

chemical formula p. 312

LESSON 3 Ionic and Metallic Bonds

- **Ionic bonds** form when valence electrons move from a metal atom to a nonmetal atom.

- An ionic compound is held together by ionic bonds, which are attractions between positively and negatively charged **ions**.

- A **metallic bond** forms when valence electrons are pooled among many metal atoms.

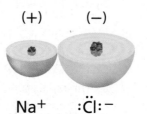

(+)   (−)

Na⁺   :C̈l:⁻

ion p. 316

ionic bond p. 318

metallic bond p. 319

FOLDABLES **Chapter Project**

Assemble your lesson Foldables as shown to make a Chapter Project. Use the project to review what you have learned in this chapter.

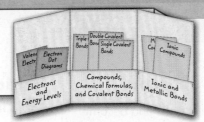

Use Vocabulary

1 The force that holds atoms together is called a(n) _____ .

2 You can predict the number of bonds an atom can form by drawing its _____ .

3 The nitrogen and hydrogen atoms that make up ammonia (NH3) are held together by a(n) _____ because the atoms share valence electrons unequally.

4 Two hydrogen atoms and one oxygen atom together are a _____ of water.

5 A positively charged sodium ion and a negatively charged chlorine ion are joined by a(n) _____ to form the compound sodium chloride.

Link Vocabulary and Key Concepts

Use vocabulary terms from the previous page and other terms from the chapter to complete the concept map.

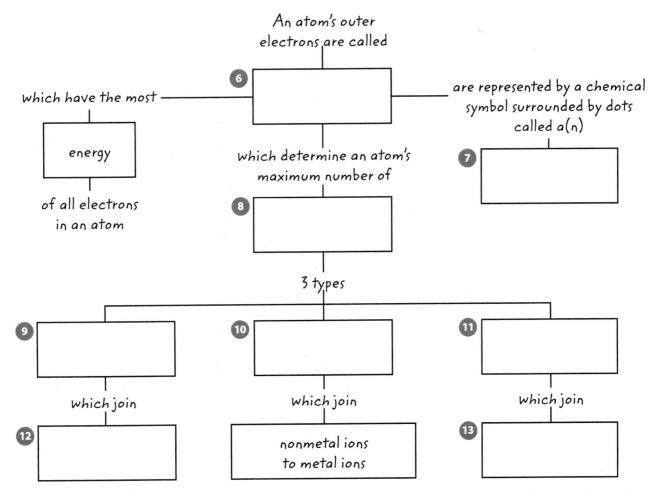

🔑 Understand Key Concepts

1. Atoms lose, gain, or share electrons and become as chemically stable as SC.8.P.8.5
   - (A) an electron.
   - (B) an ion.
   - (C) a metal.
   - (D) a noble gas.

2. Which is the correct electron dot diagram for boron, one of the group 13 elements? SC.8.P.8.5

   (A) B·

   (B) ·B:

   (C) :B:

   (D) ·B·

3. If an electron transfers from one atom to another atom, what type of bond will most likely form? SC.8.P.8.5
   - (A) covalent
   - (B) ionic
   - (C) metallic
   - (D) polar

4. What change would make an atom represented by this diagram have the same electron arrangement as a noble gas? SC.8.P.8.5

   - (A) gaining two electrons
   - (B) gaining four electrons
   - (C) losing two electrons
   - (D) losing four electrons

5. What would make bromine, a group 17 element, more similar to a noble gas? SC.8.P.8.6
   - (A) gaining one electron
   - (B) gaining two electrons
   - (C) losing one electron
   - (D) losing two electrons

Critical Thinking

6. **Classify** Use the periodic table to classify the elements potassium (K), bromine (Br), and argon (Ar) according to how likely their atoms are to do the following: SC.8.P.8.6

   a. lose electrons to form positive ions

   _____

   b. gain electrons to form negative ions

   _____

   c. neither gain nor lose electrons

   _____

7. **Describe** the change that is shown in this illustration. How does this change affect the stability of the atom? SC.8.P.8.5

   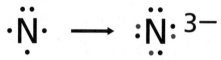

   _____

   _____

   _____

8. **Analyze** One of your classmates draws an electron dot diagram for a helium atom with two dots. He tells you that these dots mean each helium atom has two unpaired electrons and can gain, lose, or share electrons to have four pairs of valence electrons and become stable. What is wrong with your classmate's argument? SC.8.P.8.5

   _____

   _____

   _____

   _____

9. **Explain** why the hydrogen atoms in a hydrogen gas molecule (H_2) form nonpolar covalent bonds but the oxygen and hydrogen atoms in water molecules (H_2O) form polar covalent bonds. SC.8.P.8.5

   _____

   _____

   _____

   _____

10 **Contrast** Why is it possible for an oxygen atom to form a double covalent bond, but it is not possible for a chlorine atom to form a double covalent bond? SC.8.P.8.5

_____

_____

_____

_____

_____

_____

Writing in Science

11 **Compose** On a separate piece of paper, write a poem at least ten lines long that explains ionic bonding, covalent bonding, and metallic bonding. LA.8.2.2.3

Big Idea Review

12 Which types of atoms pool their valence electrons to form a "sea of electrons"? SC.8.P.8.5

_____

_____

13 Describe a way in which elements joining together to form chemical compounds is similar to the way the letters on a computer keyboard join together to form words. SC.8.P.8.5

_____

_____

_____

_____

_____

Math Skills   MA.6.A.3.6

Element	Atomic Radius	Ionic Radius
Potassium (K)	227 pm	133 pm
Iodine (I)	133 pm	216 pm

14 What is the percent change when an iodine atom (I) becomes an ion (I-)?

15 What is the percent change when a potassium atom (K) becomes an ion (K^+)?

Record your answers on the answer sheet provided by your teacher or on a sheet of paper.

Multiple Choice

1 When at atom is chemically stable, how many electrons are in its outer energy level? SC.8.P.8.7

Ⓐ 8

Ⓑ 7

Ⓒ 4

Ⓓ 1

Use the diagram below to answer question 2.

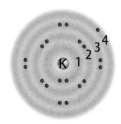

2 The diagram above shows a potassium atom. Which is the second-highest energy level? SC.8.P.8.7

Ⓕ 1

Ⓖ 2

Ⓗ 3

Ⓘ 4

3 On the periodic table, which groups of elements tend to form positive ions? SC.8.P.8.6

Ⓐ Group 1 and Group 2

Ⓑ Group 16 and Group 17

Ⓒ Group 1 and Group 16

Ⓓ Group 2 and Group 17

4 What is the electron diagram for the ionic compound sodium fluoride (NaF)? SC.8.P.8.6

Ⓕ $\left[\text{Na} \right]^+ \left[\text{F} \right]^-$

Ⓖ $\left[\text{Na} \right]^+ \left[:\text{F}: \right]^-$

Ⓗ $\left[\text{Na:} \right]^+ \left[\text{F:} \right]^-$

Ⓘ $\left[\text{Na:} \right]^+ \left[\text{F:} \right]^-$

Use the diagram below to answer question 5.

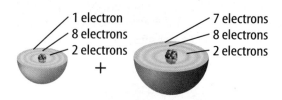

5 The atoms in the diagram above are forming a bond. Which represents that bond? SC.8.P.8.5

Ⓐ
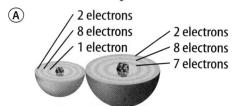

Sodium   Chloride

Ⓑ
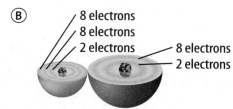

Sodium   Chloride

Ⓒ
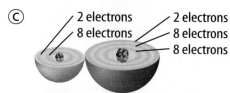

Sodium   Chloride

Ⓓ
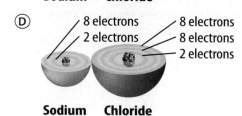

Sodium   Chloride

6 Covalent bonds typically form between the atoms of elements that share SC.8.P.8.5

Ⓕ nuclei.

Ⓖ oppositely charged ions.

Ⓗ protons.

Ⓘ valence electrons.

Use the diagram below to answer question 7.

Water Molecule

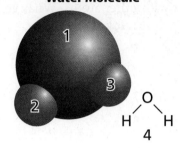

7 In the diagram above, which shows an atom with a partial negative charge? SC.8.P.8.7

Ⓐ 1

Ⓑ 2

Ⓒ 3

Ⓓ 4

8 Which compound is formed by the attraction between negatively and positively charged ions? SC.8.P.8.5

Ⓕ bipolar

Ⓖ covalent

Ⓗ ionic

Ⓘ nonpolar

9 The atoms of noble gases do NOT bond easily with other atoms because their valence electrons are SC.8.P.8.5

Ⓐ absent.

Ⓑ moving.

Ⓒ neutral.

Ⓓ stable.

Use the diagram below to answer question 10.

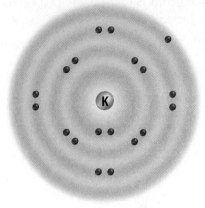

10 How many electrons does potassium, shown above, need to gain or lose to become stable? SC.8.P.8.6

Ⓕ gain 1

Ⓖ gain 2

Ⓗ lose 1

Ⓘ lose 2

11 What is the number of the group in which the elements have a stable outer energy level? SC.8.P.8.6

Ⓐ 1

Ⓑ 13

Ⓒ 16

Ⓓ 18

12 In what ways are the elements in a group similar? SC.8.P.8.6

Ⓕ atomic numbers

Ⓖ atomic masses

Ⓗ chemical properties

Ⓘ symbols

Need Extra Help?

If You Missed Question...	1	2	3	4	5	6	7	8	9	10	11	12
Go to Lesson...	2	1	3	1	3	2	2	3	1	1	1	1

Multiple Choice *Bubble the correct answer.*

Use the table below to answer questions 1 and 2.

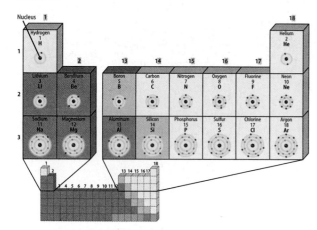

1. According to the table, how many valence electrons does phosphorus have? **SC.8.P.8.6**

   (A) 3

   (B) 4

   (C) 5

   (D) 6

2. Which element listed in the table is likely to be shiny and able to conduct electricity and heat? **SC.8.P.8.6**

   (F) neon

   (G) sulfur

   (H) beryllium

   (I) nitrogen

3. The element selenium is a nonmetal. It is an unstable atom that can form two bonds with other atoms. What group is selenium in? **SC.8.P.8.6**

   (A) group 15

   (B) group 16

   (C) group 17

   (D) group 18

4. The image above shows an electron dot diagram for silicon. How many total bonds can one atom of this metalloid make with other atoms? **SC.8.P.8.5**

   (F) 1

   (G) 2

   (H) 3

   (I) 4

Benchmark Mini-Assessment | Chapter 8 • Lesson 2

Multiple Choice *Bubble the correct answer.*

1. Which electron dot diagram shows how hydrogen and oxygen are bonded together in the compound H_2O? **SC.8.P.8.5**

(A) $:\ddot{O}:H:H$

(B) $H:\ddot{O}:H$

(C) $H::O::H$

(D) $H\cdot\ddot{O}\cdot H$

2. Which has the strongest covalent bonds? **SC.8.P.8.5**

(F) CO_2

(G) H_2O

(H) H_2

(I) N_2

3. Which compound contains four atoms of hydrogen for every one atom of carbon? **SC.8.P.8.5**

(A) $2CH_4$

(B) $4CH$

(C) C_4H

(D) CH_4

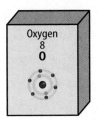

4. The diagram above shows the valence electrons in one atom of an element. If two atoms of this element joined together to create a molecule, they **SC.8.P.8.5**

(F) would form a double covalent bond.

(G) would form a quadruple covalent bond.

(H) would form a single covalent bond.

(I) would form a triple covalent bond.

Multiple Choice *Bubble the correct answer.*

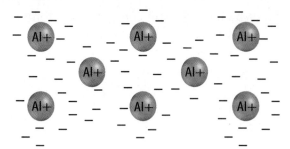

1. The atoms in the diagram above are **SC.8.P.8.5**

(A) losing valence electrons.

(B) pooling valence electrons.

(C) sharing pairs of valence electrons.

(D) transferring valence electrons.

2. How does an atom become an ion? **SC.8.P.8.5**

(F) An atom forms a compound with another atom.

(G) An atom is joined to another atom with a covalent bond.

(H) An atom loses or gains a valence electron.

(I) An atom shares a pair of valence electrons with another atom.

3. Which two elements are likely to form a compound that has ionic bonds? **SC.8.P.8.5**

(A) carbon and oxygen

(B) nitrogen and oxygen

(C) sodium and chlorine

(D) sulfur and hydrogen

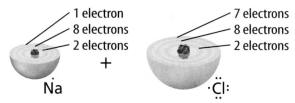

4. In the image above, if sodium loses one electron and chlorine gains one electron, how many valance electrons do sodium and chlorine now have? **SC.8.P.8.5**

(F) Sodium and chlorine each have 8 valence electrons.

(G) Sodium and chlorine each have 10 valence electrons.

(H) Sodium has 7 valence electrons, and chlorine has 9 valence electrons.

(I) Sodium has 11 valence electrons, and chlorine has 17 valence electrons.

Name _____ Date _____

Notes

Is it a mixture?

All around you are things that are made of matter. Some of these things are made of a type of matter called a mixture. Put an *X* next to each thing that you think is a mixture.

____ milk	____ soil	____ oxygen
____ salt water	____ iron	____ air
____ salt (sodium chloride)	____ shampoo	____ ginger ale
____ paint	____ sugar	____ granite rock
____ blood	____ sun tan lotion	____ carbon dioxide
____ bread	____ catsup	____ copper

Explain your thinking. Describe your rule or reasoning for deciding whether something is a mixture.

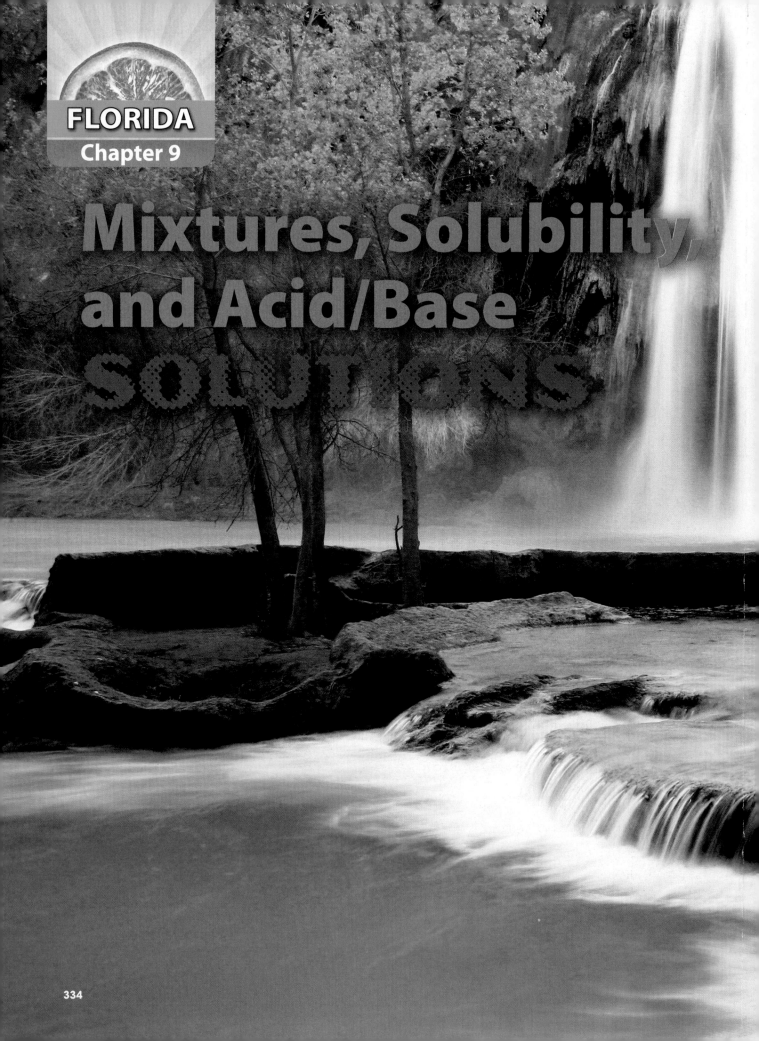

Mixtures, Solubility, and Acid/Base Solutions

Think About It!

What are solutions, and how are they described?

Havasu Falls is located in northern Arizona near the Grand Canyon. The creek that feeds these falls runs through a type of limestone called travertine. Small amounts of travertine are mixed evenly in the water, giving the water its unique blue-green color.

1 Can you think of other examples of something mixed evenly in water?

_____

_____

_____

2 What do you think of when you hear the word *solution*?

_____

_____

_____

3 How would you describe a solution?

_____

_____

_____

Get Ready to Read

What do you think about mixtures and solutions?

Before you read, decide if you agree or disagree with each of these statements. As you read this chapter, see if you change your mind about any of the statements.

	AGREE	DISAGREE
1 You can identify a mixture by looking at it without magnification.	☐	☐
2 A solution is another name for a homogeneous mixture.	☐	☐
3 Solutions can be solids, liquids, or gases.	☐	☐
4 A teaspoon of soup is less concentrated than a cup of the same soup.	☐	☐
5 Acids are found in many foods.	☐	☐
6 You can determine the exact pH of a solution by using pH paper.	☐	☐

There's More Online!
Video • Audio • Review • ⓘLab Station • WebQuest • Assessment • Concepts in Motion • Multilingual eGlossary   335

Substances and MIXTURES

ESSENTIAL QUESTIONS

How do substances and mixtures differ?

How do solutions compare and contrast with heterogeneous mixtures?

In what three ways do compounds differ from mixtures?

Vocabulary

substance p. 338

mixture p. 338

heterogeneous mixture p. 339

homogeneous mixture p. 339

solution p. 339

Florida NGSSS

LA.8.2.2.3 The student will organize information to show understanding or relationships among facts, ideas, and events (e.g., representing key points within text through charting, mapping, paraphrasing, summarizing, or comparing/contrasting);

LA.8.4.2.2 The student will record information (e.g., observations, notes, lists, charts, legends) related to a topic, including visual aids to organize and record information, as appropriate, and attribute sources of information;

SC.8.P.8.8 Identify basic examples of and compare and classify the properties of compounds, including acids, bases, and salts.

SC.8.P.8.9 Distinguish among mixtures (including solutions) and pure substances.

SC.8.N.1.1 Define a problem from the eighth grade curriculum using appropriate reference materials to support scientific understanding, plan and carry out scientific investigations of various types, such as systematic observations or experiments, identify variables, collect and organize data, interpret data in charts, tables, and graphics, analyze information, make predictions, and defend conclusions.

SC.8.N.1.6 Understand that scientific investigations involve the collection of relevant empirical evidence, the use of logical reasoning, and the application of imagination in devising hypotheses, predictions, explanations and models to make sense of the collected evidence.

 Launch Lab SC.8.P.8.9, SC.8.N.1.6

15 minutes

What makes black ink black?

Many of the products we use every day are mixtures. How can you tell if something is a mixture?

Procedure

1. Read and complete a lab safety form.

2. Lay a **coffee filter** on your table.

3. Find the center of the coffee filter and mark it lightly with a **pencil.**

4. Use a **permanent marker** to draw a circle with a diameter of 5 cm around the center of the coffee filter. Do not fill this circle in.

5. Pour **rubbing alcohol** to a depth of 1 cm into a **beaker.**

6. Using the eraser end of your pencil, push the center of the coffee filter down into the beaker until the center, but not the ink, touches the liquid. Keep the ink above the surface of the liquid.

7. Observe the liquid in the bottom of the beaker and the circle on the coffee filter.

8. Dispose of rubbing alcohol and used coffee filters as instructed by your teacher.

Think About This

1. What happened to the black ink circle on the coffee filter?

2. What was the purpose of the rubbing alcohol?

3. What do you think you would see if you used green ink instead?

4. **Key Concept** How do you think this shows that black ink is a mixture?

Inquiry What's in the water?

1. The water where these fish live is so clear that you might think it is pure water (H_2O). But is it? What do you think could be in this water that fish need in order to survive?

_____

_____

_____

_____

Active Reading

FOLDABLES® LA.8.2.2.3

Make horizontal two- and four-tab books. Assemble, staple, and label as shown. Use it to organize your notes on matter.

Two Types of Matter

Substances | Mixtures

Elements | Compounds | Heterogeneous | Homogeneous

Matter: Substances and Mixtures

Think about the journey you take to get to school. How many different types of matter do you see? You might see metal, plastic, rocks, concrete, bricks, plants, fabric, water, skin, and hair. So many different types of matter exist around you that it's hard to imagine them all. You might notice that you can group types of matter into categories. For example, keys, coins, and paper clips are all made of metal. Grouping matter into categories helps you understand how some things are similar to each other, but different from other things.

You might be surprised to know that nearly all types of matter can be sorted into just two major categories—substances and mixtures. What are substances and mixtures, and how are they different from each other?

Active Reading

2. Label Look around your classroom and list five things that you think are mixtures and five things that you think are substances. At the end of the lesson, come back and see if you are correct.

Matter	Examples
Mixtures	1.
	2.
	3.
	4.
	5.
Substances	1.
	2.
	3.
	4.
	5.

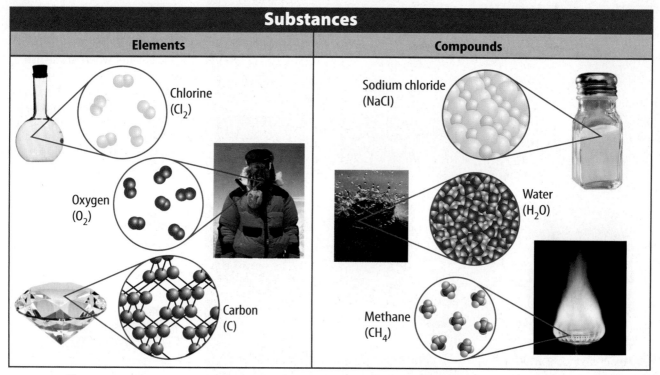

Substances

Elements	Compounds

Chlorine (Cl$_2$)

Oxygen (O$_2$)

Carbon (C)

Sodium chloride (NaCl)

Water (H$_2$O)

Methane (CH$_4$)

Figure 1 🔑 Elements are substances made of only one type of atom. Compounds are made of atoms of two or more elements bonded together.

Active Reading 3. **Explain** How do substances and mixtures differ?

_____

_____

_____

_____

What is a substance?

A **substance** *is matter that is always made up of the same combination of atoms.* Some substances are shown in **Figure 1.** There are two types of substances—elements and compounds. Recall that an element is matter made of only one type of atom, such as oxygen. A **compound** is matter made of atoms of two or more elements chemically bonded together, such as water (H$_2$O). Because the compositions of elements and compounds do not change, all elements and compounds are substances.

What is a mixture?

A **mixture** *is two or more substances that are physically blended but are not chemically bonded together.* The amounts of each substance in a mixture can vary. Granite, a type of rock, is a mixture. If you look at a piece of granite, you can see bits of white, black, and other colors. Another piece of granite will have different amounts of each color. The composition of rocks varies.

Air is a mixture, too. Air contains about 78 percent nitrogen, 21 percent oxygen, and 1 percent other substances. However, this composition varies. Air in a scuba tank can have more than 21 percent oxygen and less of the other substances.

It's not always easy to identify a mixture. A rock looks like a mixture, but air does not. Rocks and air are examples of the two different types of mixtures—heterogeneous (he tuh roh JEE nee us) and homogeneous (hoh muh JEE nee us).

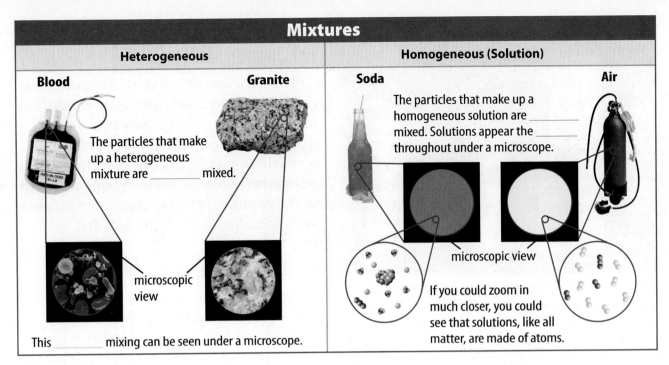

Mixtures

Heterogeneous		Homogeneous (Solution)	
Blood	**Granite**	**Soda**	**Air**

The particles that make up a heterogeneous mixture are _____ mixed.

This _____ mixing can be seen under a microscope.

microscopic view

The particles that make up a homogeneous solution are _____ mixed. Solutions appear the _____ throughout under a microscope.

microscopic view

If you could zoom in much closer, you could see that solutions, like all matter, are made of atoms.

Figure 2 🔑 Heterogeneous and homogeneous mixtures look different under the microscope.

Heterogeneous Mixtures *A* **heterogeneous mixture** *is a mixture in which substances are not evenly mixed.* For example, the substances that make up granite, a heterogeneous mixture, are unevenly mixed. When you look at a piece of granite, you can easily see the different parts. Often you can see the different substances and parts of a heterogeneous mixture with unaided eyes, but sometimes you can see them only with a microscope. For example, blood looks evenly mixed—its color and texture are the same throughout. But when you view blood with a microscope, as shown in **Figure 2,** you can see areas with more of one component and less of another.

Solutions—Homogeneous Mixtures Many mixtures look evenly mixed even when you view them with a powerful microscope. These mixtures are homogeneous. *A* **homogeneous mixture** *is a mixture in which two or more substances are evenly mixed on the atomic level but not bonded together.* The individual atoms or compounds of each substance are mixed. The mixture looks the same throughout under a microscope because individual atoms and compounds are too small to see.

Air is a homogeneous mixture. If you view air under a microscope, you can't see the individual substances that make it up. This is shown in **Figure 2.** *Another name for a homogeneous mixture is* **solution.** As you read about solutions in this chapter, remember that the term *solution* means "homogeneous mixture."

Active Reading **4. Label** Complete the table above by filling in the properties of heterogeneous and homogeneous mixtures.

Active Reading **5. Describe** What would you see if you were to look at a homogeneous mixture under an ordinary microscope?

_____

_____

WORD ORIGIN

heterogeneous
from Greek *heteros,* means "different"; and *genos,* means "kind"

homogeneous
from Greek *homos,* means "same"; and *genos,* means "kind"

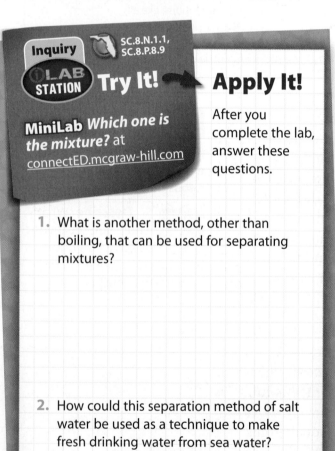

Sugar
• sweet
• solid

Mixture
• sweet
• colorless
• liquid

Water
• tasteless
• colorless
• liquid

Figure 3 Some properties of both the sugar and the water are observed in the mixture.

Inquiry SC.8.N.1.1, SC.8.P.8.9

LAB STATION Try It!

MiniLab *Which one is the mixture?* at connectED.mcgraw-hill.com

Apply It!

After you complete the lab, answer these questions.

1. What is another method, other than boiling, that can be used for separating mixtures?

2. How could this separation method of salt water be used as a technique to make fresh drinking water from sea water?

How do compounds and mixtures differ?

You have read that a compound contains two or more elements chemically bonded together. In contrast, the substances that make up a mixture are not chemically bonded. So, mixing is a physical change. The substances that exist before mixing still exist in the mixture. This leads to two important differences between compounds and mixtures.

Substances keep their properties.

Because substances that make up a mixture are not changed chemically, some of their properties are observed in the mixture. Sugar water, shown in **Figure 3,** is a mixture of two compounds—sugar and water. After the sugar is mixed in, you can't see the sugar in the water, but you can still taste it. Some properties of the water, such as its liquid state, are also observed in the mixture.

In contrast, the properties of a compound can be different from the properties of the elements that make it up. Sodium and chlorine bond to form table salt. Sodium is a soft, opaque, silvery metal. Chlorine is a greenish, poisonous gas. None of these properties are observed in table salt.

Mixtures can be separated.

Because the substances that make up a mixture are not bonded together, they can be separated from each other using physical methods. The physical properties of one substance are different from those of another. These differences can be used to separate the substances. In contrast, the only way compounds can be separated is by a chemical change that breaks the bonds between the elements. **Figure 4** summarizes the characteristics of substances and mixtures.

 6. **NGSSS Check Identify** Highlight the three ways that compounds differ from mixtures. SC.8.P.8.9

Figure 4 This organizational chart shows how different types of matter are classified. All matter can be classified as either a substance or a mixture.

Active Reading **7. Label** Fill in the blanks below to complete the figure.

Matter
- anything that has mass and takes up space
- Most matter on Earth is made up of atoms.

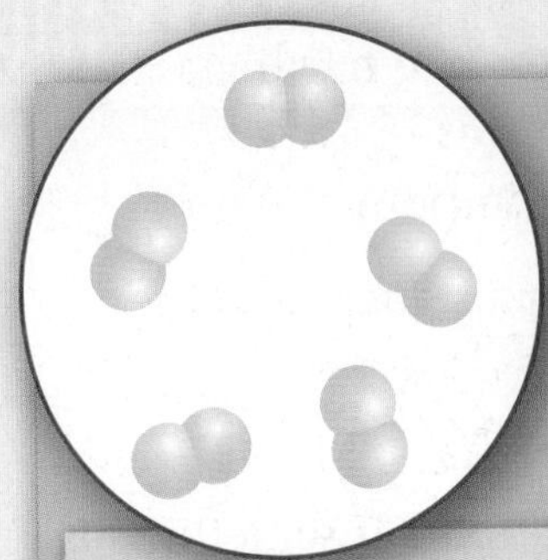

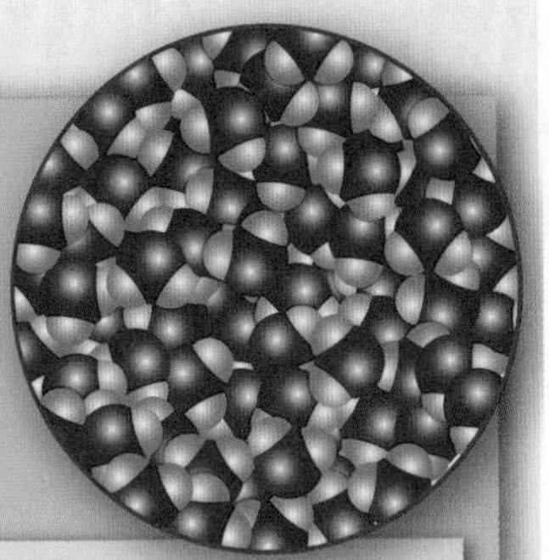

Substances
- matter with a composition that is always the same
- two types of substances: elements and compounds

- consist of just one type of atom
- organized on the periodic table
- Elements can exist as single atoms or as diatomic molecules—two atoms bonded together.

changes

- two or more types of atoms bonded together
- can't be separated by physical methods
- Properties of a compound are different from the properties of the elements that make it up.
- two types: ionic and covalent

Separating mixtures
- filtering
- boiling
- using a magnet

changes

Combining substances
- mixing
- dissolving

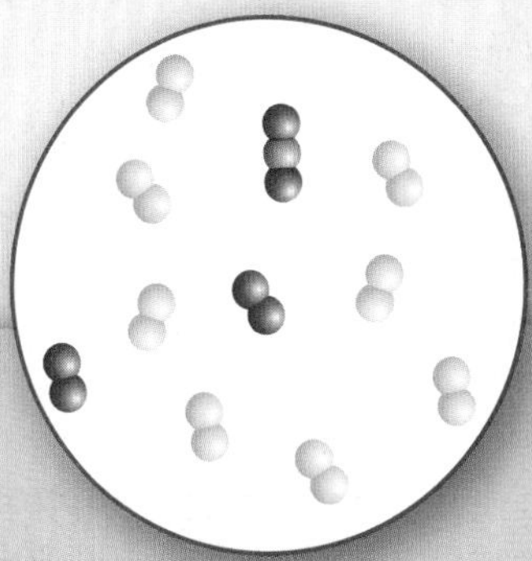

Mixtures
- matter that can vary in composition
- made of two or more substances mixed but not bonded together
- can be separated into substances by physical methods
- two types of mixtures: heterogeneous and homogeneous

_____ **mixtures**
- two or more substances unevenly mixed
- Uneven mixing is visible with unaided eyes or a microscope.

_____ **mixtures (solutions)**
- two or more substances evenly mixed
- Homogeneous mixtures appear uniform under a microscope.

Visual Summary

Substance   Mixtures
Substances have a composition that does not change. The composition of mixtures can vary.

Solutions (homogeneous mixtures) are mixed at the atomic level.

Mixtures contain parts that are not bonded together. These parts can be separated using physical means.

Use Vocabulary

1 **Identify** What is another name for a homogeneous mixture?

_____

2 **Contrast** homogeneous and heterogeneous mixtures.

_____

_____

Understand Key Concepts 🔑

3 **Explain** why a compound is classified as a substance. SC.8.P.8.9

_____

_____

4 **Describe** two tests that you can run to determine if something is a substance or a mixture. SC.8.P.8.9

_____

_____

Interpret Graphics

5 **Compare and contrast** Use the graphic organizer to compare and contrast heterogeneous mixtures and solutions. LA.8.2.2.3

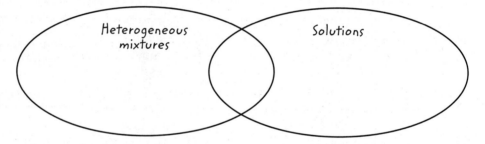

Heterogeneous mixtures   Solutions

Critical Thinking

6 **Explain** the following statement: All compounds are substances, but not all substances are compounds. SC.8.P.8.9

_____

_____

7 **Suppose** you have found an unknown substance in a laboratory. It has the formula H_2O_2 written on the bottle. Is it water? How do you know?

_____

_____

Sports Drinks and Your Body

The Importance of Electrolytes

When you exercise, you sweat. When you're thirsty, do you drink water or a sports drink? Chances are you have seen ads that claim sports drinks are better for you than water because they contain electrolytes. What are electrolytes, and how are they used in your body?

What are electrolytes?

An electrolyte is a charged particle, scientifically known as an ion. Electrolytes can conduct electric charges. Pure water cannot conduct electric charges, but water containing electrolytes can.

Electrolytes in Your Body

Why does your body need electrolytes? Solutions of water and electrolytes surround all of the cells in your body. Electrolytes enable these solutions to carry nerve impulses from one cell to another. Your body's voluntary movements, such as walking, and involuntary movements, such as your heart beating, are caused by nerve impulses. Without electrolytes these nerve impulses cannot move normally.

Replenishing Fluids

Sweat is a solution of water and electrolytes, including sodium. Sports drinks can replace water and electrolytes. Some foods, such as bananas and oranges, also contain electrolytes. However many sports drinks contain ingredients your body doesn't need, such as caffeine and sugar. Unless you are sweating for an extended period of time, you don't need to replace electrolytes, but replacing water is always essential.

Origin of the Sports Drink

Did you know that the first sports drink was created at the University of Florida in 1965? The assistant coach of the football team wondered why players lost so much weight and rarely felt the need to urinate, even though they were eating and drinking. Scientists began researching the effects of heat on the human body. Soon they understood that all the sweat was taking the players' energy, strength, and endurance. The researchers speculated that the electrolytes—primarily sodium and potassium—the players were losing through their sweat were upsetting the body's delicate chemical balance. After careful research and testing, the now famous sports drink was created and had great success. Over the next five years, only one player was taken to the hospital—and he didn't drink any of the sports drink!

Heat

Sweat

It's Your Turn

RESEARCH Study three different sports drinks. Create a bar graph that compares ingredients such as water, sugar, electrolytes, and caffeine in each brand. Draw a conclusion about which type of drink is best for your body. LA.8.4.2.2

Properties of SOLUTIONS

Vocabulary

solvent p. 345

solute p. 345

polar molecule p. 346

concentration p. 348

solubility p. 350

saturated solution p. 350

unsaturated solution p. 350

 Florida NGSSS

SC.8.P.8.4 Classify and compare substances on the basis of characteristic physical properties that can be demonstrated or measured; for example, density, thermal or electrical conductivity, solubility, magnetic properties, melting and boiling points, and know that these properties are independent of the amount of the sample.

SC.8.P.8.8 Identify basic examples of and compare and classify the properties of compounds, including acids, bases, and salts.

SC.8.P.8.9 Distinguish among mixtures (including solutions) and pure substances.

SC.8.N.1.1 Define a problem from the eighth grade curriculum using appropriate reference materials to support scientific understanding, plan and carry out scientific investigations of various types, such as systematic observations or experiments, identify variables, collect and organize data, interpret data in charts, tables, and graphics, analyze information, make predictions, and defend conclusions.

SC.8.N.1.2 Design and conduct a study using repeated trials and replication.

SC.8.N.1.6 Understand that scientific investigations involve the collection of relevant empirical evidence, the use of logical reasoning, and the application of imagination in devising hypotheses, predictions, explanations and models to make sense of the collected evidence.

Also covers: LA.8.2.2.3, MA.6.A.3.6

 Launch Lab

 SC.8.P.8.9, SC.8.N.1.6

15 minutes

How are they different?

If you have ever looked at a bottle of Italian salad dressing, you know that some substances do not easily form solutions. The oil and vinegar do not mix, and the spices sink to the bottom. How can you describe the difference quantitatively?

Procedure

1. Read and complete a lab safety form.

2. Label one **beaker** A and another **beaker** B.

3. Measure 100 mL of water and pour it into beaker A.

4. Measure 100 mL of water and pour it into beaker B.

5. Add 10 g of **baking soda** to beaker A, and stir with a **plastic spoon** for 2 min or until all the baking soda dissolves, whichever happens first.

6. Add 25 g of **sugar** to beaker B and stir with a plastic spoon for 2 min or until all of the sugar dissolves, whichever happens first.

7. Observe the mixtures in each beaker.

Think About This

1. Predict what would happen if you were to use 200 mL of water instead of 100 mL.

2. Do you think more baking soda might dissolve if you stirred the solution longer?

3. **Key Concept** Why do you think one substance dissolved more easily in water than the other substance? What factors do you think contribute to this difference?

Inquiry Large Fingers?

1. These stalactites and stalagmites, found at the Florida Caverns State Park in Marianna, Florida, were formed over thousands of years. What do you think causes these structures to form? What substances might be involved?

Parts of Solutions

You've read that a solution is a homogeneous mixture. Recall that in a solution, substances are evenly mixed on the atomic level. How does this mixing occur? Dissolving is the process of mixing one substance into another to form a solution. Scientists use two terms to refer to the substances that make up a solution. Generally, the **solvent** *is the substance that exists in the greatest quantity in a solution. All other substances in a solution are* **solutes**. Recall that air is a solution of 78 percent nitrogen, 21 percent oxygen, and 1 percent other substances. Which substance is the solvent? In air, nitrogen exists in the greatest quantity. Therefore, it is the solvent. The oxygen and other substances are solutes. In this lesson, you will read the terms *solute* and *solute* often. Refer back to this page if you forget what these terms mean.

Active Reading   **2. Contrast** How do a solute and a solvent differ?

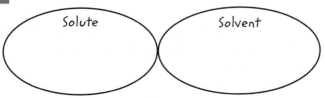

Solute   Solvent

Active Reading

FOLDABLES®   LA.8.2.2.3

Make a four-tab shutterfold. Label it as shown. Collect information about which solvents dissolve which solutes.

Polar solvents dissolve:	Nonpolar solvents dissolve:
Like Dissolves Like	
Polar solvents do not dissolve:	Nonpolar solvents do not dissolve:

Table 1 Types of Solutions

State of Solution	Solvent Is:	Solute Can Be:
Solid	solid	**gas or solid (called alloys)** This saxophone is a solid solution of solid copper and solid zinc.
Liquid	liquid	**solid, liquid, and/or gas** Soda is a liquid solution of liquid water, gaseous carbon dioxide, and solid sugar and other flavorings.
Gas	gas	**gas** This lighted sign contains a gaseous mixture of gaseous argon and gaseous mercury.

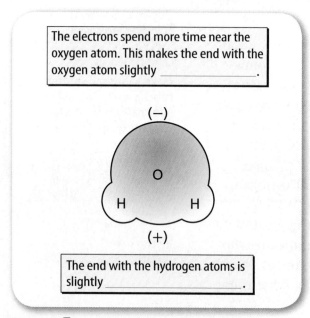

The electrons spend more time near the oxygen atom. This makes the end with the oxygen atom slightly _____.

(−)

O

H   H

(+)

The end with the hydrogen atoms is slightly _____.

Figure 5 🔑 Water is a polar molecule.

Active Reading 3. **Recall** Label the figure above indicating the charges of the atom.

Types of Solutions

When you think of a solution, you might think of a liquid. However, solutions can exist in all three states of matter—solid, liquid, or gas. The state of the solvent, because it exists in the greatest quantity, determines the state of the solution. **Table 1** contrasts solid, liquid, and gaseous solutions.

Water as a Solvent

Did you know that over 75 percent of your brain and almost 90 percent of your lungs are made of water? Water is one of the few substances on Earth that exists naturally in all three states—solid, liquid, and gas. However, much of this water is not pure water. In nature, water almost always exists as a solution; it contains dissolved solutes. All of Florida's 11,761 miles of water area contain solutes such as sodium chloride ($NaCl$). Why does nearly all water on Earth contain dissolved solutes? The answer has to do with the structure of the water molecule.

The Polarity of Water

A water molecule, such as the one shown in **Figure 5,** is a covalent compound. Recall that atoms are held together with covalent bonds when sharing electrons. In a water molecule, one oxygen atom shares electrons with two hydrogen atoms. However, these electrons are not shared equally. The electrons in the oxygen-hydrogen bonds more often are closer to the oxygen atom than they are to the hydrogen atoms. This unequal sharing of electrons gives the end with the oxygen atom a slightly negative charge. And it gives the end with the hydrogen atoms a slightly positive charge. Because of the unequal sharing of electrons, a water molecule is said to be polar. *A* **polar molecule** *is a molecule with a slightly negative end and a slightly positive end.* Nonpolar molecules have an even distribution of charge. Solutes and solvents can be polar or nonpolar.

Like Dissolves Like

Water is often called the universal solvent because it dissolves many different substances. But water can't dissolve everything. Why does water dissolve some substances but not others? Water is a polar solvent. Polar solvents dissolve polar solutes easily. Nonpolar solvents dissolve nonpolar solutes easily. This is summarized by the phrase "like dissolves like." Because water is a polar solvent, it dissolves most polar and ionic solutes.

Active Reading **4. Diagram** Why do some substances dissolve in water and some do not?

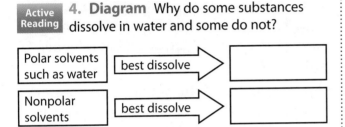

| Polar solvents such as water | best dissolve → | |
| Nonpolar solvents | best dissolve → | |

Polar Solvents and Polar Molecules

Because water molecules are polar, water dissolves groups of other polar molecules. **Figure 6** shows what rubbing alcohol, a substance used as a disinfectant, looks like when it is in a solution with water. Molecules of rubbing alcohol also are polar. Therefore, when rubbing alcohol and water mix, the positive ends of the water molecules are attracted to the negative ends of the alcohol molecules. Similarly, the negative ends of the water molecules are attracted to the positive ends of the alcohol. In this way, alcohol molecules dissolve in the solvent.

Polar Solvents and Ionic Compounds

Many ionic compounds are also soluble in water. Recall that ionic compounds are composed of alternating positive and negative ions. Sodium chloride (NaCl) is an ionic compound composed of sodium ions (Na^+) and chloride ions (Cl^-). When sodium chloride dissolves, these ions are pulled apart by the water molecules. This is shown in **Figure 7.** The negative ends of the water molecules attract the positive sodium ions. The positive ends of the water molecules attract the negative chloride ions.

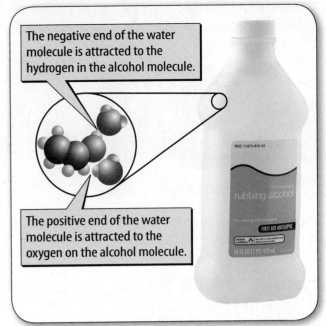

The negative end of the water molecule is attracted to the hydrogen in the alcohol molecule.

The positive end of the water molecule is attracted to the oxygen on the alcohol molecule.

Figure 6 When a polar solute, such as rubbing alcohol, dissolves in a polar solvent, such as water, the poles of the solvent are attracted to the oppositely charged poles of the solute.

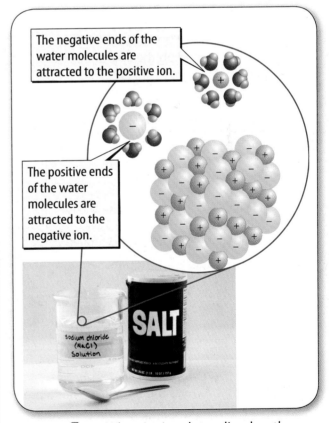

The negative ends of the water molecules are attracted to the positive ion.

The positive ends of the water molecules are attracted to the negative ion.

Figure 7 When ionic solutes dissolve, the positive poles of the solvent are attracted to the negative ions. The negative poles of the solvent are attracted to the positive ions.

More solute   Less solute

Equal
amounts
of water

Concentrated   **Dilute**

Figure 8 The volumes of both drinks are the same, but the glass on the left contains more solute than the solution on the right.

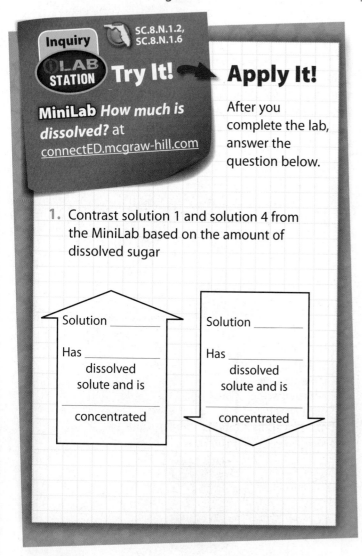

Inquiry

SC.8.N.1.2,
SC.8.N.1.6

LAB STATION **Try It!**   **Apply It!**

MiniLab *How much is dissolved?* at connectED.mcgraw-hill.com

After you complete the lab, answer the question below.

1. Contrast solution 1 and solution 4 from the MiniLab based on the amount of dissolved sugar

Solution _____
Has _____
  dissolved
  solute and is
_____
  concentrated

Solution _____
Has _____
  dissolved
  solute and is
_____
  concentrated

Active Reading   **5. Infer** Why is the term *dilute* not a precise way to describe concentration?

_____

_____

_____

_____

Concentration—How much is dissolved?

Have you ever tasted a spoonful of soup and wished it had more salt in it? In a way, your taste buds were measuring the amount, or concentration, of salt in the soup. **Concentration** *is the amount of a particular solute in a given amount of solution.* In the soup, salt is a solute. Saltier soup has a higher concentration of salt. Soup with less salt has a lower concentration of salt. Look at the glasses of fruit drink in **Figure 8.** Which drink has a higher concentration of solute? The darker blue drink has a higher concentration of solute.

Concentrated and Dilute Solutions

One way to describe the saltier soup is to say that it is more concentrated. The less-salty soup is more dilute. The terms *concentrated* and *dilute* are one way to describe how much solute is dissolved in a solution. However, these terms don't state the exact amount of solute dissolved. What one person thinks is concentrated might be what another person thinks is dilute. Soup that tastes too salty to you might be perfect for someone else. How can concentration be described more precisely?

Describing Concentration Using Quantity

A more precise way to describe concentration is to state the quantity of solute in a given quantity of solution. When a solution is made of a solid dissolved in a liquid, such as salt in water, concentration is the mass of solute in a given volume of solution. Mass usually is stated in grams, and volume usually is stated in liters. For example, concentration can be stated as grams of solute per 1 L of solution. However, concentration can be stated using any units of mass or volume.

Calculating Concentration—Mass per Volume

One way that concentration can be calculated is by the following equation:

$$\text{Concentration } (C) = \frac{\text{mass of solute } (m)}{\text{volume of solution } (V)}$$

To calculate concentration, you must know both the mass of solute and the volume of solution that contains this mass. Then divide the mass of solute by the volume of solution.

Concentration—Percent by Volume

Not all solutions are made of a solid dissolved in a liquid. If a solution contains only liquids or gases, its concentration is stated as the volume of solute in a given volume of solution. In this case, the units of volume must be the same—usually mL or L. Because the units match, the concentration can be stated as a percentage. Percent by volume is calculated by dividing the volume of solute by the total volume of solution and then multiplying the quotient by 100. For example, if a container of orange drink contains 3 mL of acetic acid in a 1,000-mL container, the concentration is 0.3 percent.

Math Skills   MA.6.A.3.6

Solve for Concentration

Suppose you want to calculate the concentration of salt in a **0.4 L** can of soup. The back of the can says it contains **1.6 g** of salt. What is its concentration in g/L? In other words, how much salt would be contained in 1 L of soup?

1. This is what you know: mass: **1.6 g**
                                               volume: **0.4 L**

2. This is what you need to find: concentration: C

3. Use this formula: $C = \dfrac{m}{V}$

4. Substitute: $C = \dfrac{1.6\text{ g}}{0.4\text{ L}} = 4\text{ g/L}$
the values for m and v
into the formula and divide.

Answer: The concentration is 4 g/L. As you might expect, 0.4 L of soup contains less salt (1.6 g) than 1 L of soup (4 g). However, the concentration of both amounts of soup is the same—4 g/L.

6. What is the concentration of 5 g of sugar in 0.2 L of solution?

For more Concentration Calculation practice, check out the Math Skill Review on page 353.

Active Reading

7. Contrast How do concentration and solubility differ?

> Concentration

> Solubility

Active Reading

8. Explain Using arrows, show what happens to the concentration of a solution as the amount of solvent or solute changes.

↑ Solvent	Concentration
Solvent	Concentration
Solute	↓ Concentration
Solute	Concentration

Solubility—How much can dissolve?

Have you ever put too much sugar into a glass of iced tea? What happens? Not all of the sugar dissolves. You stir and stir, but there is still sugar at the bottom of the glass. That is because there is a limit to how much solute (sugar) can be dissolved in a solvent (water). **Solubility** (sahl yuh BIH luh tee) *is the maximum amount of solute that can dissolve in a given amount of solvent at a given temperature and pressure.* If a substance has a high solubility, more of it can dissolve in a given solvent.

Saturated and Unsaturated Solutions

If you add water to a dry sponge, the sponge absorbs the water. If you keep adding water, the sponge becomes saturated. It can't hold any more water. This is analogous (uh NA luh gus), or similar, to what happens when you stir too much sugar into iced tea. Some sugar dissolves, but the excess sugar does not dissolve. The solution is saturated. *A* **saturated solution** *is a solution that contains the maximum amount of solute the solution can hold at a given temperature and pressure. An* **unsaturated solution** *is a solution that can still dissolve more solute at a given temperature and pressure.*

Factors that Affect How Much Can Dissolve

Can you change the amount of a particular solute that can dissolve in a solvent? Yes. Recall the definition of solubility—the maximum amount of solute that can dissolve in a given amount of solvent at a given temperature and pressure. Changing either temperature or pressure changes how much solute can dissolve in a solvent.

Figure 9 ⌐ Some solids are more soluble in warmer liquids than cooler ones. Other solids are less soluble in warmer liquids than cooler ones.

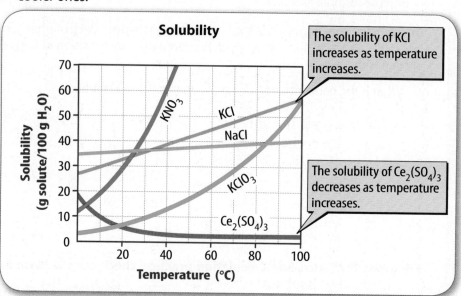

9. Visual Check
Extrapolate How many grams of KNO_3 will dissolve in 100 g of water at 10°C?

Effect of Temperature Have you noticed that more sugar dissolves in hot tea than in iced tea? The solubility of sugar in water increases with temperature. This is true for many solid solutes, as shown in **Figure 9.** Notice that some solutes become less soluble when temperature is increased.

How does temperature affect the solubility of a gas in a liquid? Recall that soda, or soft drinks, contains carbon dioxide, a gaseous solute, dissolved in liquid water. The bubbles you see in soda are made of undissolved carbon dioxide. Have you ever noticed that more carbon dioxide bubbles out when you open a warm can of soda than when you open a cold can? This is because the solubility of a gas in a liquid decreases when the temperature of the solution increases.

Effect of Pressure What keeps carbon dioxide dissolved in an unopened can of soda? In a can, the carbon dioxide in the space above the liquid soda is under pressure. This causes the gas to move to an area of lower pressure—the solvent. The gas moves into the solvent, and a solution is formed. When the can is opened, as shown in **Figure 10,** this pressure is released, and the carbon dioxide gas leaves the solution. Pressure does not affect the solubility of a solid solute in a liquid.

Active Reading

10. Describe How can the solubility of a solute be changed?

_____

_____

How Fast a Solute Dissolves

Temperature and pressure can affect how much solute dissolves. If solute and solvent particles come into contact more often, the solute dissolves faster. **Figure 11** shows three ways to increase how often solute particles contact solvent particles. Each of these methods will make a solute dissolve faster. However, it is important to note that stirring the solution or crushing the solute will not make more solute dissolve.

Figure 10 When the pressure of a gas is increased, it becomes more soluble in a liquid. When the can is opened, this pressure is lowered and the gas leaves the solution.

Stirring the solution

Crushing the solute

Increasing the temperature

Figure 11 Several factors can affect how quickly a solute will dissolve in a solution. However, dissolving more quickly won't necessarily make more solute dissolve.

Visual Summary

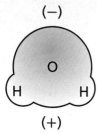

Polar molecule

Substances dissolve in other substances that have similar polarity. In other words, like dissolves like.

Concentration is the amount of substance that is dissolved. Solubility is the maximum amount that can dissolve.

Both temperature and pressure affect the solubility of solutes in solutions.

Inquiry
SC.8.N.1.1, SC.8.N.1.6, SC.8.P.8.4, SC.8.P.8.8

LAB STATION Try It!

Skills Lab *How does a solute affect the conductivity of a solution?* at connectED.mcgraw-hill.com

Use Vocabulary

1. **Define** *polar molecule* in your own words.

Understand Key Concepts 🔑

2. **Explain** how you could use the solubility of a substance to make a saturated solution.

3. **Predict** whether an ionic compound will dissolve in a nonpolar solvent.

Interpret Graphics

4. **Organize** Use the graphic organizer below to organize three factors that increase the speed a solute dissolves in a liquid. LA.8.2.2.3

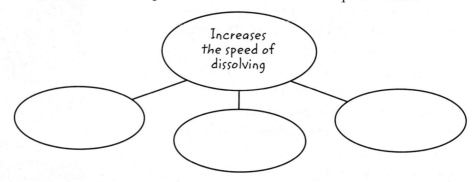

Increases the speed of dissolving

Critical Thinking

5. **Explain** A student wants to increase the maximum amount of sugar that can dissolve in water. She crushes the sugar and then stirs it into the water. Does this work? Why or why not?

Math Skills   MA.6.A.3.6

6. Use ratios to explain how a tablespoon of soup and a cup of the same soup have the same concentration.

Math Skills

Calculate Concentration Practice

1. How many grams of salt are in 5 L of a solution with a concentration of 3 g/L?

2. Suppose you add water to a 6 g of sugar to make a solution with a concentration of 3 g/L. What is the total volume of the solution?

3. What is the concentration of KCl if 10 g are dissolved in enough water to make 12 L?

4. How many grams of KMnO4 would you get if you evaporated the water from 85.75 mL of 1.27 g/L solution?

5. Your teacher created a 2 g/L salt water solution (solution A) and a 2.5 g/L salt water solution (solution B), both in 500 mL of water. Using the concentration equation, which solution is more concentrated? How do you know?

Math Skills

Difference in pH Practice

After learning about pH in Lesson 3, come back and practice calculating pH strength.

6. Suppose you have two solutions that have a pH of 1 and a pH of 4. How much more acidic is the solution with a pH of 1 than the solution with a pH of 4?

7. What is the difference in basicity between drain cleaner (pH 13) and ammonia (pH 11.9)?

8. Which is more acidic—milk of magnesia or lemon juice? What is the difference in acidity between the two? (milk of magnesia = pH 10.5, lemon juice = pH 2.5)

Acid and Base SOLUTIONS

 What happens when acids and bases dissolve in water?

 How does the concentration of hydronium ions affect pH?

 What methods can be used to measure pH?

Vocabulary

acid p. 356

hydronium ion p. 356

base p. 356

pH p. 358

indicator p. 360

 Florida NGSSS

LA.8.2.2.3 The student will organize information to show understanding or relationships among facts, ideas, and events (e.g., representing key points within text through charting, mapping, paraphrasing, summarizing, or comparing/contrasting);

MA.6.A.3.6 Construct and analyze tables, graphs, and equations to describe linear functions and other simple relations using both common language and algebraic notation.

MA.6.S.6.2 Select and analyze the measures of central tendency or variability to represent, describe, analyze, and/or summarize a data set for the purposes of answering questions appropriately.

SC.8.P.8.8 Identify basic examples of and compare and classify the properties of compounds, including acids, bases, and salts.

SC.8.N.1.1 Define a problem from the eighth grade curriculum using appropriate reference materials to support scientific understanding, plan and carry out scientific investigations of various types, such as systematic observations or experiments, identify variables, collect and organize data, interpret data in charts, tables, and graphics, analyze information, make predictions, and defend conclusions.

SC.8.N.1.2 Design and conduct a study using repeated trials and replication.

SC.8.N.1.6 Understand that scientific investigations involve the collection of relevant empirical evidence, the use of logical reasoning, and the application of imagination in devising hypotheses, predictions, explanations and models to make sense of the collected evidence.

 Inquiry Launch Lab

SC.8.N.1.6, SC.8.P.8.8

15 minutes

What color is it?

Did you know that all rain is naturally acidic? As raindrops fall through the air, they pick up molecules of carbon dioxide. An acid called carbonic acid is formed when the water molecules react with the carbon dioxide molecules. An indicator is a substance that can be used to tell if a solution is acidic, basic, or neutral.

Procedure

1. Read and complete a lab safety form.

2. Half fill a **beaker** with the **colored solution**.

3. Place one end of a **straw** into the solution.
   ⚠ **Caution:** *Do not suck liquid through the straw.*

4. Blow through the straw, making bubbles in the solution. Continue blowing, and count how many times you have to blow bubbles until you observe a change.

Think About This

1. Describe what change you saw take place.

2. What do you think made this change occur?

3. How do you think the results would have been different if you had held your breath for several seconds before blowing through the straw?

4. **Key Concept** Using the terms *acidic* and *basic,* explain what you have learned about the colored solution being used.

What's eating her?

1. When this statue was first carved, it didn't have any of these odd-shaped marks on its surface. What do you think caused these marks? What would happen to the statue if this process were to continue?

_____

_____

_____

_____

Active Reading **2. Predict** What do you already know about acids and bases? Place an *A* before properties of acids and a *B* before properties of bases. Note that some properties apply to both.

_____ Provide sour taste in food

_____ Found in saliva

_____ Can damage skin and eyes

_____ Slippery

_____ Provide bitter taste in food

_____ Found in milk

_____ Helps plants grow

What are acids and bases?

Would someone ever drink an acid? At first thought, you might answer no. After all, when people think of acids, they often think of acids such as those found in batteries or in acid rain. However, acids are found in other items, including milk, vinegar, fruits, and green leafy vegetables. Some examples of acids that you might eat are shown in **Figure 12**. Along with the word *acid*, you might have heard the word *base*. Like acids, you can also find bases in your home. Detergent, antacids, and baking soda are examples of items that contain bases. But acids and bases are found in more than just household goods. As you will learn in this lesson, they are necessities for our daily life.

Figure 12 You might be surprised to learn that acids are common in the foods you eat.

3. Differentiate What happens when acids and bases dissolve in water?

Acids

Bases

4. Visual Check
Infer How is dissolving an acid, shown above on the right similar to dissolving ammonia, shown below?

_____

_____

_____

Acids in Water

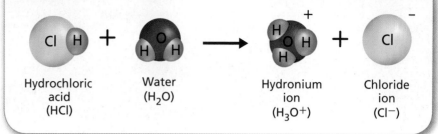

Hydrochloric acid (HCl)   Water (H_2O)   Hydronium ion (H_3O^+)   Chloride ion (Cl^-)

Figure 13 🔑 Acids, such as hydrochloric acid, produce hydronium ions when they dissolve in water.

Acids

Have you ever tasted the sourness of a lemon or a grapefruit? This sour taste is due to the acid in the fruit. *An **acid** is a substance that produces a hydronium ion (H_3O^+) when dissolved in water.* Nearly all acid molecules contain one or more hydrogen atoms (H). When an acid mixes with water, this hydrogen atom separates from the acid. It quickly combines with a water molecule, resulting in a hydronium ion. This process is shown in **Figure 13**. *A **hydronium ion**, H_3O^+, is a positively charged ion formed when an acid dissolves in water.*

Bases

*A **base** is a substance that produces hydroxide ions ($OH-$) when dissolved in water.* When a hydroxide compound such as sodium hydroxide (NaOH) mixes with water, hydroxide ions separate from the base and form hydroxide ions ($OH-$) in water. Some bases, such as ammonia ($NH3$), do not contain hydroxide ions. These bases produce hydroxide ions by taking hydrogen atoms away from water, leaving hydroxide ions ($OH-$). This process is shown in **Figure 14**. Some properties and uses of acids and bases are shown in **Table 2.**

Figure 14 🔑 Bases, such as sodium hydroxide and ammonia, produce hydroxide ions when they dissolve in water.

Bases in Water

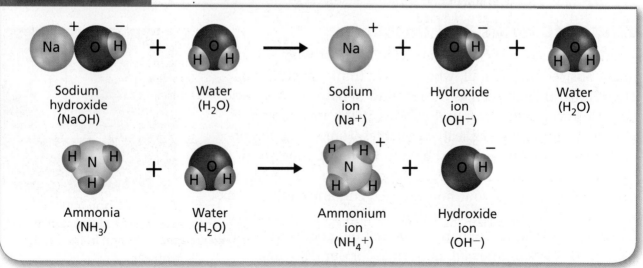

Sodium hydroxide (NaOH)   Water (H_2O)   Sodium ion (Na^+)   Hydroxide ion (OH^-)   Water (H_2O)

Ammonia (NH_3)   Water (H_2O)   Ammonium ion (NH_4^+)   Hydroxide ion (OH^-)

Table 2 Properties and Uses of Acids and Bases

	Acids	Bases
Ions produced	Acids produce H_3O^+ in water.	Bases produce OH^- ions in water.
Examples	• hydrochloric acid, HCl • acetic acid, CH_3COOH • citric acid, $H_3C_6H_5O_7$ • lactic acid, $C_3H_6O_3$	• sodium hydroxide, NaOH • ammonia, NH_3 • sodium carbonate, Na_2CO_3 • calcium hydroxide, $Ca(OH)_2$
Some properties	• Acids provide the sour taste in food (never taste acids in the laboratory). • Most can damage skin and eyes. • Acids react with some metals to produce hydrogen gas. • H_3O^+ ions can conduct electricity in water. • Acids react with bases to form neutral solutions.	• Bases provide the bitter taste in food (never taste bases in the laboratory). • Most can damage skin and eyes. • Bases are slippery when mixed with water. • OH^- ions can conduct electricity in water. • Bases react with acids to form neutral solutions.
Some uses	• Acids are responsible for for natural and artificial flavoring in foods, such as fruits. • Lactic acid is found in milk. • Acid in your stomach breaks down food. • Blueberries, strawberries, and many vegetable crops grow better in acidic soil. • Acids are used to make products such as fertilizers, detergents, and plastics.	• Bases are found in natural and artificial flavorings in food, such as cocoa beans. • Antacids neutralize stomach acid, alleviating heartburn. • Bases are found in cleaners such as shampoo, dish detergent, and window cleaner. • Many flowers grow better in basic soil. • Bases are used to make products such as rayon and paper.

What is pH?

Have you ever seen someone test the water in a swimming pool? It is likely that the person was testing the pH of the water. Swimming pool water should have a pH around 7.4. If the pH of the water is higher or lower than 7.4, the water might become cloudy, burn swimmers' eyes, or contain too many bacteria. What does a pH of 7.4 mean?

Hydronium Ions

The **pH** *is an inverse measure of the concentration of hydronium ions (H_3O^+) in a solution.* What does *inverse* mean? It means that as one thing increases, another thing decreases. In this case, as the concentration of hydronium ions increases, pH decreases. A solution with a lower pH is more acidic. As the concentration of hydronium ions decreases, the pH increases. A solution with a higher pH is more basic. This relationship is shown in **Figure 15.**

Balance of Hydronium and Hydroxide Ions

All acid and base solutions contain both hydronium and hydroxide ions. Acids have a greater concentration of hydronium ions (H_3O^+) than hydroxide ions (OH^-). Bases have a greater concentration of hydroxide ions than hydronium ions. Brackets around a chemical formula mean *concentration*.

Acids	$[H_3O^+] > [OH^-]$
Neutral	$[H_3O^+] = [OH^-]$
Bases	$[H_3O^+] < [OH^-]$

Active Reading **5. Model** How does the concentration of hydronium ions affect pH?

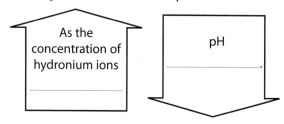

Salt Formation

A salt is a compound formed when the negative ions from an acid combine with the positive ions from a base. In the reaction between HCl and NaOH, a salt, sodium chloride (NaCl), is formed.

pH Scale

Figure 15 🔑 Notice that as hydronium concentration increases, the pH decreases.

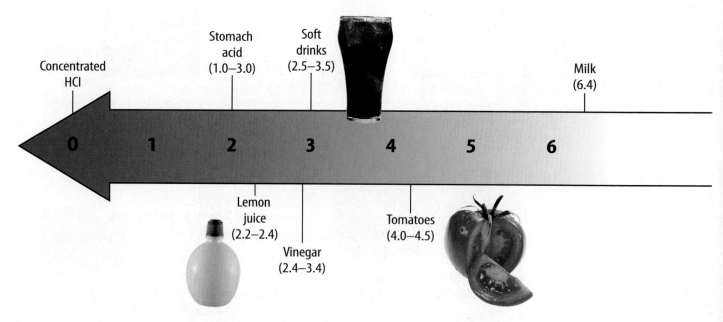

The pH Scale

The pH scale is used to indicate how acidic or basic a solution is. Notice in **Figure 15** that the pH scale contains values that range from below 0 to above 14. Acids have a pH below 7. Bases have a pH above 7. Solutions that are neutral have a pH of 7—they are neither acidic nor basic.

You might be wondering what the numbers on the pH scale mean. How is the concentration of hydronium ions different in a solution with a pH of 1 from the concentration in a solution with a pH of 2? A change in one pH unit represents a tenfold change in the acidity or basicity of a solution. For example, if one solution has a pH of 1 and a second solution has a pH of 2, then the first solution is not twice as acidic as the second solution; it is ten times more acidic.

The difference in acidity or basicity between two solutions is represented by 10^n, where n is the difference between the two pH values. For example, how much more acidic is a solution with a pH of 1 than a solution with pH of 3? First, calculate the difference, n, between the two pH values: $n = 3 - 1 = 2$. Then use the formula, 10^n, to calculate the difference in acidity: $10^2 = 100$. A solution with a pH of 1 is 100 times more acidic than a solution with a pH of 3.

To practice differences in pH go to page 353.

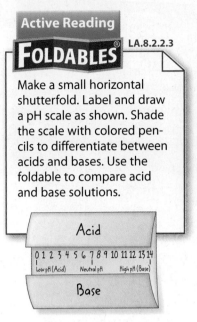

Active Reading

FOLDABLES® LA.8.2.2.3

Make a small horizontal shutterfold. Label and draw a pH scale as shown. Shade the scale with colored pencils to differentiate between acids and bases. Use the foldable to compare acid and base solutions.

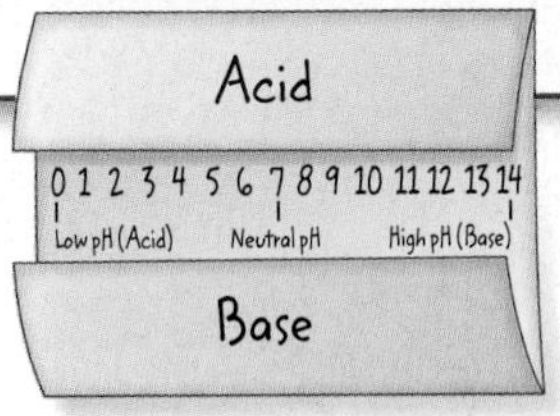
Acid
0 1 2 3 4 5 6 7 8 9 10 11 12 13 14
Low pH (Acid)
Neutral pH
High pH (Base)
Base

6. Visual Check Review Is a tomato more or less acidic than detergent?

What is the difference in acidity?

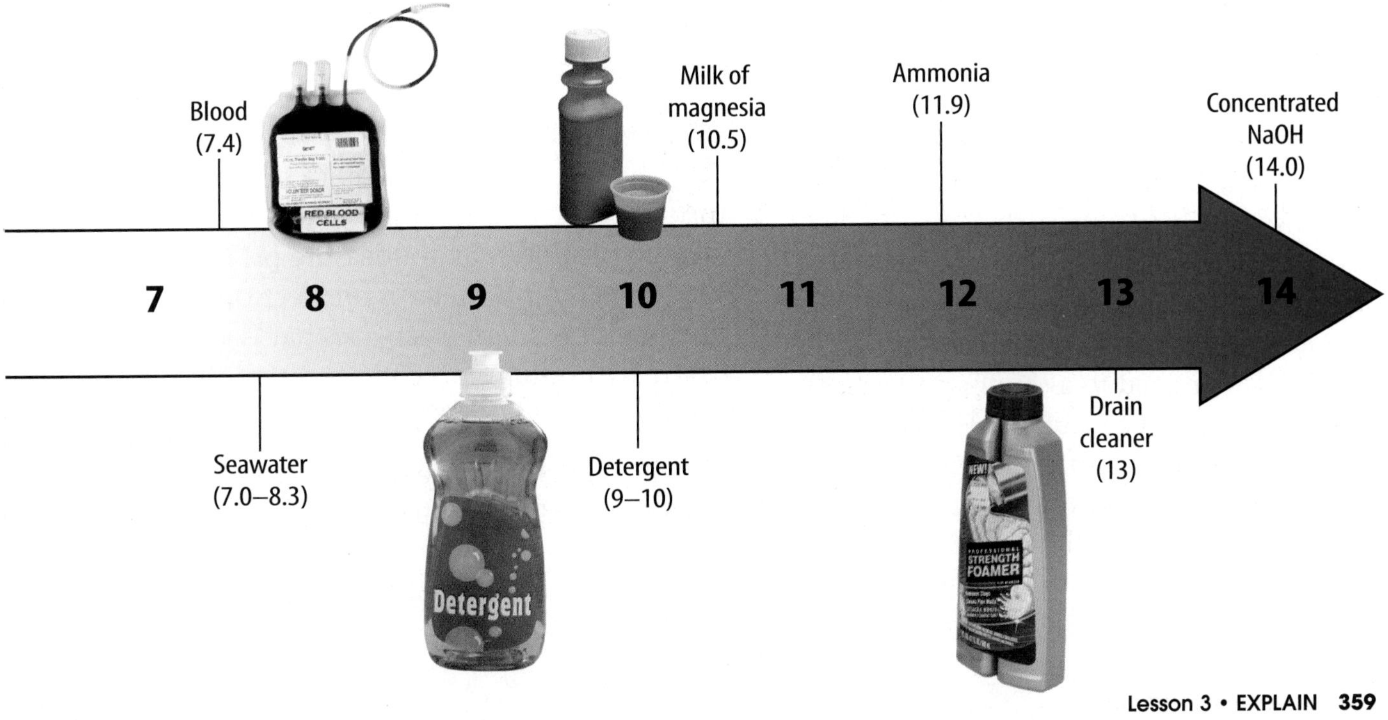
Blood
(7.4)
Milk of
magnesia
(10.5)
Ammonia
(11.9)
Concentrated
NaOH
(14.0)
7
8
9
10
11
12
13
14
Seawater
(7.0–8.3)
Detergent
(9–10)
Drain
cleaner
(13)
RED BLOOD
CELLS
Detergent
STRENGTH
FOAMER

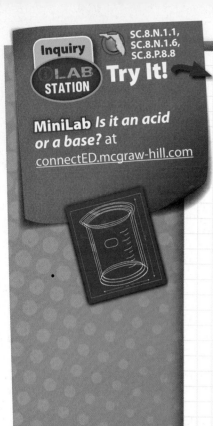

Inquiry LAB STATION **Try It!**

SC.8.N.1.1,
SC.8.N.1.6,
SC.8.P.8.8

MiniLab *Is it an acid or a base?* at
connectED.mcgraw-hill.com

Apply It! After you complete the lab, answer these questions below.

1. Choose two acids that you tested in the MiniLab and calculate how much more acidic one is from the other.

2. Why would the pH of soils be important to Florida farmers who grow orange crops?

How is pH measured?

How is the pH of a solution, such as swimming pool water, measured? Water test kits contain chemicals that change color when an acid or a base is added to them. These chemicals are called indicators.

pH Indicators

Indicators can be used to measure the approximate pH of a solution. *An **indicator** is a compound that changes color at different pH values when it reacts with acidic or basic solutions.* The pH of a solution is measured by adding a drop or two of the indicator to the solution. When the solution changes color, this color is matched to a set of standard colors that correspond to certain pH values. There are many different indicators—each indicator changes color over a specific range of pH values. For example, bromthymol blue is an indicator that changes from yellow to green to blue between pH 6 and pH 7.6.

pH Testing Strips

pH also can be measured using pH testing strips. The strips contain an indicator that changes to a variety of colors over a range of pH values. To use pH strips, dip the strip into the solution. Then match the resulting color to the list of standard colors that represent specific pH values.

pH Meters

Although pH strips are quick and easy, they provide only an approximate pH value. A more accurate way to measure pH is to use a pH meter. A pH meter is an electronic instrument with an electrode that is sensitive to the hydronium ion concentration in solution.

Active Reading

7. **List** What are two methods that can be used to measure the pH of a solution?

_____

_____

_____

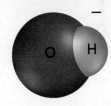

Acids contain hydrogen ions that are released and form hydronium ions in water. Bases are substances that form hydroxide ions when dissolved in water.

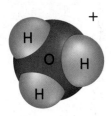

Hydronium ion concentration changes inversely with pH. This means that as hydronium ion concentration increases, the pH decreases.

pH can be measured using indicators or digital pH meters.

Inquiry
Try It!

SC.8.N.1.1,
SC.8.N.1.6,
SC.8.P.8.8,
MA.6.A.3.6,
MA.6.S.6.2

Inquiry Lab *Can the pH of a solution be changed?* at connectED.mcgraw-hill.com

Use Vocabulary

1 A measure of the concentration of hydronium ions (H_3O^+) in a solution is _____ .

2 A(n) _____ is used to determine the approximate pH of a solution.

Understand Key Concepts 🔑

3 **Describe** What happens to a hydrogen atom in an acid when the acid is dissolved in water?

_____

_____

4 **Explain** How does pH vary with hydronium ion and hydroxide ion concentrations in water?

_____

_____

5 **Show** Does an acidic solution contain hydroxide ions? Explain your answer with a diagram.

Interpret Graphics

6 **Contrast** Use the graphic organizer to describe and contrast three ways to measure pH. In the organizer, describe which methods are most and least accurate. **LA.8.2.2.3**

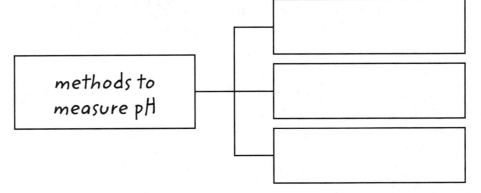

Critical Thinking

7 **Describe** the concentration of hydronium ions and hydroxide ions when a base is added slowly to a white vinegar solution. The pH of white vinegar is 3.1. **SC.8.P.8.8**

_____

_____

_____

Chapter 9 — Study Guide

  Think About It! Mixtures and substances are the two main classifications of matter. A solution is a type of mixture. Solutions can be described by the concentration and type of solute they contain.

🔑 Key Concepts Summary

Vocabulary

LESSON 1 Substances and Mixtures

- **Substances** have a fixed composition. The composition of **mixtures** can vary.
- **Solutions** and **heterogeneous mixtures** are both types of mixtures. Solutions are mixed at the atomic level.
- Mixtures contain parts that are not bonded together. These parts can be separated using physical means, and their properties can be seen in the solution.

substance p. 338

mixture p. 338

heterogeneous mixture p. 339

homogeneous mixture p. 339

solution p. 339

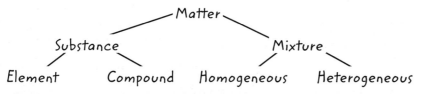

LESSON 2 Properties of Solutions

- Substances dissolve other substances that have a similar polarity. In other words, like dissolves like.
- **Concentration** is the amount of a **solute** that is dissolved. **Solubility** is the maximum amount of a solute that can dissolve.
- Both temperature and pressure affect the solubility of solutes in solutions.

solvent p. 345

solute p. 345

polar molecule p. 346

concentration p. 348

solubility p. 350

saturated solution p. 350

unsaturated solution p. 350

LESSON 3 Acid and Base Solutions

- **Acids** contain hydrogen ions that are released and form **hydronium ions** in water. **Bases** are substances that form hydroxide ions when dissolved in water.
- Hydronium ion concentration changes inversely with **pH.** This means that as hydronium ion concentration increases, the pH decreases.
- pH can be measured using **indicators** or digital pH meters.

acid p. 356

hydronium ion p. 356

base p. 356

pH p. 358

indicator p. 360

acidic   basic

0 1 2 3 4 5 6 7 8 9 10 11 12 13 14

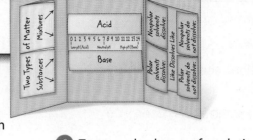
Active Reading

FOLDABLES® **Chapter Project**

Assemble your lesson Foldables as shown to make a Chapter Project. Use the project to review what you have learned in this chapter.

Use Vocabulary

1 The parts of a _____ can be seen with unaided eyes or with a microscope.

2 It is impossible to tell the difference between a solution and a _____ just by looking at them.

3 Water dissolves other _____ easily.

4 Two equal volumes of a solution that contain different amounts of the same solute have a different _____.

5 As _____ concentration decreases, pH increases.

6 A(n) _____ can be added to milk to neutralize it.

Link Vocabulary and Key Concepts

Use vocabulary terms from the previous page to complete the concept map.

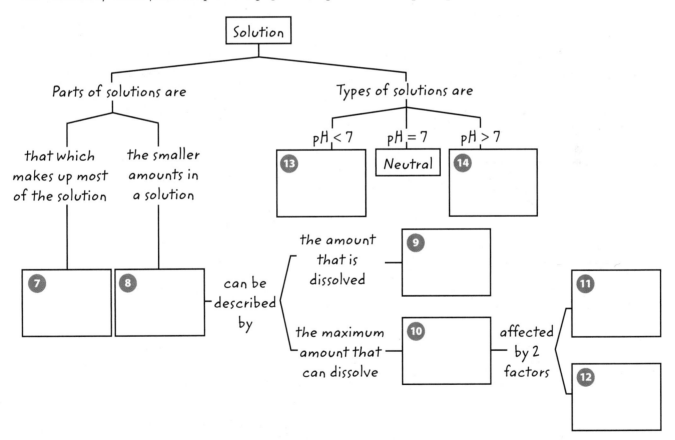

Fill in the correct answer choice.

🔑 Understand Key Concepts

1 Which is a solution? SC.8.P.8.9

Ⓐ copper
Ⓑ vinegar
Ⓒ pure water
Ⓓ a raisin cookie

2 The graph below shows the solubility of sodium chloride (NaCl) in water. MA.6.A.3.6

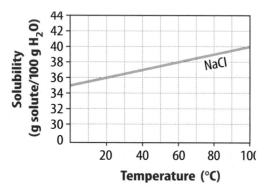

What mass of sodium chloride is needed to form a saturated solution at 80°C?

Ⓐ 30 g
Ⓑ 40 g
Ⓒ 50 g
Ⓓ 60 g

3 What would you add to a solution with a pH of 1.5 to obtain a solution with a pH of 7? SC.8.P.8.8

Ⓐ milk (pH 6.4)
Ⓑ vinegar (pH 3.0)
Ⓒ lye (pH 13.0)
Ⓓ coffee (pH 5.0)

4 Which can change the solubility of a solid in a liquid? SC.8.P.8.4

Ⓐ crushing the solute
Ⓑ stirring the solute
Ⓒ increasing the pressure of the solution
Ⓓ increasing the temperature of the solution

Critical Thinking

5 **Infer** How can you tell which component in a solution is the solvent? SC.8.P.8.4

_____

_____

_____

6 **Predict** The graph below shows the solubility of potassium chloride (KCl) in water. MA.6.A.3.6

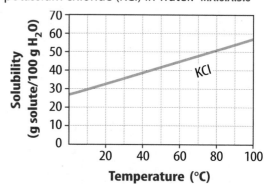

Imagine you have made a solution that contains 50 g of potassium chloride (KCl) in 100 g of solution. Predict what you would observe as you gradually increased the temperature from 0°C to 100°C.

_____

_____

_____

7 **Organize** The pH of three solutions is shown below. SC.8.P.8.8

Milk (pH 6.7)
Coffee (pH 5)
Ammonia (pH 11.6)

Place these solutions in order of

Ⓐ most acidic to least acidic

_____

Ⓑ most basic to least basic

_____

Ⓒ highest OH− concentration to lowest OH− concentration

_____

8 **Explain** The pH of a solution is inversely related to the concentration of hydronium ions in solution. Explain what this means. **MA.6.A.3.6**

_____

_____

_____

9 **Design** a method to determine the solubility of an unknown substance at 50°C. **SC.8.P.8.4**

_____

_____

_____

_____

Writing in Science

10 **Compose** A haiku is a poem containing three lines of five, seven, and five syllables, respectively. Write a haiku on a separate sheet of paper describing what happens when an acid is dissolved in water. **LA.8.2.2.3**

Big Idea Review

11 What are solutions? List at least three ways a solution can be described. **SC.8.P.8.9**

_____

_____

_____

_____

12 How do solutions differ from other types of matter? **SC.8.P.8.9**

_____

_____

_____

_____

Math Skills   MA.6.A.3.6

Calculate Concentration

13 Calculate the concentration of sugar in g/L in a solution that contains 40 g of sugar in 100 mL of solution. There are 1,000 mL in 1 L.

14 There are many ways to make a solution of a given concentration. What are two ways you could make a sugar solution with a concentration of 100 g/L?

15 A salt solution has a concentration of 200 g/L. How many grams of salt are contained in 500 mL of this solution? How many grams of salt would be contained in 2 L of this solution?

Fill in the correct answer choice.

Multiple Choice

Use the figures below to answer question 1.

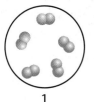 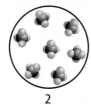

1   2

1 Which statement describes the two figures? SC.8.P.8.9

(A) Both 1 and 2 are mixtures.

(B) Both 1 and 2 are substances.

(C) 1 is a mixture and 2 is a substance.

(D) 1 is a substance and 2 is a mixture.

2 Which statement is an accurate comparison of solutions and homogeneous mixtures? SC.8.P.8.9

(F) They are the same.

(G) They are opposites.

(H) Solutions are more evenly mixed than homogeneous mixtures.

(I) Homogeneous mixtures are more evenly mixed than solutions.

3 A worker uses a magnet to remove bits of iron from a powdered sample. Which describes the sample before the worker used the magnet to remove the iron? SC.8.P.8.9

(A) The sample is a compound because the iron was removed using a physical method.

(B) The sample is a compound because the iron was removed using a chemical change.

(C) The sample is a mixture because the iron was removed using a chemical change.

(D) The sample is a mixture because the iron was removed using a physical method.

4 A beaker contains a mixture of sand and small pebbles. What kind of mixture is this? SC.8.P.8.9

(F) compound

(G) heterogeneous

(H) homogeneous

(I) solution

5 What ions must be present in the greatest amount in a solution with a pH of 8.5? SC.8.P.8.8

(A) hydrogen ions

(B) oxygen ions

(C) hydronium ions

(D) hydroxide ions

6 According to the pH scale, which pH measurement is basic? SC.8.P.8.8

pH Scale

0 1 2 3 4 5 6 7 8 9 10 11 12 13 14

(F) 7.0

(G) 9.5

(H) 5.5

(I) 1.2

7 Which is a property of an acidic solution? SC.8.P.8.8

(A) It tastes sour.

(B) It feels slippery.

(C) It is used in many cleaning products.

(D) It tastes bitter.

8 The neutralization reaction between sodium hydroxide (NaOH) and hydrogen chloride (HCl) forms a salt. What is the correct chemical formula? SC.8.P.8.8

(F) NaCl

(G) ClNa

(H) Na$_2$Cl

(I) Cl$_2$Na

Use the table below to answer question 9.

Sample solution	Change in blue litmus	Change in red litmus
1	turns red	no change
2	no change	turns blue
3	turns red	no change
4	no change	no change

9 A scientist collects the data above using litmus paper. Blue litmus paper is a type of pH indicator that turns red when placed in an acidic solution. Red litmus paper is an indicator that turns blue when placed in a basic solution. Neutral solutions cause no change in either color of litmus paper. Which sample solution must be a base? SC.8.P.8.8

(A) solution 1

(B) solution 2

(C) solution 3

(D) solution 4

10 Which statement is an accurate description for bases? SC.8.P.8.8

(F) They decrease the concentration of hydroxide ions when dissolved in water.

(G) They increase the concentration of hydroxide ions when dissolved in water.

(H) They increase the concentration of hydronium ions when dissolved in water.

(I) They have no effect on the concentration of hydroxide ions when dissolved in water.

11 Which product has a bitter taste, is slippery when mixed with water, and can damage skin and eyes? SC.8.P.8.8

(A) sodium hydroxide

(B) citric acid

(C) lactic acid

(D) acetic acid

12 How does a solution with a pH of 2 compare to a solution with a pH of 1? SC.8.P.8.8

(F) The pH 2 solution is two times more acidic than that with a pH of 1.

(G) The pH 1 solution is ten times more acidic than that with a pH of 2.

(H) The pH 1 solution is two times more acidic than that with a pH of 2.

(I) The pH 2 solution is ten times more acidic than that with a pH of 1.

NEED EXTRA HELP?

If You Missed Question...	1	2	3	4	5	6	7	8	9	10	11	12
Go to Lesson...	1	1	1	1	3	3	3	3	3	3	3	3

Multiple Choice *Bubble the correct answer.*

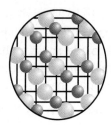

1. The image above represents a(n) **SC.8.P.8.9**

   (A) compound.

   (B) element.

   (C) heterogeneous mixture.

   (D) homogeneous mixture.

2. Which is another name for a homogeneous mixture? **SC.8.P.8.9**

   (F) compound

   (G) element

   (H) solution

   (I) substance

3. Which is a homogeneous mixture? **SC.8.P.8.9**

   (A) blood

   (B) granite

   (C) lemonade

   (D) pizza

**Blood
Samples**

4. Which is true about the substances shown in the figure above? **SC.8.P.8.9**

   (F) The different substances in each sample are bonded together.

   (G) The different substances in each sample are evenly mixed at the atomic level.

   (H) The different substances in each sample are not evenly mixed.

   (I) The different substances in each sample cannot be separated from each other.

Multiple Choice *Bubble the correct answer.*

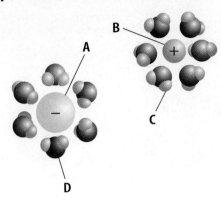

1. The image above shows sodium and chloride ions in solution. Which label in the diagram represents the chloride ion? **SC.8.P.8.9**

   (A) Label A

   (B) Label B

   (C) Label C

   (D) Label D

2. Which does NOT describe a water molecule? **SC.8.P.8.9**

   (F) The molecule has a bent shape.

   (G) The molecule is an ionic compound.

   (H) The molecule dissolves most polar solutes.

   (I) The oxygen end of the molecule has a slightly negative charge.

3. Rosalinda is thirsty and wants to make a glass of lemonade as quickly as possible. She needs to add sugar to lemon juice and water to make the lemonade. Which type of sugar should she use? **SC.8.P.8.4**

   (A) granulated sugar with fine grains

   (B) raw sugar with coarse grains

   (C) sugar from rock candy

   (D) sugar formed into cubes

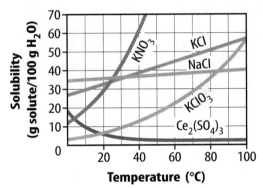

4. The solubility of which substance in the figure above increases MOST quickly as temperature increases? **SC.8.P.8.4**

   (F) KCl

   (G) $KClO_3$

   (H) KNO_3

   (I) $NaCl$

369

Benchmark Mini-Assessment   Chapter 9 • Lesson 3

Multiple Choice *Bubble the correct answer.*

Use the images below to answer questions 1–4.

Reactants and Products in Two Reactions

Reaction A

Reaction B

1. One of the products of Reaction A is
SC.8.P.8.8

   Ⓐ an ammonium ion.

   Ⓑ an oxygen ion.

   Ⓒ a hydronium ion.

   Ⓓ a sodium ion.

2. Which product could contain a substance formed by Reaction B? **SC.8.P.8.8**

   Ⓕ antacid

   Ⓖ milk

   Ⓗ baking soda

   Ⓘ dark chocolate

3. Reaction B could produce a substance with a pH of **SC.8.P.8.8**

   Ⓐ 4.

   Ⓑ 7.

   Ⓒ 10.

   Ⓓ 14.

4. Which is one of the reactants in Reaction A?
SC.8.P.8.8

   Ⓕ copper

   Ⓖ silver

   Ⓗ sodium

   Ⓘ water

Notes

Notes

Chemical Reaction

When you mix two different substances together, sometimes a chemical reaction occurs that forms a new substance. Which best describes what happens during a chemical reaction when two different substances form a new substance?

A. Some of the atoms are destroyed and replaced by new atoms.

B. All of the atoms are destroyed and all new atoms form.

C. None of the atoms are destroyed but some new atoms are added.

D. None of the atoms are destroyed and no new atoms form.

Explain your thinking. Describe what you think happens to the atoms during a chemical reaction.

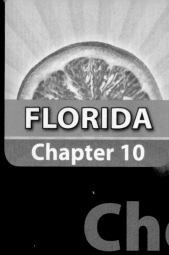

Chemical Reactions and EQUATIONS

FLORIDA BIG IDEAS

1 The Practice of Science

3 The Role of Theories, Laws, Hypotheses, and Models

9 Changes in Matter

Think About It!

What happens to atoms and energy during a chemical reaction?

An air bag deploys in less than the blink of an eye. How does the bag open so fast? At the moment of impact, a sensor triggers a chemical reaction between two chemicals. This reaction quickly produces a large amount of nitrogen gas. This gas inflates the bag with a pop.

1 A chemical reaction can produce a gas. How is this different from a gas produced when a liquid boils?

_____

_____

_____

2 What do you think happens to atoms and energy during a chemical reaction?

_____

_____

_____

Get Ready to Read

What do you think about chemical reactions?

Before you read, decide if you agree or disagree with each of these statements. As you read this chapter, see if you change your mind about any of the statements.

	AGREE	DISAGREE
1 If a substance bubbles, you know a chemical reaction is occurring.	☐	☐
2 During a chemical reaction, some atoms are destroyed and new atoms are made.	☐	☐
3 Reactions always start with two or more substances that react with each other.	☐	☐
4 Water can be broken down into simpler substances.	☐	☐
5 Reactions that release energy require energy to get started.	☐	☐
6 Energy can be created in a chemical reaction.	☐	☐

 **Connect ED**

There's More Online!
Video • Audio • Review • ⓘLab Station • WebQuest • Assessment • Concepts in Motion • Multilingual eGlossary

Understanding Chemical REACTIONS

Vocabulary

chemical reaction p. 377

chemical equation p. 380

reactant p. 381

product p. 381

law of conservation of mass p. 382

coefficient p. 384

 Florida NGSSS

SC.8.N.1.1 Define a problem from the eighth grade curriculum using appropriate reference materials to support scientific understanding, plan and carry out scientific investigations of various types, such as systematic observations or experiments, identify variables, collect and organize data, interpret data in charts, tables, and graphics, analyze information, make predictions, and defend conclusions.

SC.8.N.3.1 Select models useful in relating the results of their own investigations.

SC.8.P.8.5 Recognize that there are a finite number of elements and that their atoms combine in a multitude of ways to produce compounds that make up all of the living and nonliving things that we encounter.

SC.8.P.9.1 Explore the Law of Conservation of Mass by demonstrating and concluding that mass is conserved when substances undergo physical and chemical changes.

SC.8.P.9.2 Differentiate between physical changes and chemical changes.

LA.8.2.2.3 The student will organize information to show understanding or relationships among facts, ideas, and events (e.g., representing key points within text through charting, mapping, paraphrasing, summarizing, or comparing/contrasting);

 SC.8.P.9.1

Inquiry Launch Lab

15–20 minutes

Where did it come from?

Does a boiled egg have more mass than a raw egg? What happens when liquids change to a solid?

Procedure

1. Read and complete a lab safety form.

2. Use a **graduated cylinder** to add 25 mL of **solution A** to a **self-sealing plastic bag.** Place a **stoppered test tube** containing **solution B** into the bag. Be careful not to dislodge the stopper.

3. Seal the bag completely, and wipe off any moisture on the outside with a **paper towel.** Place the bag on the **balance.** Record the total mass.

4. Without opening the bag, remove the stopper from the test tube and allow the liquids to mix. Observe and record what happens.

5. Place the sealed bag and its contents back on the balance. Read and record the mass.

Data and Observations

Think About This

1. **Express** What did you observe when the liquids mixed? How would you account for this observation?

2. **Evaluate** Did the mass of the bag's contents change? If so, could the change have been due to the precision of the balance, or did the matter in the bag change its mass? Explain.

3. 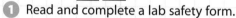 **Key Concept** **Assess** Do you think matter was gained or lost in the bag? How can you tell?

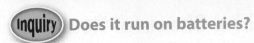

1. From Florida to Maine, east of the Mississippi River fireflies dot summer evening skies with flickering spots of light. But fireflies don't use batteries. Their light is produced using a two-step chemical process called bioluminescence (bi oh lew muh NE sents). In this process, chemicals in a firefly's body combine to make new chemicals and light. How do chemicals produce light? What are some other common chemical processes?

_____

_____

_____

_____

_____

_____

_____

Chemical Changes

Recall that during a chemical change, one or more substances change into new substances. The starting substances and the substances produced have different physical and chemical properties. For example, when brownie batter bakes, a chemical change occurs. Many of the substances in the baked brownies are different from the substances in the batter. As a result, baked brownies have physical and chemical properties that are different from those of brownie batter.

A chemical change also is called a chemical reaction. These terms mean the same thing. _A **chemical reaction** is a process in which atoms of one or more substances rearrange to form one or more new substances._ In this lesson, you will read what happens to atoms during a reaction and how these changes can be described using equations.

Active Reading **2. Identify** What types of properties change during a chemical reaction?

_____

_____

_____

Changes in Matter

When you put liquid water in a freezer, it changes to solid water, or ice. When you pour brownie batter into a pan and bake it, the liquid batter changes to a solid, too. In both cases, a liquid changes to a solid. Are these changes the same?

Physical Changes

Recall that matter can undergo two types of changes—chemical or physical. A physical change does not produce new substances. The substances that exist before and after the change are the same, although they might have different physical properties. This is what happens when liquid water freezes. Its physical properties change from a liquid to a solid, but the water, H_2O, does not change into a different substance. Water molecules are always made up of two hydrogen atoms bonded to one oxygen atom regardless of whether they are solid, liquid, or gas.

Signs of a Chemical Reaction

How can you tell if a chemical reaction has taken place? You have read that the substances before and after a reaction have different properties. You might think that you could look for changes in properties as a sign that a reaction occurred. In fact, changes in the physical properties of color, state of matter, and odor are all signs that a chemical reaction might have occurred. Another sign of a chemical reaction is a change in energy. If substances get warmer or cooler or if they give off light or sound, it is likely that a reaction has occurred. Some signs that a chemical reaction might have occurred are shown in **Figure 1**.

However, these signs are not proof of a chemical change. For example, bubbles appear when water boils. But, bubbles also appear when baking soda and vinegar react and form carbon dioxide gas. How can you be sure that a chemical reaction has taken place? The only way to know is to study the chemical properties of the substances before and after the change. If they have different chemical properties, then the substances have undergone a chemical reaction.

3. NGSSS Check **Recognize** How can you know for sure that a chemical reaction has occurred? <mark>Highlight</mark> your answer in the text.

Figure 1 🔑 You can detect a chemical reaction by looking for changes in properties and changes in energy of the substances that reacted.

Change in Properties

Change in color
Bright copper changes to green when the copper reacts with certain gases in the air.

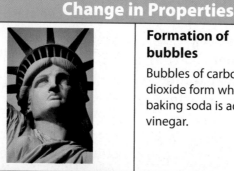

Formation of bubbles
Bubbles of carbon dioxide form when baking soda is added to vinegar.

Change in odor
When food burns or rots, a change in odor is a sign of chemical change.

Formation of a precipitate
A precipitate is a solid formed when two liquids react.

Change in Energy

Warming or cooling
Thermal energy is either given off or absorbed during a chemical change.

Release of light
A firefly gives off light as the result of a chemical change.

What happens in a chemical reaction?

During a chemical reaction, one or more substances react and form one or more new substances. How are these new substances formed?

Atoms Rearrange and Form New Substances

To understand what happens in a reaction, first review substances. Recall that there are two types of substances—elements and compounds. Substances have a fixed arrangement of atoms. For example, in a single drop of water, there are trillions of oxygen and hydrogen atoms. However, all of these atoms are arranged in the same way—two atoms of hydrogen are bonded to one atom of oxygen. If this arrangement changes, the substance is no longer water. Instead, a different substance forms with different physical and chemical properties. This is what happens during a chemical reaction. Atoms of elements or compounds rearrange and form different elements or compounds.

Bonds Break and Bonds Form

How does the rearrangement of atoms happen? Atoms rearrange when **chemical bonds** between atoms break. Recall that constantly moving particles make up all substances, including solids. As particles move, they collide with one another. If the particles collide with enough energy, the bonds between atoms can break. The atoms separate and rearrange, and new bonds can form. The reaction that forms hydrogen and oxygen from water is shown in **Figure 2**. Adding electric energy to water molecules can cause this reaction. The added energy causes bonds between the hydrogen atoms and the oxygen atoms to break. After the bonds between the atoms in water molecules break, new bonds can form between pairs of hydrogen atoms and between pairs of oxygen atoms.

Active Reading 4. Sequence
Fill in the graphic organizer below to list the changes in substances during a chemical reaction.

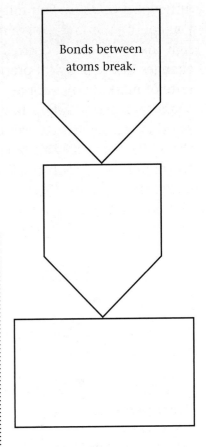

Bonds between atoms break.

REVIEW VOCABULARY

chemical bond
an attraction between atoms when electrons are shared, transferred, or pooled

Figure 2 🗝 Notice that no new atoms are created in a chemical reaction. The existing atoms rearrange and form new substances.

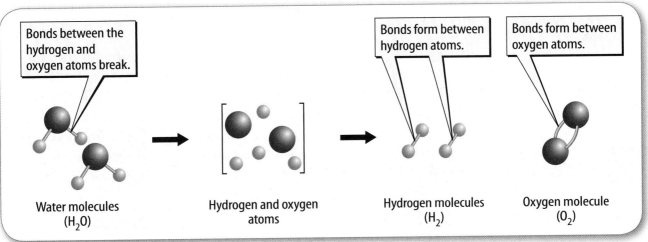

Bonds between the hydrogen and oxygen atoms break.

Bonds form between hydrogen atoms.

Bonds form between oxygen atoms.

Water molecules (H_2O)

Hydrogen and oxygen atoms

Hydrogen molecules (H_2)

Oxygen molecule (O_2)

Table 1 Symbols and Formulas of Some Elements and Compounds

Substance		Formula	# of atoms
Carbon		C	C: 1
Copper		Cu	Cu: ___
Cobalt		Co	Co: ___
Oxygen		O_2	O: 2
Hydrogen		H_2	H: ___
Chlorine		Cl_2	Cl: ___
Carbon dioxide		CO_2	C: 1 O: 2
Carbon monoxide		CO	C: ___ O: ___
Water		H_2O	H: ___ O: ___
Hydrogen peroxide		H_2O_2	H: 2 O: 2
Glucose		$C_6H_{12}O_6$	C: ___ H: ___ O: ___
Sodium chloride		NaCl	Na: 1 Cl: 1
Magnesium hydroxide		$Mg(OH)_2$	Mg: 1 O: 2 H: 2

Table 1 Symbols and subscripts describe the type and number of atoms in an element or a compound.

Active Reading 5. **Write** In **Table 1**, fill in the correct number of atoms of each element.

Chemical Equations

Suppose your teacher asks you to produce a specific reaction in your science laboratory. How might your teacher describe the reaction to you? He or she might say something such as "react baking soda and vinegar to form sodium acetate, water, and carbon dioxide." It is more likely that your teacher will describe the reaction in the form of a chemical equation. A **chemical equation** *is a description of a reaction using element symbols and chemical formulas.* Element symbols represent elements. Chemical formulas represent compounds.

Element Symbols

Recall that symbols of elements are shown in the periodic table. For example, the symbol for carbon is C. The symbol for copper is Cu. Each element can exist as just one atom. However, some elements exist in nature as diatomic molecules—two atoms of the same element bonded together. A formula for one of these diatomic elements includes the element's symbol and the subscript *2*. A subscript describes the number of atoms of an element in a compound. Oxygen (O_2) and hydrogen (H_2) are examples of diatomic molecules. Some element symbols are shown above the blue line in **Table 1.**

Chemical Formulas

When atoms of two or more different elements bond, they form a compound. Recall that a chemical formula uses elements' symbols and subscripts to describe the number of atoms in a compound. If an element's symbol does not have a subscript, the compound contains only one atom of that element. For example, carbon dioxide (CO_2) is made up of one carbon atom and two oxygen atoms. Remember that two different formulas, no matter how similar, represent different substances. Some chemical formulas are shown below the blue line in **Table 1.**

Writing Chemical Equations

A chemical equation includes both the substances that react and the substances that are formed in a chemical reaction. *The starting substances in a chemical reaction are* **reactants**. *The substances produced by the chemical reaction are* **products**. **Figure 3** shows how a chemical equation is written. Chemical formulas are used to describe the reactants and the products. The reactants are written to the left of an arrow, and the products are written to the right of the arrow. Two or more reactants or products are separated by a plus sign. The general structure for an equation is:

reactant + reactant → product + product

When writing chemical equations, it is important to use correct chemical formulas for the reactants and the products. For example, suppose a certain chemical reaction produces carbon dioxide and water. The product carbon dioxide would be written as CO_2 and not as CO. CO is the formula for carbon monoxide, which is not the same compound as CO_2. Water would be written as H_2O and not H_2O_2, the formula for hydrogen peroxide.

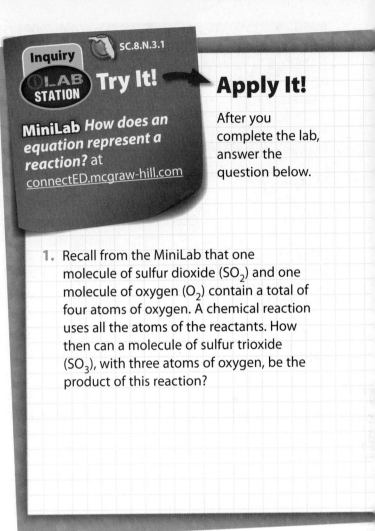

Inquiry SC.8.N.3.1

Try It!

MiniLab *How does an equation represent a reaction?* at connectED.mcgraw-hill.com

Apply It!

After you complete the lab, answer the question below.

1. Recall from the MiniLab that one molecule of sulfur dioxide (SO_2) and one molecule of oxygen (O_2) contain a total of four atoms of oxygen. A chemical reaction uses all the atoms of the reactants. How then can a molecule of sulfur trioxide (SO_3), with three atoms of oxygen, be the product of this reaction?

Figure 3 An equation is read much like a sentence. This equation is read as "carbon plus oxygen produces carbon dioxide."

Active Reading 6. **Recognize** Fill in the blanks in **Figure 3** below.

Parts of an Equation

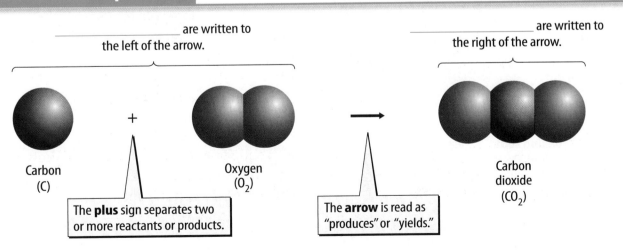

_____ are written to the left of the arrow.

_____ are written to the right of the arrow.

Carbon (C) + Oxygen (O_2) → Carbon dioxide (CO_2)

The **plus** sign separates two or more reactants or products.

The **arrow** is read as "produces" or "yields."

Conservation of Mass

A French chemist named Antoine Lavoisier (AN twan · luh VWAH see ay) (1743–1794) discovered something interesting about chemical reactions. In a series of experiments, Lavoisier measured the masses of substances before and after a chemical reaction inside a closed container. He found that the total mass of the reactants always equaled the total mass of the products. Lavoisier's results led to the law of conservation of mass. *The **law of conservation of mass** states that the total mass of the reactants before a chemical reaction is the same as the total mass of the products after the chemical reaction.*

Atoms are conserved.

The discovery of atoms provided an explanation for Lavoisier's observations. Mass is conserved in a reaction because atoms are conserved. Recall that during a chemical reaction, bonds break and new bonds form. However, atoms are not destroyed, and no new atoms form. All atoms at the start of a chemical reaction are present at the end of the reaction. **Figure 4** shows that mass is conserved in the reaction between baking soda and vinegar.

7. NGSSS Check Describe What happens to the total mass of the reactants in a chemical reaction? SC.8.P.9.1

_____

_____

Figure 4 As this reaction takes place, the mass on the balance remains the same, showing that mass is conserved.

Active Reading **8. Draw** For the products of the reaction shown in **Figure 4**, indicate the correct number of atoms of each element. Then use colored pencils to draw the correct number of atoms. Match the colors of the atoms of the product to the colors of the atoms of reactants.

Conservation of Mass 🔑

Baking soda is contained in a balloon. The balloon is attached to a flask that contains vinegar.

When the balloon is tipped up, the baking soda pours into the vinegar. The reaction forms a gas that is collected in the balloon.

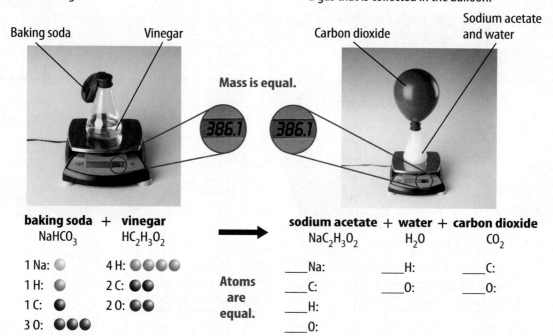

Baking soda   Vinegar

Carbon dioxide

Sodium acetate and water

Mass is equal.

386.1   386.1

baking soda	+	**vinegar**		**sodium acetate**	+	**water**	+	**carbon dioxide**
$NaHCO_3$		$HC_2H_3O_2$		$NaC_2H_3O_2$		H_2O		CO_2

1 Na: ⚪   4 H: ⚪⚪⚪⚪

1 H: ⚪   2 C: ⚫⚫

1 C: ⚫   2 O: ⚫⚫

3 O: ⚫⚫⚫

Atoms are equal.

___ Na: ___ H: ___ C:

___ C: ___ O: ___ O:

___ H:

___ O:

Is an equation balanced?

How does a chemical equation show that atoms are conserved? An equation is written so that the number of atoms of each element is the same, or balanced, on each side of the arrow. The equation showing the reaction between carbon and oxygen that produces carbon dioxide is shown below. Remember that oxygen is written as O_2 because it is a diatomic molecule. The formula for carbon dioxide is CO_2.

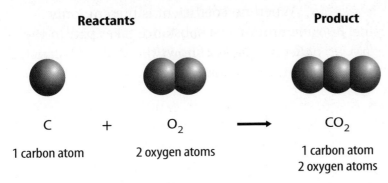

Reactants				Product
C	+	O_2	→	CO_2
1 carbon atom		2 oxygen atoms		1 carbon atom
				2 oxygen atoms

Is there the same number of carbon atoms on each side of the arrow? Yes, there is one carbon atom on the left and one on the right. Carbon is balanced. Is oxygen balanced? There are two oxygen atoms on each side of the arrow. Oxygen also is balanced. The atoms of all elements are balanced. Therefore, the equation is balanced.

You might think a balanced equation happens automatically when you write the symbols and formulas for reactants and products. However, this usually is not the case. For example, the reaction between hydrogen (H_2) and oxygen (O_2) that forms water (H_2O) is shown below.

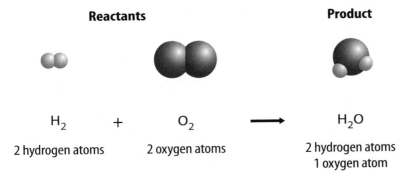

Reactants				Product
H_2	+	O_2	→	H_2O
2 hydrogen atoms		2 oxygen atoms		2 hydrogen atoms
				1 oxygen atom

Count the number of hydrogen atoms on each side of the arrow. There are two hydrogen atoms in the product and two in the reactants. They are balanced. Now count the number of oxygen atoms on each side of the arrow. Did you notice that there are two oxygen atoms in the reactants and only one in the product? Because they are not equal, this equation is not balanced. To accurately represent this reaction, the equation needs to be balanced.

Active Reading

FOLDABLES® LA.8.2.2.3

Make a vertical four-tab book. Label it as shown. Use it to study the steps of balancing equations.

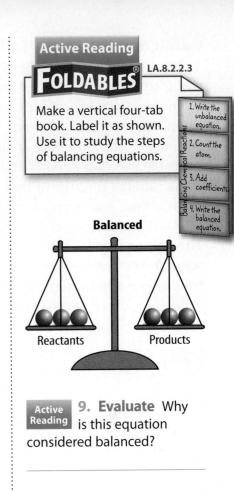

Balancing Chemical Reactions

1. Write the unbalanced equation.
2. Count the atom.
3. Add coefficients.
4. Write the balanced equation.

Balanced

Reactants   Products

Active Reading **9. Evaluate** Why is this equation considered balanced?

_____

_____

_____

Unbalanced

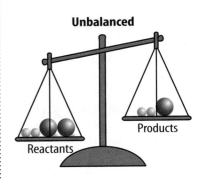

Products

Reactants

Active Reading **10. Explain** Why is this equation considered unbalanced?

_____

_____

_____

Balancing Chemical Equations

When you balance a chemical equation, you count the atoms in the reactants and the products and then add coefficients to balance the number of atoms. *A coefficient is a number placed in front of an element symbol or chemical formula in an equation.* It is the number of units of that substance in the reaction. For example, in the formula $2H_2O$, the *2* in front of H_2O is a coefficient. This means that there are two molecules of water in the reaction. Only coefficients can be changed when balancing an equation. Changing subscripts changes the identities of the substances that are in the reaction.

If one molecule of water contains two hydrogen atoms and one oxygen atom, how many H and O atoms are in two molecules of water ($2H_2O$)? Multiply each by 2.

> 2×2 H atoms $= 4$ H atoms
> 2×1 O atom $= 2$ O atoms

When no coefficient is present, only one unit of that substance takes part in the reaction. **Table 2** shows the steps of balancing a chemical equation.

Active Reading 11. **Write** Fill in the correct numbers and chemical formulas to complete **Table 2**.

Table 2 Balancing a Chemical Equation

1 **Write the unbalanced equation.** Make sure that all chemical formulas are correct.	H_2 + O_2 → H_2O **reactants** **products**
2 **Count atoms of each element in the reactants and in the products.** **a.** Note which, if any, elements have a balanced number of atoms on each side of the equation. Which atoms are not balanced? **b.** If all of the atoms are balanced, the equation is balanced.	H_2 + O_2 → H_2O **reactants** **products** $H = 2$ $H = $ ____ $O = 2$ $O = $ ____
3 **Add coefficients to balance the atoms.** **a.** Pick an element in the equation that is not balanced, such as oxygen. Write a coefficient in front of a reactant or a product that will balance the atoms of that element. **b.** Recount the atoms of each element in the reactants and the products. Note which atoms are not balanced. Some atoms that were balanced before might no longer be balanced. **c.** Repeat step 3 until the atoms of each element are balanced.	H_2 + O_2 → $2H_2O$ **reactants** **products** $H = 2$ $H = $ ____ $O = 2$ $O = $ ____ $2H_2$ + O_2 → $2H_2O$ **reactants** **products** $H = $ ____ $H = $ ____ $O = $ ____ $O = $ ____
4 **Write the balanced chemical equation** including the coefficients.	____ + ____ → ____

Visual Summary

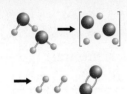

A chemical reaction is a process in which bonds break and atoms rearrange, forming new bonds.

$$2H_2 + O_2 \rightarrow 2H_2O$$

A chemical equation uses symbols to show reactants and products of a chemical reaction.

The mass and the number of each type of atom do not change during a chemical reaction. This is the law of conservation of mass.

Use Vocabulary

1 **Define** *reactants* and *products*.

_____

_____

_____

Understand Key Concepts 🔑

2 Which is a definite sign of a chemical reaction?

(A) Chemical properties change.   (C) A gas forms.

(B) Physical properties change.   (D) A solid forms.

3 **Explain** why subscripts cannot change when balancing a chemical equation. **SC.8.P.9.2**

_____

_____

4 **Infer** Is the reaction below possible? Explain why or why not.

$$H_2O + NaOH \rightarrow NaCl + H_2$$

_____

_____

Interpret Graphics

5 **Interpret** Complete the table to determine if this equation is balanced:

$$CH_4 + 2O_2 \rightarrow CO_2 + 2H_2O$$

Is this reaction balanced? Explain. **SC.8.P.9.1**

_____

Type of Atom	Number of Atoms in the Balanced Chemical Equation	
	Reactants	**Products**

Critical Thinking

6 Balance this chemical equation. Hint: Balance Al last and then use a multiple of 2 and 3.

$$Al + HCl \rightarrow AlCl_3 + H_2$$

SC.8.N.1.1, SC.8.P.8.5, SC.8.P.9.2

Inquiry

LAB STATION **Try It!**

Skill Lab *What can you learn from an experiment?* at connectED.mcgraw-hill.com

1 Fill in the graphic organizer to distinguish the parts of a chemical equation.

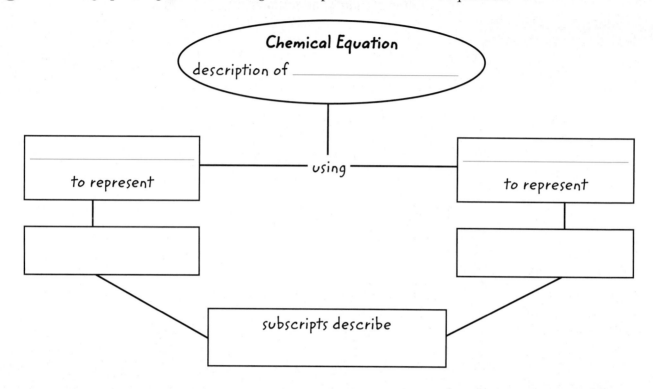

Complete the table below with information regarding the writing of chemical equations.

Define *reactant*.	2
Define *product*.	3
Write the general structure for a chemical equation.	4
How is the arrow sign read?	5
Write the equation for "carbon plus oxygen produces carbon dioxide."	6

7 Restate the law of conservation of mass.

_____

_____

_____

Types of Chemical REACTIONS

ESSENTIAL QUESTIONS

 How can you recognize the type of chemical reaction by the number or type of reactants and products?

 What are the different types of chemical reactions?

Vocabulary

synthesis p. 389

decomposition p. 389

single replacement p. 390

double replacement p. 390

combustion p. 390

 **Launch Lab**

15 minutes

What combines with what?

The reactants and the products in a chemical reaction can be elements, compounds, or both. In how many ways can these substances combine?

Procedure

1 Read and complete a lab safety form.

2 Divide a **sheet of paper** into four equal sections labeled *A*, *B*, *Y*, and *Z*. Place **red paper clips** in section A, **yellow clips** in section B, **blue clips** in section Y, and **green clips** in section Z.

3 Use another sheet of paper to copy the table shown to the right. Turn the paper so that a long edge is at the top. Print *REACTANTS → PRODUCTS* across the top, and then complete the table.

	REACTANTS → PRODUCTS
1	$AY \rightarrow A + Y$
2	$B + Z \rightarrow BZ$
3	$2A_2 + Y_2 \rightarrow 2A_2Y$
4	$A + BY \rightarrow B + AY$
5	$Z + BY \rightarrow Y + BZ$
6	$AY + BZ \rightarrow AZ + BY$

4 Using the paper clips, model the equations listed in the table. Hook the clips together to make diatomic elements or compounds. Place each clip model onto your paper over the matching written equation.

5 As you read Lesson 2, match the types of equations to your paper clip equations.

Think About This

1. **Select** the equation that represents hydrogen combining with oxygen and forming water. How did you determine which equation to choose?

2. **Key Concept** **Identify** How could you use the number and type of reactants to identify a type of chemical reaction?

Florida NGSSS

LA.8.2.2.3 The student will organize information to show understanding or relationships among facts, ideas, and events (e.g., representing key points within text through charting, mapping, paraphrasing, summarizing, or comparing/contrasting);

LA.8.4.2.2 The student will record information (e.g., observations, notes, lists, charts, legends) related to a topic, including visual aids

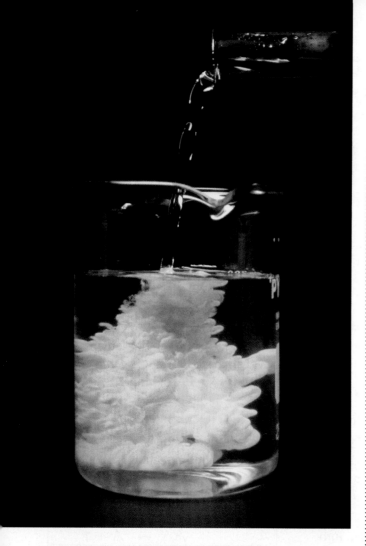

1. When lead nitrate and potassium iodide—both clear liquids—combine, a yellow solid appears instantly. Where did the solid come from? Here's a hint—the name of the solid is lead iodide. Did you guess that parts of each reactant combined and formed it? What are chemical reactants? What chemical reactants were combined in **Figure 5** to cause the explosion?

_____

_____

_____

_____

_____

_____

_____

Active Reading **2. Summarize** Fill in the graphic organizer below to describe the concept of patterns in chemical reactions.

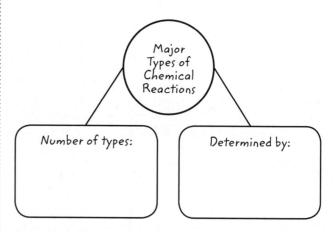

Patterns in Reactions

If you have ever used hydrogen peroxide, you might have noticed that it is stored in a dark bottle. This is because light causes hydrogen peroxide to change into other substances. Maybe you have seen a video of an explosion demolishing an old building, like in **Figure 5.** How is the reaction with hydrogen peroxide and light similar to a building demolition? In both, one reactant breaks down into two or more products.

The breakdown of one reactant into two or more products is one of four major types of chemical reactions. Each type of chemical reaction follows a unique pattern in the way atoms in reactants rearrange to form products. In this lesson, you will read how chemical reactions are classified by recognizing patterns in the way the atoms recombine.

Figure 5 When dynamite explodes, it chemically changes into several products and releases energy.

Types of Chemical Reactions

There are many different types of reactions. It would be impossible to memorize them all. However, most chemical reactions fit into four major categories. Understanding these categories of reactions can help you predict how compounds will react and what products will form.

Synthesis

A **synthesis** (SIHN thuh sus) *is a type of chemical reaction in which two or more substances combine and form one compound.* In the synthesis reaction shown in **Figure 6,** magnesium (Mg) reacts with oxygen (O_2) in the air and forms magnesium oxide (MgO). You can recognize a synthesis reaction because two or more reactants form only one product.

Decomposition

In a **decomposition** *reaction, one compound breaks down and forms two or more substances.* You can recognize a decomposition reaction because one reactant forms two or more products. For example, hydrogen peroxide (H_2O_2), shown in **Figure 6,** decomposes and forms water (H_2O) and oxygen gas (O_2). Notice that decomposition is the reverse of synthesis.

Figure 6 Synthesis and decomposition reactions are opposites of each other.

Active Reading

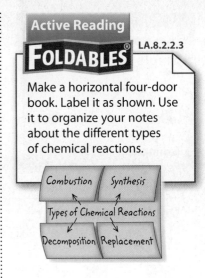

FOLDABLES LA.8.2.2.3

Make a horizontal four-door book. Label it as shown. Use it to organize your notes about the different types of chemical reactions.

Combustion | Synthesis

Types of Chemical Reactions

Decomposition | Replacement

Active Reading

3. Differentiate How can you tell the difference between synthesis and decomposition reactions?

_____

_____

_____

_____

Synthesis and Decomposition Reactions 🔑

Synthesis Reactions

Examples:
$2Na + Cl_2 \rightarrow 2NaCl$
$2H_2 + O_2 \rightarrow 2H_2O$
$H_2O + SO_3 \rightarrow H_2SO_4$

$2Mg$	+	O_2	→	$2MgO$
magnesium		oxygen		magnesium oxide

Decomposition Reactions

Examples:
$CaCO_3 \rightarrow CaO + CO_2$
$2H_2O \rightarrow 2H_2 + O_2$
$2KClO_3 \rightarrow 2KCl + 3O_2$

$2H_2O_2$	→	$2H_2O$	+	O_2
hydrogen peroxide		water		oxygen

Replacement Reactions 🔑

Single Replacement

Examples:
Fe + CuSO$_4$ → FeSO$_4$ + Cu
Zn + 2HCl + ZnCl$_2$ + H$_2$

2AgNO$_3$ + Cu → Cu(NO$_3$)$_2$ + 2Ag
silver nitrate   copper   copper nitrate   silver

Double Replacement

Examples:
NaCl + AgNO$_3$ → NaNO$_3$ + AgCl
HCl + FeS → FeCl$_2$ + H$_2$S

Pb(NO$_3$)$_2$ +   2KI   →   2KNO$_3$ +   PbI$_2$
lead nitrate   potassium iodide   potassium nitrate   lead iodide

Figure 7 In each of these reactions, an atom or group of atoms replaces another atom or group of atoms.

Combustion Reactions 🔑

substance + O$_2$ → substance(s)

C$_3$H$_8$ + 5O$_2$ → 3CO$_2$ + 4H$_2$O
propane   oxygen   carbon   water
                    dioxide

Example:
2C$_4$H$_{10}$ + 13O$_2$ → 8CO$_2$ + 10H$_2$O

Figure 8 Combustion reactions always contain oxygen (O$_2$) as a reactant and often produce carbon dioxide (CO$_2$) and water (H$_2$O).

Replacement

In a replacement reaction, an atom or group of atoms replaces part of a compound. There are two types of replacement reactions. *In a **single-replacement** reaction, one element replaces another element in a compound.* In this type of reaction, an element and a compound react and form a different element and a different compound. *In a **double-replacement** reaction, the negative ions in two compounds switch places, forming two new compounds.* In this type of reaction, two compounds react and form two new compounds. **Figure 7** describes these replacement reactions.

Combustion

Combustion *is a chemical reaction in which a substance combines with oxygen and releases energy.* This energy usually is released as thermal energy and light energy. For example, burning is a common combustion reaction. The burning of fossil fuels, such as the propane (C$_3$H$_8$) shown in **Figure 8,** produces the energy we use to cook food, power vehicles, and light cities.

Active Reading 4. **Select** What are the two types of energy typically released during a combustion reaction? (Circle) your answer in the text above.

Visual Summary

Chemical reactions are classified according to patterns seen in their reactants and products.

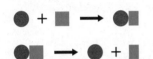

In a synthesis reaction, there are two or more reactants and one product. A decomposition reaction is the opposite of a synthesis reaction.

In replacement reactions, an element, or elements, in a compound is replaced with another element or elements.

Use Vocabulary

1 **Contrast** synthesis and decomposition reactions using a diagram.

2 A reaction in which parts of two substances switch places and make two new substances is a(n) _____.

Understand Key Concepts 🔑

3 **Classify** the following reaction: $2Na + Cl_2 \rightarrow 2NaCl$
   - (A) combustion
   - (B) decomposition
   - (C) single replacement
   - (D) synthesis

4 Write a balanced equation that produces Na and Cl_2 from NaCl. Classify this reaction.

5 **Classify** Which two types of reactions describe this reaction:

$$2SO_2 + O_2 \rightarrow 2SO_3$$

_____

_____

Interpret Graphics

6 **Complete** this table to identify four types of chemical reactions and the patterns shown by the reactants and the products. **LA.8.2.2.3**

Type of Reaction	Pattern of Reactants and Products
Synthesis	at least two reactants; one product

Critical Thinking

7 **Infer** The combustion of methane (CH_4) produces energy. Where do you think this energy comes from?

_____

_____

_____

_____

How Does a Light Stick Work?

What makes it glow?

A light stick consists of a plastic tube with a glass tube inside it. Hydrogen peroxide fills the glass tube.

A solution of phenyl oxalate ester and fluorescent dye surrounds the glass tube.

When you bend the outer plastic tube, the inner glass tube breaks, causing the hydrogen peroxide, ester, and dye to combine.

When the solutions combine, they react. Energy produced by the reaction causes the electrons in the dye to produce light.

Glowing necklaces, bracelets, or sticks—chances are you've worn or used them. Light sticks—also known as glow sticks—come in brilliant colors and provide light without electricity or batteries. Because they are lightweight, portable, and waterproof, they provide an ideal light source for campers, scuba divers, and other people participating in activities for which electricity is not readily available. Light sticks also are useful in emergency situations in which an electric current from battery-powered lights could ignite a fire.

Light sticks give off light because of a chemical reaction that happens inside the tube. During the reaction, energy is released as light. This is known as chemiluminescence (ke mee lew muh NE sunts).

It's Your Turn

RESEARCH AND REPORT Research bioluminescent organisms, such as fireflies and sea animals. How is the reaction that occurs in these organisms similar to or different from that in a glow stick? Work in small groups, and present your findings to the class. LA.8.4.4.2

Energy Changes and Chemical REACTIONS

Vocabulary

endothermic p. 395

exothermic p. 395

activation energy p. 396

catalyst p. 398

enzyme p. 398

inhibitor p. 398

 Florida NGSSS

LA.8.2.2.3 The student will organize information to show understanding or relationships among facts, ideas, and events (e.g., representing key points within text through charting, mapping, paraphrasing, summarizing, or comparing/contrasting);

MA.6.A.3.6 Construct and analyze tables, graphs, and equations to describe linear functions and other simple relations using both common language and algebraic notation.

MA.6.S.6.2 Select and analyze the measures of central tendency or variability to represent, describe, analyze, and/or summarize a data set for the purposes of answering questions appropriately.

SC.8.P.8.8 Identify basic examples of and compare and classify the properties of compounds, including acids, bases, and salts.

SC.8.P.9.3 Investigate and describe how temperature influences chemical changes.

SC.8.N.1.6 Understand that scientific investigations involve the collection of relevant empirical evidence, the use of logical reasoning, and the application of imagination in devising hypotheses, predictions, explanations and models to make sense of the collected evidence.

Inquiry Launch Lab   SC.8.P.9.3, MA.6.S.6.2

20 minutes

Where's the heat?

Does a chemical change always produce a temperature increase?

Procedure

1. Read and complete a lab safety form.
2. Copy the table on a sheet of paper.
3. Use a **graduated cylinder** to measure 25 mL of **citric acid solution** into a **foam cup**. Record the temperature with a **thermometer**.
4. Use a **spoon** to add a spoonful of **dry sodium bicarbonate** to the cup. Stir.
5. Record the temperature every 15 s until it stops changing. Record your observations during the reaction.
6. Add 25 mL of **sodium bicarbonate solution** to another cup. Record the temperature. Add a spoonful of **calcium chloride**. Repeat step 5.

Time	Temperature (°C)	
	Citric Acid Solution	Sodium Bicarbonate Solution
Starting temp.		
15 s		
30 s		
45 s		
1 min		
1 min, 15 s		
1 min, 30 s		
1 min, 45 s		
2 min		
2 min, 15 sec		

Think About This

1. **Explain** What evidence do you have that the changes in the two cups were chemical reactions?

2. **Describe** Account for the temperature changes in the two cups.

3. **Key Concept** **Infer** Based on your experiences, would a change in temperature convince you that a chemical change had taken place? Why or why not? What else could cause a temperature change?

Energy Changes

What is about 1,500 times heavier than a typical car and 300 times faster than a roller coaster? Do you need a hint? The energy it needs to move this fast comes from a chemical reaction that produces water. If you guessed a space shuttle, you are right!

It takes a large amount of energy to launch a space shuttle. The shuttle's main engines burn almost 2 million liters of liquid hydrogen and liquid oxygen. This chemical reaction produces water vapor and a large amount of energy. The energy produced heats the water vapor to high temperatures, causing it to expand rapidly. When the water expands, it pushes the shuttle into orbit. Where does all this energy come from?

Chemical Energy in Bonds

Recall that when a chemical reaction occurs, chemical bonds in the reactants break and new chemical bonds form. Chemical bonds contain a form of energy called chemical energy. Breaking a bond absorbs energy from the surroundings. The formation of a chemical bond releases energy to the surroundings. Some chemical reactions release more energy than they absorb. Some chemical reactions absorb more energy than they release. You can feel this energy change as a change in the temperature of the surroundings. Keep in mind that in all chemical reactions, energy is conserved.

 Inquiry **Energy from Bonds?**

1. On April 12, 1981, the space shuttle *Columbia* became the first space shuttle to orbit Earth. What was the source of the energy that produced the deafening roar, the blinding light, and the power to lift the 2-million-kg spacecraft? The energy stored in chemical bonds and the reactions that released that energy carried *Columbia* on its mission. What is a chemical bond? Where can you find these chemical bonds that can release so much energy?

_____

_____

_____

_____

Active Reading **2. Explain** Fill in the blanks to describe the energy changes that occur during a chemical reaction.

When bonds between atoms break, energy

_____ .

When new bonds form, energy

_____ .

Active Reading **3. Identify** What type of energy is contained in a chemical bond?

_____

Endothermic Reactions—Energy Absorbed

Have you ever heard someone say that the sidewalk was hot enough to fry an egg? To fry, the egg must absorb energy. *Chemical reactions that absorb thermal energy are* **endothermic** *reactions.* For an endothermic reaction to continue, energy must be added constantly.

$$\text{reactants} + \text{thermal energy} \rightarrow \text{products}$$

In an endothermic reaction, more energy is required to break the bonds of the reactants than is released when the products form. Therefore, the overall reaction absorbs energy. The reaction on the left in **Figure 9** is an endothermic reaction.

Exothermic Reactions—Energy Released

Most chemical reactions release energy as opposed to absorbing it. *An* **exothermic** *reaction is a chemical reaction that releases thermal energy.*

$$\text{reactants} \rightarrow \text{products} + \text{thermal energy}$$

In an exothermic reaction, more energy is released when the products form than is required to break the bonds in the reactants. Therefore, the overall reaction releases energy. The reaction shown on the right in **Figure 9** is exothermic.

Figure 9 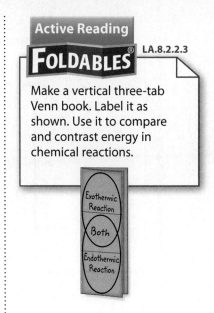 Whether a reaction is endothermic or exothermic depends on the amount of energy contained in the bonds of the reactants and in the bonds of the products.

Active Reading

FOLDABLES® LA.8.2.2.3

Make a vertical three-tab Venn book. Label it as shown. Use it to compare and contrast energy in chemical reactions.

Exothermic Reaction

Both

Endothermic Reaction

WORD ORIGIN ———

exothermic

from Greek *exo–*, means "outside"; and *therm*, means "heat"

Active Reading **4. Summarize** What happens to energy during an endothermic and an exothermic reaction? Fill in the blanks in **Figure 9**.

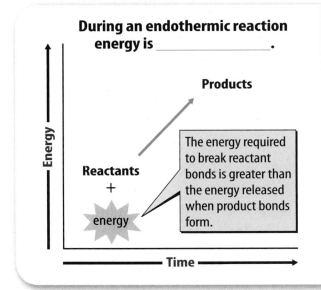

During an endothermic reaction energy is _____.

Products

The energy required to break reactant bonds is greater than the energy released when product bonds form.

Reactants +

energy

Energy

Time

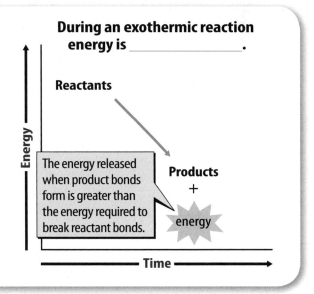

During an exothermic reaction energy is _____.

Reactants

The energy released when product bonds form is greater than the energy required to break reactant bonds.

Products +

energy

Energy

Time

Active Reading **5. Explain** Why does one orange arrow point upward and the other points downward in these diagrams?

_____

_____

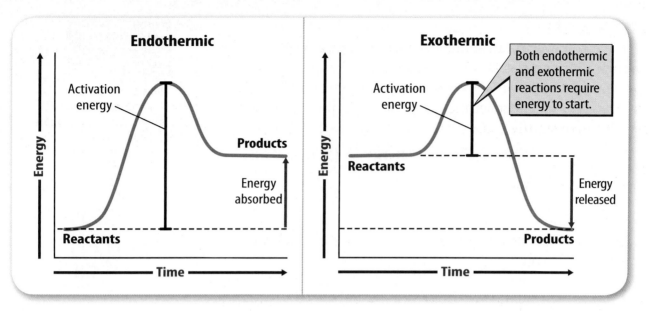

Endothermic

Activation energy

Energy

Products

Energy absorbed

Reactants

Time

Exothermic

Activation energy

Energy

Both endothermic and exothermic reactions require energy to start.

Reactants

Energy released

Products

Time

Figure 10 🔑 Both endothermic and exothermic reactions require activation energy to start the reaction.

Active Reading 6. **Explain** How can a reaction absorb energy to start, but remain exothermic?

_____

_____

_____

_____

_____

_____

Active Reading 7. **Record** On the lines below, write two factors about particle collisions that affect reaction rate.

_____

_____

_____

_____

_____

_____

Activation Energy

You might have noticed that some chemical reactions do not start by themselves. For example, a newspaper does not burn when it comes into contact with oxygen in air. However, if a flame touches the paper, it starts to burn.

All reactions require energy to start the breaking of bonds. This energy is called activation energy. **Activation energy** _is the minimum amount of energy needed to start a chemical reaction._ Different reactions have different activation energies. Some reactions, such as the rusting of iron, have low activation energy. The energy in the surroundings is enough to start these reactions. If a reaction has high activation energy, more energy is needed to start the reaction. For example, wood requires the thermal energy of a flame to start burning. Once the reaction starts, it releases enough energy to keep the reaction going. **Figure 10** shows the role activation energy plays in endothermic and exothermic reactions.

Reaction Rates

Some chemical reactions, such as the rusting of a bicycle wheel, happen slowly. Other chemical reactions, such as the explosion of fireworks, happen in less than a second. The rate of a reaction is the speed at which it occurs. What controls how fast a chemical reaction occurs? Recall that particles must collide before they can react. Chemical reactions occur faster if particles collide more often or move faster when they collide. Several factors affect how often particles collide and how fast particles move.

Surface Area

Surface area is the amount of exposed, outer area of a solid. Increased surface area increases reaction rate because more particles on the surface of a solid come into contact with the particles of another substance. For example, if you place a piece of chalk in vinegar, the chalk reacts slowly with the acid. This is because the acid contacts only the particles on the surface of the chalk. But, if you grind the chalk into powder, more chalk particles contact the acid, and the reaction occurs faster.

Temperature

Imagine a crowded hallway. If everyone in the hallway were running, they would probably collide with each other more often and with more energy than if everyone were walking. This is also true when particles move faster. At higher temperatures, the average speed of particles is greater. This speeds reactions in two ways. First, particles collide more often. Second, collisions with more energy are more likely to break chemical bonds.

Concentration and Pressure

Think of a crowded hallway again. Because the concentration of people is higher in the crowded hallway than in an empty hallway, people probably collide more often. Similarly, increasing the concentration of one or more reactants increases collisions between particles. More collisions result in a faster reaction rate. In gases, an increase in pressure pushes gas particles closer together. When particles are closer together, more collisions occur. Factors that affect reaction rate are shown in **Figure 11.**

Inquiry LAB STATION Try It!

MiniLab *Can you speed up a reaction?* at connectED.mcgraw-hill.com

Apply It! After you complete the lab, answer the question below.

1. Chemical reactions within some bacteria produce poisons that are harmful to humans. Explain why some foods are stored in a refrigerator or freezer.

Figure 11 🔑 Several factors can affect reaction rate.

Slower Reaction Rate

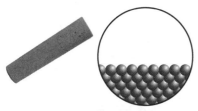

Less surface area

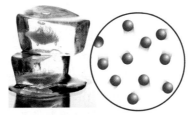

Lower temperature

Lower concentration

Faster Reaction Rate

More surface area

Higher temperature

Higher concentration

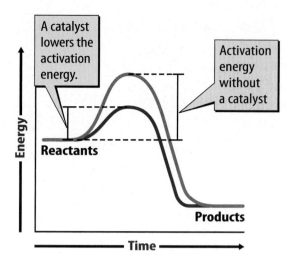

Figure 12 The blue line shows how a catalyst can increase the reaction rate.

Catalysts

A **catalyst** *is a substance that increases reaction rate by lowering the activation energy of a reaction.* One way catalysts speed reactions is by helping reactant particles contact each other more often. Look at **Figure 12.** Notice that the activation energy of the reaction is lower with a catalyst than it is without a catalyst. A catalyst isn't changed in a reaction, and it doesn't change the reactants or products. Also, a catalyst doesn't increase the amount of reactant used or the amount of product that is made. It only makes a given reaction happen faster. Therefore, catalysts are not considered reactants in a reaction.

You might be surprised to know that your body is filled with catalysts called enzymes. *An* **enzyme** *is a catalyst that speeds up chemical reactions in living cells.* For example, the enzyme protease (PROH tee ays) breaks the protein molecules in the food you eat into smaller molecules that can be absorbed by your intestine. Without enzymes, these reactions would occur too slowly for life to exist.

Inhibitors

Recall than an enzyme is a molecule that speeds reactions in organisms. However, some organisms, such as bacteria, are harmful to humans. Some medicines contain molecules that attach to enzymes in bacteria. This keeps the enzymes from working properly. If the enzymes in bacteria can't work, the bacteria die and can no longer infect a human. The active ingredients in these medicines are called inhibitors. *An* **inhibitor** *is a substance that slows, or even stops, a chemical reaction.* Inhibitors can slow or stop the reactions caused by enzymes.

Inhibitors are also important in the food industry. Preservatives in food are substances that inhibit, or slow down, food spoilage.

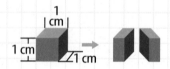

Visual Summary

Endothermic

Chemical reactions that release energy are exothermic, and those that absorb energy are endothermic.

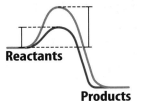

Activation energy must be added to a chemical reaction for it to proceed.

Catalysts, including enzymes, speed up chemical reactions. Inhibitors slow them down.

Use Vocabulary

1. The smallest amount of energy required by reacting particles for a chemical reaction to begin is the _____.

Understand Key Concepts

2. How does a catalyst increase reaction rate?
   - (A) by increasing the activation energy
   - (B) by increasing the amount of reactant
   - (C) by increasing the contact between particles
   - (D) by increasing the space between particles

3. **Explain** how increasing temperature affects reaction rate. SC.8.P.9.3

4. **Explain** When propane burns, heat and light are produced. Where does this energy come from?

Interpret Graphics

5. **List** Complete the graphic organizer to describe four ways to increase the rate of a reaction.

Increase reaction rate

Critical Thinking

6. **Infer** Explain why a catalyst does not increase the amount of product that can form.

Math Skills   MA.6.A.3.6

7. An object measures 1 cm × 1 cm × 3 cm.
   a. What is the surface area of the object?
   b. What is the total surface area if you cut the object into three equal cubes?

Chapter 10 Study Guide

Think About It! Matter can undergo a variety of changes. When matter is changed chemically, a rearrangement of bonds between the atoms occurs. This results in new substances with new properties.

 ## Key Concepts Summary

LESSON 1 Understanding Chemical Reactions

- There are several signs that a **chemical reaction** might have occurred, including a change in temperature, a release of light, a release of gas, a change in color or odor, and the formation of a solid from two liquids.

- In a chemical reaction, atoms of **reactants** rearrange and form **products**.

- The total mass of all the reactants is equal to the total mass of all the products in a reaction.

Reactants			Products		
1 Na:	4 H:	Atoms are equal.	1 Na:	2 H:	1 C:
1 H:	2 C:		2 C:	1 O:	2 O:
1 C:	2 O:		3 H:		
3 O:			2 O:		

chemical reaction p. 377

chemical equation p. 380

reactant p. 381

product p. 381

law of conservation of mass p. 382

coefficient p. 384

LESSON 2 Types of Chemical Reactions

- Most chemical reactions fit into one of a few main categories—synthesis, decomposition, combustion, and single- or double-replacement.

- **Synthesis** reactions create one product. **Decomposition** reactions start with one reactant. **Single-** and **double-replacement** reactions involve replacing one element or group of atoms with another element or group of atoms. **Combustion** reactions involve a reaction between one reactant and oxygen, and they release thermal energy.

synthesis p. 389

decomposition p. 389

single replacement p. 390

double replacement p. 390

combustion p. 390

LESSON 3 Energy Changes and Chemical Reactions

- Chemical reactions always involve breaking bonds, which requires energy, and forming bonds, which releases energy.

- In an **endothermic** reaction, the reactants contain less energy than the products. In an **exothermic** reaction, the reactants contain more energy than the products.

Less surface area   More surface area

- The rate of a chemical reaction can be increased by increasing the surface area, the temperature, or the concentration of the reactants or by adding a **catalyst**.

endothermic p. 395

exothermic p. 395

activation energy p. 396

catalyst p. 398

enzyme p. 398

inhibitor p. 398

Active Reading
FOLDABLES® Chapter Project

Assemble your lesson Foldables as shown to make a Chapter Project. Use the project to review what you have learned in this chapter.

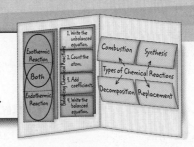

Use Vocabulary

1 When water forms from hydrogen and oxygen, water is the _____.

2 A(n) _____ uses symbols instead of words to describe a chemical reaction.

3 In a(n) _____ reaction, one element replaces another element in a compound.

4 When Na_2CO_3 is heated, it breaks down into CO_2 and Na_2O in a(n) _____ reaction.

5 The chemical reactions that keep your body warm are _____ reactions.

6 Even exothermic reactions require _____ to start.

Link Vocabulary and Key Concepts

Concepts in Motion Interactive Concept Map

Copy this concept map, and then use vocabulary terms from the previous page and other terms from the chapter to complete the concept map.

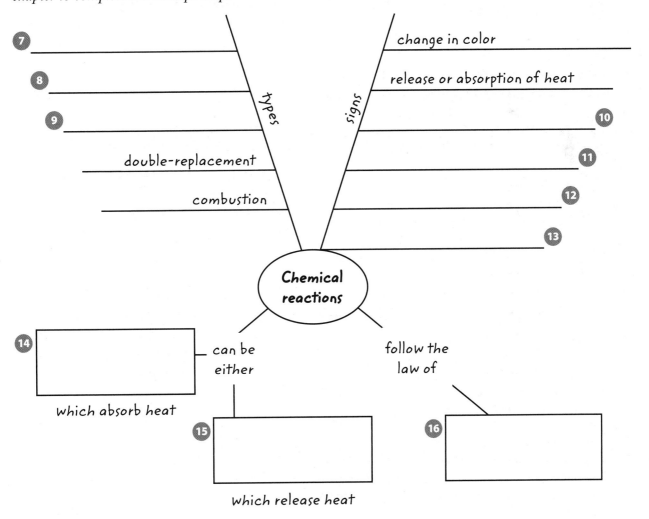

Chapter 10   Review

Fill in the correct answer choice.

Understand Key Concepts

1. How many carbon atoms react in this equation? SC.8.P.9.1

$$2C_4H_{10} + 13O_2 \rightarrow 8CO_2 + 10H_2O$$

   - (A) 2
   - (B) 4
   - (C) 6
   - (D) 8

2. The chemical equation below is unbalanced.

$$Zn + HCl \rightarrow ZnCl_2 + H_2$$

   Which is the correct balanced chemical equation? SC.8.P.9.1
   - (A) $Zn + H_2Cl_2 \rightarrow ZnCl_2 + H_2$
   - (B) $Zn + HCl \rightarrow ZnCl + H$
   - (C) $2Zn + 2HCl \rightarrow ZnCl_2 + H_2$
   - (D) $Zn + 2HCl \rightarrow ZnCl_2 + H_2$

3. When iron combines with oxygen gas and forms rust, the total mass of the products SC.8.P.9.1
   - (A) depends on the reaction conditions.
   - (B) is less than the mass of the reactants.
   - (C) is the same as the mass of the reactants.
   - (D) is greater than the mass of the reactants.

4. Potassium nitrate forms potassium oxide, nitrogen, and oxygen in certain fireworks. SC.8.P.9.1

$$4KNO_3 \rightarrow 2K_2O + 2N_2 + 5O_2$$

   This reaction is classified as a
   - (A) combustion reaction.
   - (B) decomposition reaction.
   - (C) single-replacement reaction.
   - (D) synthesis reaction.

5. Which type of reaction is the reverse of a decomposition reaction? SC.8.P.9.1
   - (A) combustion
   - (B) synthesis
   - (C) double-replacement
   - (D) single-replacement

Critical Thinking

6. **Predict** The diagram below shows two reactions—one with a catalyst (blue) and one without a catalyst (orange).

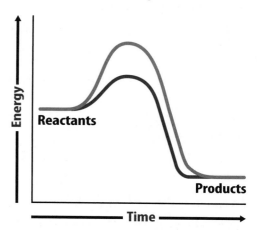

   How would the blue line change if an inhibitor were used instead of a catalyst? MA.6.A.3.6

   _____

   _____

   _____

7. **Analyze** A student observed a chemical reaction and collected the following data:

Observations before the reaction	A white powder was added to a clear liquid.
Observations during the reaction	The reactants bubbled rapidly in the open beaker.
Mass of reactants	4.2 g
Mass of products	4.0 g

   The student concludes that mass was not conserved in the reaction. Explain why this is not a valid conclusion. What might explain the difference in mass? SC.8.P.9.1

   _____

   _____

   _____

   _____

   _____

8 **Explain Observations** How did the discovery of atoms explain the observation that the mass of the products always equals the mass of the reactants in a reaction? SC.8.P.9.1

_____
_____
_____
_____
_____
_____
_____

Writing in Science

9 **Write** On a separate sheet of paper, write instructions that explain the steps in balancing a chemical equation. Use the following equation as an example. LA.8.2.2.3

$$MnO_2 + HCl \rightarrow MnCl_2 + H_2O + Cl_2$$

Big Idea Review

10 Explain how atoms and energy are conserved in a chemical reaction. SC.8.P.9.1

_____
_____
_____
_____
_____

11 When a car air bag inflates, sodium azide (NaN_3) decomposes and produces nitrogen gas (N_2) and another product. What element does the other product contain? How do you know? SC.8.P.9.1

_____
_____
_____
_____
_____

Math Skills MA.6.A.3.6

12 What is the surface area of the cube shown below? What would the total surface area be if you cut the cube into 27 equal cubes?

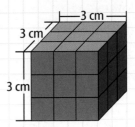

13 Suppose you have ten cubes that measure 2 cm on each side.

a. What is the total surface area of the cubes?

b. What would the surface area be if you glued the cubes together to make one object that is two cubes wide, one cube high, and five cubes long? Hint: Draw a picture of the final cube and label the length of each side.

Fill in the correct answer choice.

Multiple Choice

1 Which is NOT a chemical change? SC.8.P.9.2

Ⓐ copper turning green in air

Ⓑ baking a cake

Ⓒ drying clothes

Ⓓ exploding dynamite

Use the figure below to answer question 2.

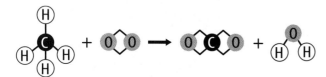

2 The figure above shows models of molecules in a chemical reaction. How would you describe this reaction? SC.8.P.9.2

Ⓕ a chemical change

Ⓖ a physical change

Ⓗ a chemical property

Ⓘ a physical property

3 The reaction between hydrogen chloride and sodium hydroxide is shown below. The name and the mass of each substance involved in the chemical reaction are shown. What mass of hydrogen chloride was used in this reaction? SC.8.P.9.1

? g	**40.0 g**	**58.5 g**	**18.0 g**
HCl +	**NaOH** →	**NaCl** +	**H₂O**
hydrogen chloride	sodium hydroxide	sodium chloride	water

Ⓐ 18.5 g

Ⓑ 36.5 g

Ⓒ 40.5 g

Ⓓ 116.5 g

4 Which occurs before new bonds can form during a chemical reaction? SC.8.P.9.1

Ⓕ The atoms in the original substances are destroyed.

Ⓖ The bonds between atoms in the original substances are broken.

Ⓗ The atoms in the original substances are no longer moving.

Ⓘ The bonds between atoms in the original substances get stronger.

5 Which can be used to speed up a reaction? SC.8.P.9.3

Ⓐ increase temperature

Ⓑ decrease concentration

Ⓒ add an inhibitor

Ⓓ decrease temperature

6 Preparing a meal involves both physical and chemical changes. Which involves a chemical change? SC.8.P.9.2

Ⓕ boiling water

Ⓖ making ice cubes

Ⓗ slicing a carrot

Ⓘ toasting bread

7 The law of conservation of mass states that in a chemical reaction, the mass of the reactants equals the mass of the products. If approximately 20 g of water reacts in the equation below, producing about 15 g of oxygen, what mass of hydrogen (H_2) is produced in this reaction? SC.8.P.9.1

$$2H_2O \rightarrow 2H_2 + O_2$$

Ⓐ 2 g

Ⓑ 5 g

Ⓒ 10 g

Ⓓ 15 g

8 Why does an oven's high temperature speed up chemical reactions? SC.8.P.9.3

(F) because heat lowers the activation energy

(G) because heat activates catalysts

(H) because heat makes molecules collide with each other

(I) because heat reduces the particle size of the reactants

9 Which is an example of a chemical change? SC.8.P.9.2

(A) boiling

(B) burning

(C) evaporation

(D) melting

10 Which statement best describes the law of conservation of mass? SC.8.P.9.1

(F) The mass of the products is always greater than the mass of the materials that react in a chemical change.

(G) The mass of the products is always less than the mass of the materials in a chemical change.

(H) A certain mass of material must be present for a reaction to occur.

(I) Matter is neither lost nor gained during a chemical change.

11 Antoine Lavoisier is credited with the discovery of the law of conservation of mass. The law states that in a chemical reaction, matter is not created or destroyed but preserved. Which equation correctly models this law? SC.8.P.9.1

(A) $H_2 + Cl_2 \rightarrow 2HCl$

(B) $H_2 + Cl_2 \rightarrow HCl$

(C) $H + Cl \rightarrow 2HCl$

(D) $2H_2 + Cl_2 \rightarrow 2HCl_2$

12 When a newspaper is left in direct sunlight for a few days, the paper begins to turn yellow? What is this change in color? SC.8.P.9.2

(F) physical property

(G) chemical property

(H) physical change

(I) chemical change

13 How does increasing the temperature affect a chemical reaction? SC.8.P.9.3

(A) The average speed of the particles speeds up causing more particle collisions.

(B) The average speed of the particles speeds up causing fewer particle collisions.

(C) The average speed of the particles slows down causing more particle collisions.

(D) The average speed of the particles slows down causing fewer particle collisions.

NEED EXTRA HELP?

If You Missed Question...	1	2	3	4	5	6	7	8	9	10	11	12	13
Go to Lesson...	1	1	1	1	2	1	1	3	1	1	1	1	2

Multiple Choice *Bubble the correct answer.*

1. The diagram above is a model of the compound **SC.8.P.8.5**

   (A) glucose, $C_6H_{12}O_6$.

   (B) hydrogen peroxide, H_2O_2.

   (C) magnesium hydroxide, $Mg(OH)_2$.

   (D) sodium chloride, $NaCl$.

2. Jennifer and Cody are mixing substances during a lab experiment. They observe the four effects listed below. Which of these effects indicates that a chemical reaction has taken place between two substances? **SC.8.P.9.2**

   (F) A clear, lemon-scented liquid continues to smell like lemon when it is mixed with water.

   (G) A piece of shiny metal begins to bubble when placed in a clear liquid.

   (H) A pink powdered substance turns water pink when they are mixed together.

   (I) A white solid disappears when mixed with water.

3. In which equation is carbon dioxide a product? **SC.8.P.8.5**

   (A) $CO_2 \rightarrow C + O_2$

   (B) $CH_4 + 2O_2 \rightarrow CO_2 + 2H_2O$

   (C) $6CO_2 + 6H_2O \rightarrow C_6H_{12}O_6 + 6O_2$

   (D) $2H_2O \rightarrow 2H_2 + O_2$

   Reactants   **Product**

   __H_2 +   __O_2 $\longrightarrow$ __H_2O

4. What is the balanced equation that would represent the reaction above? **SC.8.P.9.1**

   (F) $H_2 + O_2 = H_2O$

   (G) $H_2 + 2O_2 = 2H_2O$

   (H) $2H_2 + O_2 = 2H_2O$

   (I) $4H_2 + 4O_2 = 4H_2O$

mini BAT

Multiple Choice *Bubble the correct answer.*

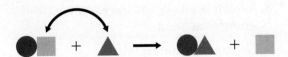

1. Which reaction could be modeled by the illustration above? **SC.8.P.9.1**

   (A) $CaCO_3 \rightarrow CaO + CO_2$

   (B) $H_2O + SO_3 \rightarrow H_2SO_4$

   (C) $NaCl + AgNO_3 \rightarrow NaNO_3 + AgCl$

   (D) $Zn + 2HCl \rightarrow ZnCl_2 + H_2$

2. Which type of reaction forms water from hydrogen and oxygen atoms? **SC.8.P.8.5**

   (F) decomposition reaction

   (G) synthesis reaction

   (H) double-replacement reaction

   (I) single-replacement reaction

3. NaCl and $AgNO_3$ react in a double replacement reaction. Which is one of the products of this reaction? **SC.8.P.8.5**

   (A) $AgClO_3$

   (B) CO_2

   (C) $NaAg$

   (D) $NaNO_3$

   $$2C_4H_{10} + 13O_2 \rightarrow 8CO_2 + 10H_2O$$

4. Which is true about the reaction shown above? **SC.8.P.8.4**

   (F) The reaction is a single replacement.

   (G) The reaction releases energy.

   (H) The reaction has negative ions in two of the compounds that switch places.

   (I) The reaction has two elements that combine to form a compound.

Multiple Choice *Bubble the correct answer.*

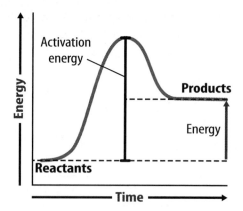

1. Which describes energy in the reaction represented by the graph above? **SC.8.P.9.3**

(A) Energy is required to start the reaction; energy is then absorbed.

(B) Energy is required to start the reaction; energy is then released.

(C) No energy is required to start the reaction; energy is then absorbed.

(D) No energy is required to start the reaction; energy is then released.

2. Which represents an endothermic reaction? **SC.8.P.9.3**

(F) burning wood

(G) exploding firecracker

(H) melting ice

(I) shining sun

3. Which reaction has the lowest activation energy? **SC.8.P.9.3**

(A) combustion of oil

(B) explosion of fireworks

(C) rotting of bananas

(D) synthesis of water

4. What is one way to increase the reaction rate of a chemical reaction?

(F) decrease surface area

(G) increase surface area

(H) lower concentration

(I) lower temperature

Notes

Unit 4

Matter and Energy Transformations

THE PLANT TAKES IN LIGHT ENERGY...

CARBON DIOXIDE...

AND WATER...

PHOTOSYNTHESIS PLANT

CO_2

H_2O

1700　　　　　　　　1800　　　　　　　　1900

1753
Swedish botanist Carl Linnaeus publishes *Species Plantarum,* a list of all plants known to him and the starting point of modern plant nomenclature.

1892
Russian botanist Dmitri Iwanowski discovers the first virus while studying tobacco mosaic disease. Iwanowski finds that the cause of the disease is small enough to pass through a filter made to trap all bacteria.

1930s
The development of commercial hybrid crops begins in the United States.

1983
Luc Montagnier's team at the Pasteur Institute in France isolates the retrovirus now called HIV.

Patterns

Have you ever seen an individual snowflake close-up? You might have heard someone say that no two snowflakes are alike. While this is true, it is also true that all snowflakes have similar patterns. A **pattern** is a consistent plan or model used as a guide for understanding or predicting things. Patterns can be created or occur naturally. The formation of snowflakes is an example of a pattern. They form as water drops in the air freeze into a six-sided crystal.

How Scientists Use Patterns

Studying and using patterns is useful to scientists because it can help explain the natural world or predict future events. A biologist might study patterns in DNA to predict what organisms will look like. A meteorologist might study cloud formation patterns to predict the weather. When doing research, scientists also try to match patterns found in their data with patterns that occur in nature. This helps to determine whether data are accurate and helps to predict outcomes.

Active Reading

1. **Review** How can patterns be useful to scientists?

_____

_____

Types of Patterns

Cyclic Patterns

A cycle, or repeated series of events, is a form of pattern. An organism's life cycle typically follows the pattern of birth, growth, and death. Scientists study an organism's life cycle to predict the life of its offspring.

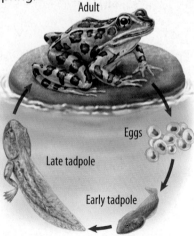

Adult

Eggs

Late tadpole

Early tadpole

Physical Patterns

Physical patterns have an artistic or decorative design. Physical patterns can occur naturally, such as the patterns in the colors on butterfly wings or flower petals, or they can be created intentionally, such as a design in a brick wall.

Patterns in Life Science

Why do police detectives or forensic scientists take fingerprints at a crime scene? They know that every fingerprint is unique. Fingerprints contain patterns that can help detectives narrow a list of suspects. The patterns on the fingerprints can then be examined more closely to identify an individual. No two humans have the same set of fingerprints, just as no two zebras have the same stripe pattern.

Patterns are an important key to understanding science and the natural world. They are found across all classifications of science. Patterns help scientists understand the genetic makeup, lifestyle, and similarities of various species of plants and animals. Patterns help determine weather, the relative age of rocks, forces of nature, and the orbits of the planets. Look around you and observe patterns in your world.

 2. Apply What patterns might a zoologist or botanist study?

_____

_____

Mathematical Patterns

Patterns are frequently applied in mathematics. Whenever you read a number, perform a mathematical operation, or describe a shape or graph, you are using patterns.

2, 5, 8, 11, ___, ___, ___

What numerals come next in this number pattern?

What will the next shape look like according to the pattern?

Inquiry LAB STATION Try It!   SC.8.N.1.6

MiniLab *Leaf Patterns* at connectED.mcgraw-hill.com

Apply It!

After you complete the lab, answer these questions.

1. **Illustrate** Design a pattern for each of the following types of patterns.

```
┌─────────────────────────┐
│                         │
│                         │
│                         │
│                         │
└─────────────────────────┘
         Cyclic Pattern
```

```
┌─────────────────────────┐
│                         │
│                         │
│                         │
│                         │
└─────────────────────────┘
        Physical Pattern
```

```
┌─────────────────────────┐
│                         │
│                         │
│                         │
│                         │
└─────────────────────────┘
       Mathematical Pattern
```

2. **Extend** What might a change in a young bird's feather pattern indicate? Support your reasoning.

Notes

Plant Processes

Four friends were comparing their ideas about plant processes. This is what they said:

Hildy: I think plants carry on photosynthesis but not cellular respiration.

Flo: I think plants carry on cellular respiration but not photosynthesis.

Al: I think plants carry on both cellular respiration and photosynthesis.

Tamir: I think plants carry on neither photosynthesis nor cellular respiration.

With whom do you agree most? Explain why you agree with that person.

Plant
PROCESSES

Think About It!

What processes enable plants to survive?

The common morning glory is native to the southeastern United States, including Florida. The tendril of this morning glory vine grows around the stem of another plant.

1. How do you think growing around another plant might help the morning glory plant survive?

_____

_____

_____

_____

_____

2. Can you think of any other processes that enable plants to survive?

_____

_____

_____

_____

_____

_____

Get Ready to Read

What do you think about plant processes?

Before you read, decide if you agree or disagree with each of these statements. As you read this chapter, see if you change your mind about any of the statements.

	AGREE	DISAGREE
1 Plants do not carry on cellular respiration.	☐	☐
2 Plants are the only organisms that carry on photosynthesis.	☐	☐
3 Plants make food in their underground roots.	☐	☐
4 Plants do not produce hormones.	☐	☐
5 Plants can respond to their environments.	☐	☐
6 All plants flower when nights are 10–12 hours long.	☐	☐

There's More Online!
Video • Audio • Review • ⓘLab Station • WebQuest • Assessment • Concepts in Motion • Multilingual eGlossary

Energy Processing in PLANTS

ESSENTIAL QUESTIONS

How do materials move inside plants?

How do plants perform photosynthesis?

What is cellular respiration?

How are photosynthesis and cellular respiration alike, and how are they different?

Vocabulary

photosynthesis p. 420

cellular respiration p. 422

Florida NGSSS

LA.8.2.2.3 The student will organize information to show understanding or relationships among facts, ideas, and events (e.g., representing key points within text through charting, mapping, paraphrasing, summarizing, or comparing/contrasting);

LA.8.4.2.2 The student will record information (e.g., observations, notes, lists, charts, legends) related to a topic, including visual aids to organize and record information, as appropriate, and attribute sources of information;

SC.8.N.1.1 Define a problem from the eighth grade curriculum using appropriate reference materials to support scientific understanding, plan and carry out scientific investigations of various types, such as systematic observations or experiments, identify variables, collect and organize data, interpret data in charts, tables, and graphics, analyze information, make predictions, and defend conclusions.

SC.8.N.1.6 Understand that scientific investigations involve the collection of relevant empirical evidence, the use of logical reasoning, and the application of imagination in devising hypotheses, predictions, explanations and models to make sense of the collected evidence.

SC.8.L.18.1 Describe and investigate the process of photosynthesis, such as the roles of light, carbon dioxide, water and chlorophyll; production of food; release of oxygen.

SC.8.L.18.2 Describe and investigate how cellular respiration breaks down food to provide energy and releases carbon dioxide.

 Launch Lab SC.8.N.1.6

20 minutes

How can you show the movement of materials inside a plant?

Most parts of plants need water. They also need a system to move water throughout the plant so cells can use it for plant processes.

Procedure

1 Read and complete a lab safety form.

2 Gently pull two stalks from the base of a bunch of **celery.** Leave one stalk complete. Use a **paring knife** to carefully cut directly across the bottom of the second stalk.

3 Pour 100 mL of water into each of two **beakers.** Place 3–4 drops of **blue food coloring** into the water. Place one celery stalk in each beaker.

4 After 20 min, observe the celery near the bottom of each stalk. Observe again after 24 h. Record your observations.

Data and Observations

Think About This

1. What happened in each celery stalk?

2. **Key Concept** What did the colored water do? Why do you think this occurred?

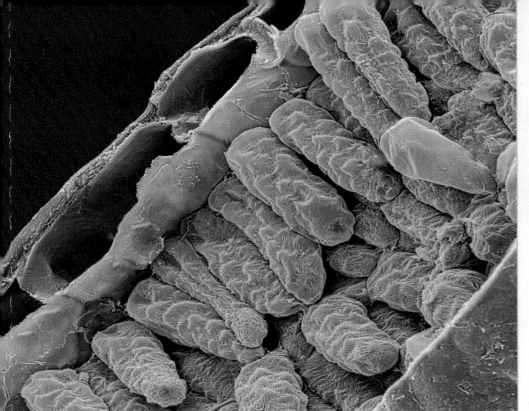

Inquiry **All Leaf Cells?**

1. You are looking at a magnified cross section of a leaf. As you can see, the cells in the middle of the leaf are different from the cells on the edges. What do you think this might have to do with the cellular processes a leaf carries out that enable a plant to survive?

_____

_____

_____

_____

_____

Materials for Plant Processes

Food, water, and oxygen are three things you need in order to survive. Some of your organ systems process these materials, and others transport them throughout your body. Like you, plants need food, water, and oxygen to survive. Unlike you, plants do not take in food. Most of them make their own.

Moving Materials Inside Plants

You might recall reading about xylem (ZI lum) and phloem (FLOH em)—the vascular tissue in most plants. These tissues transport materials throughout a plant.

After water enters a plant's roots, it moves into xylem. Water then flows inside xylem to all parts of a plant. Without enough water, plant cells wilt, as shown in **Figure 1.**

Most plants make their own food—a liquid sugar. The liquid sugar moves out of food-making cells, enters phloem, and flows to all plant cells. Cells break down the sugar and release energy. Some plant cells can store food.

Plants require oxygen and carbon dioxide to make food. Like you, plants produce water vapor as a waste product. Carbon dioxide, oxygen, and water vapor pass into and out of a plant through tiny openings in leaves.

Active Reading

3. Describe How do materials move through plants?

_____

_____

Active Reading

2. Summarize Make an outline of the information in the lesson on a separate sheet of paper. Use the main headings in the lesson as the main headings in your outline. Use your outline to review the lesson.

Figure 1 This plant wilted due to lack of water in the soil.

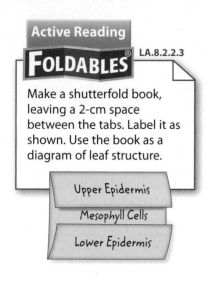
WORD ORIGIN

photosynthesis
from Greek *photo–*, means "light"; and *synthesis*, means "composition"

Active Reading **4. Identify** Which layer of cells contains vascular tissue? Place a (circle) around this structure in **Figure 2**.

Photosynthesis

Plants need food, but they cannot eat as people do. They make their own food, and leaves are the major food-producing organs of plants. This means that leaves are the sites of photosynthesis (foh toh SIHN thuh sus). **Photosynthesis** *is a series of chemical reactions that convert light energy, water, and carbon dioxide into the food-energy molecule glucose and give off oxygen.* The structure of a leaf is well-suited to its role in photosynthesis.

Leaves and Photosynthesis

As shown in **Figure 2,** leaves have many types of cells. The cells that make up the top and bottom layers of a leaf are flat, irregularly shaped cells called epidermal (eh puh DUR mul) cells. On the bottom epidermal layer of most leaves are small openings called stomata (STOH muh tuh). Carbon dioxide, water vapor, and oxygen pass through stomata. Epidermal cells can produce a waxy covering called the cuticle.

Most photosynthesis occurs in two types of mesophyll (ME zuh fil) cells inside a leaf. These cells contain chloroplasts, the organelle where photosynthesis occurs. Near the top surface of the leaf are palisade mesophyll cells. They are packed together. This arrangement exposes the most cells to light. Spongy mesophyll cells have open spaces between them. Gases needed for photosynthesis flow through the spaces between the cells.

Figure 2 Photosynthesis occurs inside the chloroplasts of mesophyll cells in most leaves.

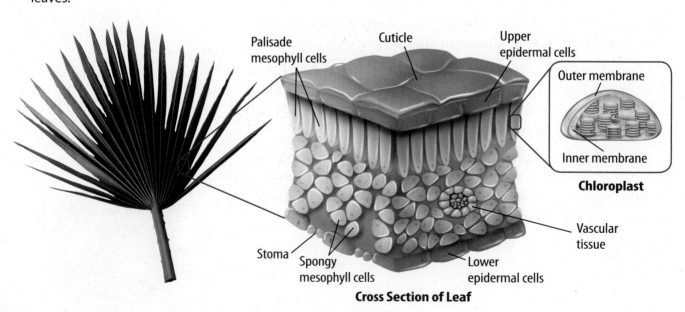

Palisade mesophyll cells

Cuticle

Upper epidermal cells

Outer membrane

Inner membrane

Chloroplast

Vascular tissue

Stoma

Spongy mesophyll cells

Lower epidermal cells

Cross Section of Leaf

Capturing Light Energy

As you read about the steps of photosynthesis, refer to **Figure 3** to help you understand the process. In the first step of photosynthesis, plants capture the energy in light. This occurs in chloroplasts. Chloroplasts contain plant pigments. Pigments are chemicals that can absorb and reflect light. Chlorophyll, the most common plant pigment, is necessary for photosynthesis. Most plants appear green because chlorophyll reflects green light. Chlorophyll absorbs other colors of light. This light energy is used during photosynthesis.

Once chlorophyll traps and stores light energy, this energy can be transferred to other molecules. During photosynthesis, water molecules are split apart. This releases oxygen into the atmosphere, as shown in **Figure 3**.

 **5. Explain** How do plants capture light energy?

_____

_____

Making Sugars

Sugars are made in the second step of photosynthesis. This step can occur without light. In chloroplasts, carbon dioxide from the air is converted into sugars by using the energy stored and trapped by chlorophyll. Carbon dioxide combines with hydrogen atoms from the splitting of water molecules and forms sugar molecules. Plants can use this sugar as an energy source, or they can store it. Potatoes and carrots are examples of plant structures where excess sugar is stored.

6. NGSSS Check Identify What are the two steps of photosynthesis? SC.8.L.18.1

_____

_____

_____

Why is photosynthesis important?

Try to imagine a world without plants. How would humans or other animals get the oxygen they need? Plants help maintain the atmosphere you breathe. Photosynthesis produces most of the oxygen in the atmosphere.

Photosynthesis 🔑

Figure 3 Photosynthesis is a series of complex chemical processes. The first step is capturing light energy. In the second step, that energy is used for making glucose, a type of sugar.

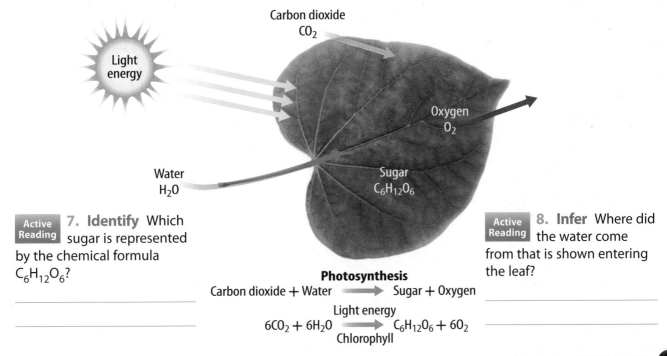

7. Identify Which sugar is represented by the chemical formula $C_6H_{12}O_6$?

_____

Photosynthesis

Carbon dioxide + Water $\longrightarrow$ Sugar + Oxygen

$$6CO_2 + 6H_2O \xrightarrow[\text{Chlorophyll}]{\text{Light energy}} C_6H_{12}O_6 + 6O_2$$

8. Infer Where did the water come from that is shown entering the leaf?

_____

Cellular Respiration

All organisms require **energy** to survive. Energy is in the chemical bonds in food **molecules**. A two-step process called cellular respiration releases energy. **Cellular respiration** *is a series of chemical reactions that convert the energy in food molecules into a usable form of energy called ATP.*

Why is cellular respiration important?

If your body did not break down the food you eat through cellular respiration, you would not have energy to do anything. Plants produce sugar, but without cellular respiration, plants could not grow, reproduce, or repair tissues.

Reactions in the Cytoplasm

The first step of cellular respiration is an anaerobic process because it does not require oxygen. During this step, glucose molecules, a type of sugar, are broken down into smaller molecules. This process produces some ATP molecules. Reactions during this step occur in the cytoplasm of the cell.

 9. NGSSS Check **Summarize** What is cellular respiration? SC.8.L.18.2

_____

_____

_____

ACADEMIC VOCABULARY

energy
(noun) usable power

REVIEW VOCABULARY

molecule
a group of atoms held together by the energy in chemical bonds

Inquiry **LAB STATION** **Try It!**

SC.8.N.1.1,
SC.8.L.18.1,
SC.8.L.18.2

MiniLab *Can you observe plant processes?* at connectED.mcgraw-hill.com

Apply It! After you complete the lab, answer these questions.

1. What gaseous element is required for cellular respiration?

2. What waste products does cellular respiration produce?

Table 1 Photosynthesis v. Cellular Respiration		
Process	Photosynthesis	Cellular Respiration
Reactants	light energy, CO_2, H_2O	glucose [sugar], O_2
Products	glucose, O_2	CO_2, H_2O, ATP
Organelle in which it occurs	chloroplasts	cytoplasm, mitochondria
Type of organism	photosynthetic organisms including plants and algae	most organisms, including plants and animals

Reactions in the Mitochondria

The second step of cellular respiration is called an aerobic process because it requires oxygen. The smaller molecules made from glucose during the first step are broken down. Large amounts of ATP, a usable energy, are produced. Cells use ATP to power all cellular processes. Two waste products—water and carbon dioxide (CO_2)—are given off during this step. More ATP is made during the second step of cellular respiration than during the first step. Reactions during this step occur in the mitochondria of the cell.

The CO_2 released by cells as a waste product is used by plants and some unicellular organisms in another process called photosynthesis.

Comparing Photosynthesis and Cellular Respiration

Photosynthesis requires light energy and the reactants—substances that react with one another during the process—carbon dioxide and water. Oxygen and the energy-rich molecule glucose are the products, or end substances, of photosynthesis. Most plants, some protists, and some bacteria carry on photosynthesis.

Cellular respiration requires the reactants glucose and oxygen, produces carbon dioxide and water, and releases energy in the form of ATP. Most organisms carry on cellular respiration. The reactants and the products of photosynthesis and cellular respiration are interrelated, as shown in **Table 1**. The products of photosynthesis—glucose and oxygen—become the reactants for cellular respiration. Some of the products of cellular respiration—carbon dioxide and water—then become the reactants for photosynthesis. Life on Earth depends on a balance of these two processes as shown in **Figure 4**.

 11. **NGSSS Check** Compare and Contrast How are photosynthesis and cellular respiration alike, and how are they different? SC.8.L.18.1, SC.8.L.18.2

_____

Table 1 🔑 The relationship between cellular respiration and photosynthesis is important for life.

✓ 10. **Visual Check** List What are the reactants of cellular respiration? What are the products?

_____

_____

Figure 4 The balance of cellular respiration and photosynthesis helps maintain the survival of the Florida marsh rabbit.

Materials that a plant requires to survive move through the plant in the vascular tissue, xylem and phloem.

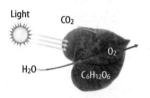

Plants can make their own food by using light energy, water, and carbon dioxide.

The products of photosynthesis are the reactants for cellular respiration.

Use Vocabulary

1 A series of chemical reactions that convert the energy in food molecules into a usable form of energy, called ATP, is called

_____. SC.8.L.18.2

2 **Define** *photosynthesis* in your own words. SC.8.L.18.1

_____

_____

Understand Key Concepts 🔑

3 Which structure moves water through plants?

Ⓐ chloroplast   Ⓒ nucleus

Ⓑ mitochondrion   Ⓓ xylem

4 **Describe** how plants use chlorophyll for photosynthesis. SC.8.L.18.1

_____

_____

_____

5 **Summarize** the process of cellular respiration. SC.8.L.18.2

_____

_____

Interpret Graphics

6 **Compare and Contrast** Fill in the table below to compare and contrast photosynthesis and cellular respiration.

Process	Similarities	Differences

Critical Thinking

7 **Predict** the effect of a plant disease that destroys all of the chloroplasts in a plant.

_____

_____

_____

8 **Evaluate** why plants perform cellular respiration. SC.8.L.18.2

_____

_____

Deforestation and Carbon Dioxide
in the Atmosphere

How does carbon dioxide affect climate?

What do you think when you hear the words *greenhouse gases?* Many people picture pollution from automobiles or factory smokestacks. It might be surprising to learn that cutting down forests affects the amount of one of the greenhouse gases in the atmosphere—carbon dioxide.

Deforestation is the term used to describe the destruction of forests. Deforestation happens because people cut down forests to use the land for other purposes, such as agriculture or building sites, or to use the trees for fuel or building materials.

Trees, like most plants, carry out photosynthesis and make their own food. Carbon dioxide from the atmosphere is one of the raw materials, or reactants, of photosynthesis. When deforestation occurs, trees are unable to remove carbon dioxide from the atmosphere. As a result, the level of carbon dioxide in the atmosphere increases.

Trees affect the amount of atmospheric carbon dioxide in other ways. Large amounts of carbon are stored in the molecules that make up trees. When trees are burned or left to rot, much of this stored carbon is released as carbon dioxide. This increases the amount of carbon dioxide in the atmosphere.

Carbon dioxide in the atmosphere has an impact on climate. Greenhouse gases, such as carbon dioxide, increase the amount of the Sun's energy that is absorbed by the atmosphere. They also reduce the ability of heat to escape back into space. So, when levels of carbon dioxide in the atmosphere increase, more heat is trapped in Earth's atmosphere. This can lead to climate change.

These cattle are grazing on land that once was part of a forest in Brazil.

In a process called slash-and-burn, forest trees are cut down and burned to clear land for agriculture.

It's Your Turn

RESEARCH AND REPORT How can we lower the rate of deforestation? What are some actions you can take that could help slow the rate of deforestation? Research to find out how you can make a difference. Make a poster to share what you learn.

LA.8.4.2.2

Plant RESPONSES

Vocabulary

stimulus p. 427

tropism p. 428

photoperiodism p. 430

plant hormone p. 431

 Florida NGSSS

LA.8.2.2.3 The student will organize information to show understanding or relationships among facts, ideas, and events (e.g., representing key points within text through charting, mapping, paraphrasing, summarizing, or comparing/contrasting);

MA.6.A.3.6 Construct and analyze tables, graphs, and equations to describe linear functions and other simple relations using both common language and algebraic notation.

MA.6.S.6.2 Select and analyze the measures of central tendency or variability to represent, describe, analyze, and/or summarize a data set for the purposes of answering questions appropriately.

SC.8.N.1.6 Understand that scientific investigations involve the collection of relevant empirical evidence, the use of logical reasoning, and the application of imagination in devising hypotheses, predictions, explanations and models to make sense of the collected evidence.

 Inquiry Launch Lab SC.8.N.1.6

15 minutes

How do plants respond to stimuli?

Plants use light energy and make their own food during photosynthesis. How else do plants respond to light in their environment?

Procedure

1. Read and complete a lab safety form.
2. Choose a **pot of young radish seedlings.**
3. Place **toothpicks** parallel to a few of the seedlings in the pot in the direction of growth.
4. Place the pot near a **light source**, such as a gooseneck lamp or next to a window. The light source should be to one side of the pot, not directly above the plants.
5. Check the position of the seedlings in relation to the toothpicks after 30 min. Record your observations.
6. Observe the seedlings when you come to class the next day.

Think About This

1. What happened to the position of the seedlings after the first 30 min? What is your evidence of change?

2. What happened to the position of the seedlings after a day?

3. **Key Concept** Why do you think the position of the seedlings changed?

Inquiry A Meat-Eating Plant?

1. The sundew is a carnivorous plant that is native to Florida. It has glistening drops of mucilage at the tip of each tentacle. The drops resemble morning dew. This plant can lure, capture, and digest its prey. When stimulated, the plant will wrap its tentacles around the trapped prey. To what other stimuli do you think plants might respond?

_____

_____

_____

_____

Stimuli and Plant Responses

Have you ever been in a dark room when someone suddenly turned on the light? You might have reacted by quickly shutting or covering your eyes. **Stimuli** (STIM yuh li; singular, stimulus) *are any changes in an organism's environment that cause a response.*

Often a plant's response to a stimulus might be so slow that it is hard to see it happen. The response might occur gradually over a period of hours or days. Light is a stimulus. A plant responds to light by growing toward it, as shown in **Figure 5.** This response occurs over several hours.

In some cases, the response to a stimulus is quick, such as a Venus flytrap's response to touch. When stimulated by an insect's touch, the two sides of the trap snap shut immediately, trapping the insect inside.

Active Reading **3. Express** Why is it sometimes hard to see a plant's response to a stimulus?

_____

_____

Active Reading **2. Identify** Write a phrase beside each paragraph that summarizes the main point of the paragraph. Use the phrases to review the lesson.

Figure 5 The light is the stimulus, and the seedlings have responded by growing toward the light.

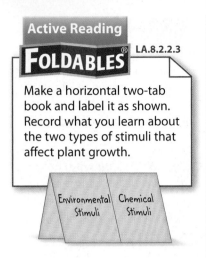
WORD ORIGIN

tropism
from Greek *tropos*, means "turn" or "turning"

Environmental Stimuli

When it is cold outside, you probably wear a sweatshirt or a coat. Plants cannot put on warm clothes, but they do respond to their environments in a variety of ways. You might have seen trees flower in the spring or drop their leaves in the fall. Both are plant responses to environmental stimuli.

Growth Responses

Plants respond to a number of different environmental stimuli. These include light, touch, and gravity. *A* **tropism** (TROH pih zum) *is a response that results in plant growth toward or away from a stimulus.* When the growth is toward a stimulus, the tropism is called positive. A plant bending toward light is a positive tropism. Growth away from a stimulus is considered negative. A plant's stem growing upward against gravity is a negative tropism.

Light The growth of a plant toward or away from light is a tropism called phototropism. A plant has a light-sensing chemical that helps it detect light. Leaves and stems tend to grow in the direction of light, as shown in **Figure 6.** This response maximizes the amount of light the plant's leaves receive. Roots generally grow away from light. This usually means that the roots grow down into the soil and help anchor the plant.

Active Reading 4. **Explain** How is phototropism beneficial to a plant?

_____

_____

Response to Light 🔑

Figure 6 As a plant's leaves turn toward the light, the amount of light that the leaves can absorb increases.

Active Reading 5. **Locate** (Circle) the plant that is responding to the light through phototropism.

Touch The response of a plant to touch is called a thigmotropism (thihg MAH truh pih zum). You might have seen vines growing up the side of a building or on a fence. This happens because the plant has special structures that respond to touch. These structures, called tendrils, can wrap around or cling to objects, as shown in **Figure 7.** A tendril wrapping around an object is an example of positive thigmotropism. Roots display negative thigmotropism. They grow away from objects in soil, enabling them to follow the easiest path through the soil.

Gravity The response of a plant to gravity is called gravitropism. Stems grow away from gravity, and roots grow toward gravity. The seedlings in **Figure 8** are exhibiting both responses. No matter how a seed lands on soil, when it starts to grow, its roots grow down into the soil. The stem grows up. This happens even when a seed is grown in a dark chamber, indicating that these responses can occur independently of light.

Response to Touch 🔑

Figure 7 The tendrils of the vine respond to touch and coil around the blade of grass.

 Active Reading

6. **List** What types of environmental stimuli do plants respond to? Give three examples.

_____

_____

_____

Response to Gravity 🔑

Figure 8 Both of these plant stems are growing away from gravity. The upward growth of a plant's stem is negative gravitropism, and the downward growth of its roots is positive gravitropism.

✔ 7. **Visual Check Identify** How is the plant on the left responding to the pot being placed on its side?

_____

Flowering Responses

You might think all plants respond to light, but in some plants, flowering is actually a response to darkness! **Photoperiodism** *is a plant's response to the number of hours of darkness in its environment.* Scientists once hypothesized that photoperiodism was a response to light. Therefore, these flowering responses are called long-day, short-day, and day-neutral and relate to the number of hours of daylight in a plant's environment.

Long-Day Plants Plants that flower when exposed to less than 10–12 hours of darkness are called long-day plants. The carnations shown in **Figure 9** are examples of long-day plants. This plant usually produces flowers in summer, when the number of hours of daylight is greater than the number of hours of darkness.

Short-Day Plants Short-day plants require 12 or more hours of darkness for flowering to begin. An example of a short-day plant is the poinsettia, shown in **Figure 9.** Poinsettias tend to flower in late summer or early fall when the number of hours of daylight is decreasing and the number of hours of darkness is increasing.

Day-Neutral Plants The flowering of some plants doesn't seem to be affected by the number of hours of darkness. Day-neutral plants flower when they reach maturity and the environmental conditions are right. Plants such as the roses in **Figure 9** are day-neutral plants.

Active Reading

8. **Infer** How is the flowering of day-neutral plants affected by exposure to hours of darkness?

_____

_____

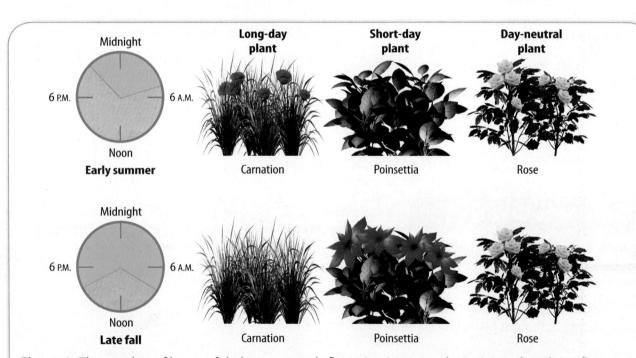

Figure 9 The number of hours of darkness controls flowering in many plants. Long-day plants flower when there are more hours of daylight than darkness, and short-day plants flower when there are more hours of darkness than daylight.

9. **Visual Check Identify** According to **Figure 9,** what time of year receives more darkness?

Active Reading

10. **Choose** (Circle) the type of plant that produces flowers during the late fall season.

Chemical Stimuli

Plants respond to chemical stimuli as well as environmental stimuli. **Plant hormones** *are substances that act as chemical messengers within plants.* These chemicals are produced in tiny amounts. They are called messengers because they usually are produced in one part and affect another part of a plant.

Auxins

One of the first plant hormones discovered was auxin (AWK sun). There are many different kinds of auxins. Auxins generally cause increased plant growth. They are responsible for phototropism, the growth of a plant toward light. Auxins concentrate on the dark side of a plant's stem, and these cells grow longer. This causes the stem of the plant to grow toward the light, as shown in **Figure 10.**

Ethylene

The plant hormone ethylene helps stimulate the ripening of fruit. Ethylene is a gas that can be produced by fruits, seeds, flowers, and leaves. You might have heard someone say that one rotten apple spoils the whole barrel. This is based on the fact that rotting fruits release ethylene. This can cause other fruits nearby to ripen and possibly rot. Ethylene also can cause plants to drop their leaves.

Light

• Auxin

Figure 10 🔑 Auxin on the left side of the seedling causes more growth and makes the seedling bend to the right.

Active Reading

11. **Explain** How do plants respond to the chemical stimuli, or hormones, auxin and ethylene?

_____

_____

_____

Inquiry

LAB STATION **Try It!**

SC.8.N.1.6, LA.8.2.2.3

MiniLab *When will plants flower?* at connectED.mcgraw-hill.com

Apply It! After you complete the lab, answer these questions.

1. In your own words, what is photoperiodism?

2. How do you think plants that flower well with any amount of darkness might have developed?

Figure 11 🔑 The grapes on the left were treated with gibberellins, and the grapes on the right were not treated.

Active Reading **12. Locate** <u>Underline</u> the text that tells the function of gibberellins.

Math Skills   MA.6.A.3.6

Use Percentages

A percentage is a ratio that compares a number to 100. For example, if a plant grows 2 cm per day with no chemical stimulus and 3 cm per day with a chemical stimulus, what is the percentage increase in growth?

Subtract the original value from the final value.

3 cm − 2 cm = 1 cm

Set up a ratio between the difference and the original value. Find the decimal equivalent.

$$\frac{1 \text{ cm}}{2 \text{ cm}} = 0.5$$

Multiply by 100 and add a percent sign.

0.5 × 100 = 50%

Practice

13. Without gibberellins, pea seedlings grew to 2 cm in 3 days. With gibberellins, the seedlings grew to 4 cm in 3 days. What was the percentage increase in growth?

Gibberellins and Cytokinins

Rapidly growing areas of a plant, such as roots and stems, produce gibberellins (jih buh REL unz). These hormones increase the rate of cell division and cell elongation. This results in increased growth of stems and leaves. Gibberellins also can be applied to the outside of plants. As shown in **Figure 11,** applying gibberellins to the outside of plants can have a dramatic effect.

Root tips produce most of the cytokinins (si tuh KI nunz), another type of hormone. Xylem carries cytokinins to other parts of a plant. Cytokinins increase the rate of cell division, and in some plants, cytokinins slow the aging process of flowers and fruits.

Summary of Plant Hormones

Plants produce many different hormones. The hormones you have just read about are groups of similar compounds. Often, two or more hormones interact and produce a plant response. Scientists continue to discover new information about plant hormones.

Humans and Plant Responses

Humans depend on plants for food, fuel, shelter, and clothing. Humans make plants more productive by using plant hormones. Some crops now are easier to grow because humans understand how they respond to hormones.

Active Reading **14. Infer** In what ways are humans dependent on plants?

_____

_____

_____

Visual Summary

Plants respond to stimuli in their environments in many ways.

Carnation

Poinsettia

Photoperiodism occurs in long-day plants and short-day plants. Day-neutral plants are not affected by the number of hours of darkness.

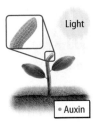

Light

• Auxin

Plant hormones are internal chemical stimuli that produce different responses in plants.

SC.8.N.1.6, MA.6.S.6.2, MA.6.A.3.6, LA.8.2.2.3

Inquiry **iLAB STATION** Try It!

Inquiry Lab *Design a Stimulating Environment for Plants* at connectED.mcgraw-hill.com

Use Vocabulary

1 **Define** *plant hormone* in your own words.

_____

2 The response of an organism to the number of hours of darkness in its environment is called _____.

Understand Key Concepts 🔑

3 **Describe** an example of a plant responding to environmental stimuli.

_____

_____

4 Which is NOT likely to cause a plant response?
- (A) changing the amount of daylight
- (B) moving plants away from each other
- (C) treating with plant hormones
- (D) turning a plant on its side

Interpret Graphics

5 **Identify** List the plant hormones mentioned in this lesson. Describe the effect of each on plants. LA.8.2.2.3

Hormone	Effect on Plants

Critical Thinking

6 **Infer** why the plant shown to the right is growing at an angle. SC.8.N.1.6

_____

Math Skills   MA.6.A.3.6

7 When sprayed with gibberellins, the diameter of mature grapes increased from 1.0 cm to 1.75 cm. What was the percent increase in size?

Chapter 11 Study Guide

Think About It! Plants survive by maintaining homeostasis and responding to stimuli. In addition, they acquire the energy they need for life processes through photosynthesis and cellular respiration.

 Key Concepts Summary

Vocabulary

LESSON 1 Energy Processing in Plants

- The vascular tissues in most plants, xylem and phloem, move materials throughout plants.

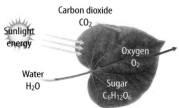

- In **photosynthesis,** plants convert light energy, water, and carbon dioxide into the food-energy molecule glucose through a series of chemical reactions. The process gives off oxygen.

- **Cellular respiration** is a series of chemical reactions that convert the energy in food molecules into a usable form of energy called ATP.

- Photosynthesis and cellular respiration can be considered opposite processes of each other.

photosynthesis p. 420
cellular respiration p. 422

LESSON 2 Plant Responses

- Although plants cannot move from one place to another, they do respond to **stimuli,** or changes in their environments. Plants respond to stimuli in different ways. **Tropisms** are growth responses toward or away from stimuli such as light, touch, and gravity. **Photoperiodism** is a plant's response to the number of hours of darkness in its environment.

- Plants respond to chemical stimuli, or **plant hormones,** such as auxins, ethylene, gibberellins, and cytokinins. Different hormones have different effects on plants.

stimulus p. 427
tropism p. 428
photoperiodism p. 430
plant hormone p. 431

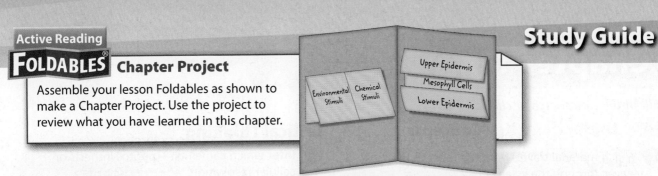

Active Reading

FOLDABLES® Chapter Project

Assemble your lesson Foldables as shown to make a Chapter Project. Use the project to review what you have learned in this chapter.

Use Vocabulary

1 Long-day and short-day plants are examples of plants that respond to

_____ .

2 The process that uses oxygen and produces carbon dioxide is

_____ .

3 Any change in an environment that causes an organism to respond is called a(n)

_____ .

4 Food is produced for plants through the process of

_____ .

Link Vocabulary and Key Concepts

Use vocabulary terms from the previous page to complete the concept map.

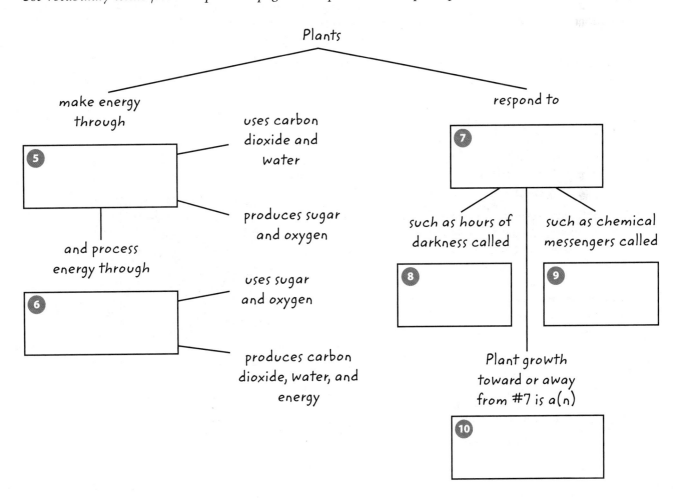

Fill in the correct answer choice.

🗝 Understand Key Concepts

1. Which material travels from the roots to the leaves through the xylem? LA.8.2.2.3
   - (A) oxygen
   - (B) sugar
   - (C) sunlight
   - (D) water

2. Which organelle is the site of photosynthesis? SC.8.L.18.1
   - (A) chloroplast
   - (B) mitochondria
   - (C) nucleus
   - (D) ribosome

3. Which is a product of cellular respiration? SC.8.L.18.2
   - (A) ATP
   - (B) light
   - (C) oxygen
   - (D) sugar

Use the image below to answer questions 4 and 5.

4. What type of plant-growth response is shown in the photo above? LA.8.2.2.3
   - (A) flowering
   - (B) gravitropism
   - (C) photoperiodism
   - (D) thigmotropism

5. Which stimulus is responsible for this type of growth? LA.8.2.2.3
   - (A) gravity
   - (B) light
   - (C) nutrients
   - (D) touch

Critical Thinking

6. **Infer** which came first—photosynthesis or cellular respiration. SC.8.L.18.1, SC.8.L.18.2

7. **Assess** the importance of material transport in plants. LA.8.2.2.3

8. **Evaluate** the internal structure of a leaf as a location for photosynthesis. SC.8.L.18.1

9. **Predict** what would happen if a short-day plant were exposed to more hours of daylight. SC.8.N.1.6

10. **Infer** from the photo below where the light source is in relation to the plant. LA.8.2.2.3

11. **Compare and contrast** the functions of xylem and phloem. LA.8.2.2.3

12. **Evaluate** the importance of auxins. LA.8.2.2.3

13 **Predict** the effect of an atmosphere with no gravity on a plant growing from a seed. SC.8.N.1.6

Writing in Science

14 **Write** On a separate piece of paper write a five-sentence paragraph about the importance of plants in your life. Include a main idea, supporting details, and a concluding sentence. SC.8.L.18.1, SC.8.L.18.2

Big Idea Review

15 What plant processes have you learned about in this chapter? Make a list. SC.8.L.18.1, SC.8.L.18.2

---

16 How do these processes, such as the one shown below, help a plant survive? SC.8.L.18.1, SC.8.L.18.2

Math Skills   MA.6.A.3.6

Use Percentages

17 Without treatment with gibberellins, 500 out of 1,000 grass seeds germinated. When sprayed with gibberellins, 875 of the seeds germinated. What was the percentage increase?

18 A bunch of bananas ripens (turns from green to yellow) in 42 hours. When the bananas are placed in a bag with an apple, which releases ethylene, the bananas ripen in 21 hours. What is the percentage change in ripening time?

Fill in the correct answer choice.

Multiple Choice

1 Why is photosynthesis important to humans? SC.8.L.18.1

Ⓐ It produces our only source of sugar.

Ⓑ It produces most of the oxygen in the atmosphere.

Ⓒ It produces most of the carbon dioxide in the atmosphere.

Ⓓ It produces our only source of water.

2 Which product of photosynthesis is used as a reactant in cellular respiration?
SC.8.L.18.1, SC.8.L.18.2

Ⓕ light

Ⓖ carbon dioxide

Ⓗ ATP

Ⓘ oxygen

Use the diagram below to answer question 3.

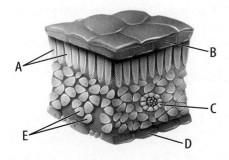

3 Where does most of the photosynthesis in the leaf above take place? SC.8.L.18.1

Ⓐ cells A and B

Ⓑ cells A and E

Ⓒ cells B and C

Ⓓ cells D and E

4 In the first step of photosynthesis, plants capture the energy in light. In which part of the plant does this occur? SC.8.L.18.1

Ⓕ chloroplast

Ⓖ stomata

Ⓗ epidermal cells

Ⓘ cuticle

5 Which is a product of photosynthesis? SC.8.L.18.1

Ⓐ ATP

Ⓑ glucose

Ⓒ light

Ⓓ water

Use the image below to answer question 6.

6 Which cellular process occurs within the organelle shown above? SC.8.L.18.2

Ⓕ cellular respiration

Ⓖ photosynthesis

Ⓗ transport of phloem

Ⓘ transport of xylem

7 What are the reactants necessary for cellular respiration to occur? SC.8.L.18.2

Ⓐ glucose and carbon dioxide

Ⓑ oxygen and carbon dioxide

Ⓒ ATP and glucose

Ⓓ glucose and oxygen

8 How is cellular respiration related to photosynthesis? SC.8.L.18.2

   (F) Through cellular respiration animals produce sugars that are broken down by plants through photosynthesis.

   (G) Animals use cellular respiration, and plants use photosynthesis.

   (H) Cellular respiration produces sugars, which are stored through photosynthesis.

   (I) Photosynthesis produces sugars, which are broken down in cellular respiration.

Use the diagram below to answer question 9.

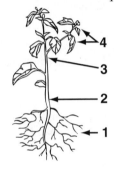

9 In which part of the plant does photosynthesis occur? SC.8.L.18.1

   (A) 1

   (B) 2

   (C) 3

   (D) 4

Use the figure below to answer question 10.

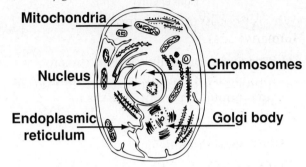

10 Where in the cell does respiration occur? SC.8.L.18.2

   (F) endoplasmic reticulum

   (G) mitochondria

   (H) Golgi body

   (I) nucleus

11 Which common plant pigment is necessary for photosynthesis? SC.8.L.18.1

   (A) stomata

   (B) chloroplasts

   (C) chlorophyll

   (D) mesophyll

12 Which process is a series of chemical reactions that convert the energy in food molecules into a usable form of energy called ATP? SC.8.L.18.2

   (F) photosynthesis

   (G) stomata

   (H) chlorophyll

   (I) cellular respiration

NEED EXTRA HELP?

If You Missed Question...	1	2	3	4	5	6	7	8	9	10	11	12
Go to Lesson...	1	1	1	1	1	1	1	1	1	1	1	1

Multiple Choice *Bubble the correct answer.*

Light energy
$6CO_2 + 6H_2O \xrightarrow{\text{Chlorophyll}} \text{Product A} + 6O_2$

1. The formula above represents the process of photosynthesis. What does Product A in the formula represent? **SC.8.L.18.1**

   (A) carbon dioxide

   (B) carbon monoxide

   (C) sugar

   (D) water

2. Which color of light does chlorophyll reflect and use for photosynthesis? **SC.8.L.18.1**

   (F) blue

   (G) green

   (H) red

   (I) violet

3. Which are the products of cellular respiration? **SC.8.L.18.2**

   (A) carbon dioxide and water

   (B) sunlight and carbon dioxide

   (C) sugar and oxygen

   (D) sugar and water

4. Which is produced in the organelle shown above? **SC.8.L.18.2**

   (F) ATP

   (G) chlorophyll

   (H) oxygen

   (I) sugar

Multiple Choice *Bubble the correct answer.*

1. Which plant hormone has caused the response shown by the plant in the picture above? **SC.8.L.18.1**

(A) auxin

(B) cytokinin

(C) ethylene

(D) gibberellin

2. A florist wants to prolong the life of cut flowers. What could he add to the water that the flowers are in? **SC.8.L.18.1**

(F) auxin

(G) cytokinin

(H) ethylene

(I) gibberellin

3. Which is an example of thigmotropism? **SC.8.L.18.1**

(A) flowers blooming in the summer

(B) roots growing down into the ground

(C) stems of ivy climbing a wall

(D) sunflowers pointing toward the Sun

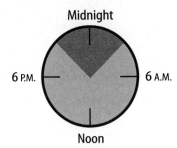

Midnight

6 P.M.   6 A.M.

Noon

Early summer

4. Sam is walking in a garden on a day represented by the figure above. Which plants in the garden could be blooming on this day? **SC.8.L.18.1**

(F) day-neutral plants only

(G) long-day and day-neutral plants

(H) short-day and day-neutral plants

(I) short-day, long-day, and day-neutral plants

Notes

Cycling of Matter

Three friends were talking about carbon dioxide and oxygen in the ecosystem. They each had different ideas. This is what they said:

Flynn: I think animals take in oxygen and breathe out carbon dioxide. Plants then take in the carbon dioxide and release oxygen, and the cycle continues.

Jervis: I think both plants and animals take in oxygen and release carbon dioxide; but only the plants take in the carbon dioxide and release oxygen, and the cycle continues.

Melody: I think both plants and animals take in oxygen and release carbon dioxide. The oxygen is used up and carbon dioxide is not cycled again by living things.

Circle the name of the friend you agree with the most. Explain why you agree. Describe your ideas about the cycling of matter.

Matter and Energy in the ENVIRONMENT

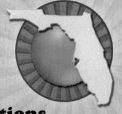

The Big Idea

FLORIDA BIG IDEAS

1   The Practice of Science

3   The Role of Theories, Laws, Hypotheses, and Models

18 Matter and Energy Transformations

Think About It!

How do living things and the nonliving parts of the environment interact?

The turtle needs food, air, water, and shelter to survive. The environment provides the turtle with all that it needs to survive.

1 How do you think the turtle depends on the nonliving things in the photo?

_____

_____

2 How might the turtle interact with living things in its environment?

_____

_____

_____

Get Ready to Read

What do you think about the environment?

Before you read, decide if you agree or disagree with each of these statements. As you read this chapter, see if you change your mind about any of the statements.

	AGREE	DISAGREE
1 The air you breathe is mostly oxygen.	☐	☐
2 Living things are made mostly of water.	☐	☐
3 Carbon, nitrogen, and other types of matter are used by living things over and over again.	☐	☐
4 Clouds are made of water vapor.	☐	☐
5 The Sun is the source for all energy used by living things on Earth.	☐	☐
6 All living things get their energy from eating other living things.	☐	☐

 **Connect ED**   **There's More Online!**
Video • Audio • Review • ⓘLab Station • WebQuest • Assessment • Concepts in Motion • Multilingual eGlossary

Abiotic FACTORS

Vocabulary

ecosystem p. 447

biotic factor p. 447

abiotic factor p. 447

climate p. 448

atmosphere p. 449

Inquiry **Launch Lab** SC.8.N.1.6

10 minutes

Is it living or nonliving?

You are surrounded by living and nonliving things, but it is sometimes difficult to tell what is alive. Some nonliving things may appear to be alive at first glance. Others are alive or were once living, but seem nonliving. In this lab, you will explore which items are alive and which are not.

Procedure

1 Your teacher will provide you with a list of items. Decide if each item is living or nonliving, then write its name in the appropriate column in the table to the right.

Living	Nonliving

Think About This

1. What are some characteristics that the items in the *Living* column share?

2. **Key Concept** How might the nonliving items be a part of your environment?

 Florida NGSSS

LA.8.2.2.3 The student will organize information to show understanding or relationships among facts, ideas, and events (e.g., representing key points within text through charting, mapping, paraphrasing, summarizing, or comparing/contrasting);

LA.8.4.2.2 The student will record information (e.g., observations, notes, lists, charts, legends) related to a topic, including visual aids to organize and record information, as appropriate, and attribute sources of information;

SC.8.N.1.6 Understand that scientific investigations involve the collection of relevant empirical evidence, the use of logical reasoning, and the application of imagination in devising hypotheses, predictions, explanations and models to make sense of the collected evidence.

Inquiry Why so Blue?

1. Have you ever seen a picture of a bright blue ocean like this one in Miami Beach, Florida? The water looks so colorful in part because of nonliving factors, such as matter in the water and the gases surrounding Earth. These nonliving things change the way you see light from the Sun, another nonliving part of the environment.

What is an ecosystem?

Have you ever watched a bee fly from flower to flower? Certain flowers and bees depend on each other. Bees help flowering plants reproduce. In return, flowers provide the nectar that bees use to make honey. Flowers also need nonliving things to survive, such as sunlight and water. For example, if plants don't get enough water, they can die. The bees might die, too, because they feed on the plants. All organisms need both living and nonliving things to survive.

An **ecosystem** *is all the living things and nonliving things in a given area.* Ecosystems vary in size. An entire forest can be an ecosystem, and so can a rotting log on the forest floor. Other examples of ecosystems include a pond, a desert, an ocean, and your neighborhood.

Biotic (bi AH tihk) **factors** *are the living things in an ecosystem.* **Abiotic** (ay bi AH tihk) **factors** *are the nonliving things in an ecosystem, such as sunlight and water.* Biotic factors and abiotic factors depend on each other. If just one factor—either abiotic or biotic—is disturbed, other parts of the ecosystem are affected. For example, severe droughts, or periods of water shortages, occurred in South Florida in 2001. Many fish in rivers and lakes died. Animals that fed on the fish had to find food elsewhere. A lack of water, an abiotic factor, affected biotic factors in this ecosystem, such as the fish and the animals that fed on the fish.

Active Reading **2. Write** What are three examples of biotic factors that you interact with daily?

_____

_____

_____

WORD ORIGIN

biotic

from Greek *biotikos*, means "fit for life"

Figure 1 Abiotic factors include sunlight, water, atmosphere, soil, temperature, and climate.

What are the nonliving parts of ecosystems?

Some abiotic factors in an ecosystem are shown in **Figure 1.** Think about how these factors might affect you. You need sunlight for warmth and air to breathe. You would have no food without water and soil. These nonliving parts of the environment affect all living things.

The Sun

The source of almost all energy on Earth is the Sun. It provides warmth and light. In addition, many plants use sunlight and make food, as you'll read in Lesson 3. The Sun also affects two other abiotic factors—climate and temperature.

3. NGSSS Check **Explain** How do living things use the Sun's energy? LA.8.2.2.3

_____

_____

Climate

Alligator snapping turtles live in the Florida Panhandle. This area has a warm, moist climate. **Climate** *describes average weather conditions in an area over time.* These weather conditions include temperature, moisture, and wind.

Climate influences where organisms can live. A desert climate, for example, is dry and often hot. A plant that needs a lot of water could not survive in a desert. In contrast, a cactus is well adapted to a dry climate because it can survive with little water.

Temperature

Is it hot or cold where you live? Temperatures on Earth vary greatly. Temperature is another abiotic factor that influences where organisms can survive. Some organisms, such as alligators, thrive in hot conditions. Others, such as polar bears, are well adapted to the cold. Alligators don't live in cold ecosystems, and polar bears don't live in warm ecosystems.

Water

All life on Earth requires water. In fact, most organisms are made mostly of water. All organisms need water for important life processes, such as growing and reproducing. Every ecosystem must contain some water to support life.

Gases in Atmosphere

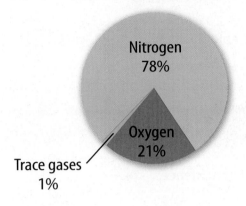

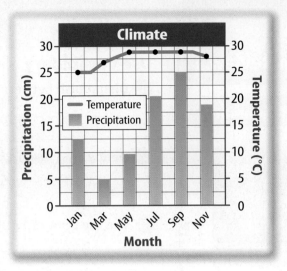

4. **Visual Check** How does the jaguar interact with the abiotic factors in its ecosystem?

_____

_____

_____

Atmosphere

Every time you take a breath you are interacting with another abiotic factor that is necessary for life—the atmosphere. *The* **atmosphere** (AT muh sfir) *is the layer of gases that surrounds Earth.* The atmosphere is mostly nitrogen and oxygen with trace amounts of other gases, also shown in **Figure 1.** Besides providing living things with oxygen, the atmosphere also protects them from certain harmful rays from the Sun.

Soil

Bits of rocks, water, air, minerals, and the remains of once-living things make up soil. When you think about soil, you might picture a farmer growing crops. Soil provides water and nutrients for the plants we eat. However, it is also a home for many organisms, such as insects, bacteria, and fungi.

Factors such as water, soil texture, and the amount of available nutrients affect the types of organisms that can live in soil. Bacteria break down dead plants and animals, returning nutrients to the soil. Earthworms and insects make small tunnels in the soil, allowing air and water to move through it. Even very dry soil, like that in the desert, is home to living things.

Active Reading 5. **List** What are the nonliving things in ecosystems?

_____

_____

Ecosystems include all the biotic and abiotic factors in an area.

Biotic factors are the living things in ecosystems.

Gases in Atmosphere

Nitrogen 78%

Oxygen 21%

Trace gases 1%

Abiotic factors are the nonliving things in ecosystems, including water, sunlight, temperature, climate, air, and soil.

Use Vocabulary

1. **Distinguish** between biotic and abiotic factors.

2. **Define** *ecosystem* in your own words.

3. **Use the term** *climate* in a complete sentence.

Understand Key Concepts 🔑

4. What role do bacteria play in soil ecosystems?
   - (A) They add air to soil.
   - (B) They break down rocks.
   - (C) They return nutrients to soil.
   - (D) They tunnel through soil.

5. **Explain** How would a forest ecosystem change if no sunlight were available to it?

Interpret Graphics

6. **Organize** Fill in each oval with an abiotic factor. LA.8.2.2.3

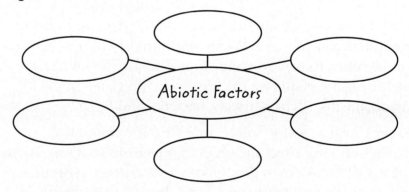

Abiotic Factors

Critical Thinking

7. **Predict** Imagine that the soil in an area is carried away by wind and water, leaving only rocks behind. How would this affect the living things in that area?

Saving Florida's Citrus Crops

Florida citrus growers are the leading citrus producers in the United States. They produce about 75 percent of the U.S. orange crop and about 80 percent of the U.S. grapefruit crop. Florida growers also produce lemons, limes, and tangerines.

Growing citrus fruit might seem simple, but the citrus growers deal with many issues including how to provide water and nutrients to their trees. The process of transporting water to fields for the purpose of helping crops grow is irrigation. Growers supply nutrients by applying fertilizers to the soil beneath the trees. Without fertilizers, the soil in which the citrus trees are grown would become depleted and the trees would not produce as much fruit.

Because Florida has a subtropical climate, it has high rainfall and high humidity. This environment is ideal for growing citrus. Unfortunately, these same conditions are a welcoming habitat for citrus diseases and pests—the two largest problems Florida citrus growers face. A habitat provides the food, shelter, moisture, temperature, and other factors necessary for an organism to survive.

Citrus trees are affected by pests such as fungi, bacteria, viruses, and insects. Infected trees can be treated with chemicals that kill the harmful organisms. Growers and researchers must take care to use treatments that will keep the trees healthy, but will not harm humans or the environment.

Citrus root weevils can damage leaves and harm plants.

It's Your Turn

RESEARCH Investigate one pest or disease that is harmful to Florida citrus crops and its method of treatment. Create a poster to share your findings.

Cycles of MATTER

Vocabulary

evaporation p. 454

condensation p. 454

precipitation p. 454

nitrogen fixation p. 456

 Florida NGSSS

LA.8.2.2.3 The student will organize information to show understanding or relationships among facts, ideas, and events (e.g., representing key points within text through charting, mapping, paraphrasing, summarizing, or comparing/contrasting);

MA.6.S.6.2 Select and analyze the measures of central tendency or variability to represent, describe, analyze, and/or summarize a data set for the purposes of answering questions appropriately.

SC.8.L.18.3 Construct a scientific model of the carbon cycle to show how matter and energy are continuously transferred within and between organisms and their physical environment.

SC.8.L.18.4 Cite evidence that living systems follow the Laws of Conservation of Mass and Energy.

SC.8.N.1.2 Design and conduct a study using repeated trials and replication.

SC.8.N.1.3 Use phrases such as "results support" or "fail to support" in science, understanding that science does not offer conclusive 'proof' of a knowledge claim.

SC.8.N.1.6 Understand that scientific investigations involve the collection of relevant empirical evidence, the use of logical reasoning, and the application of imagination in devising hypotheses, predictions, explanations and models to make sense of the collected evidence.

SC.8.N.3.1 Select models useful in relating the results of their own investigations.

 **Inquiry Launch Lab**

SC.8.N.1.6,
SC.8.N.3.1,
SC.8.L.18.4

15 minutes

How can you model raindrops?

Like all matter on Earth, water is recycled. It constantly moves between Earth and its atmosphere. You could be drinking the same water that a *Tyrannosaurus rex* drank 65 million years ago!

Procedure

1 Read and complete a lab safety form.

2 Half-fill a **plastic cup** with warm water.

3 Cover the cup with **plastic wrap.** Secure the plastic with a **rubber band.**

4 Place an **ice cube** on the plastic wrap. Observe the cup for several minutes.

Think About This

1. What did you observe on the underside of the plastic wrap? Why do you think this happened?

2. How does this activity model the formation of raindrops?

3. **Key Concept** Do you think other matter moves through the environment? Explain your answer.

Inquiry **Where does the water go?**

1. All water, including the water in this waterfall in Florida's Falling Waters State Park, can move throughout an ecosystem in a cycle. It also can change forms. What other forms do you think water takes as it move through an ecosystem?

_____

_____

How does matter move in ecosystems?

The water that you used to wash your hands this morning might have once traveled through the roots of a tree in Africa or even have been part of an Antarctic glacier. How can this be? Water moves continuously through ecosystems. It is used over and over again. Like water, other types of matter, such as carbon, oxygen, and nitrogen, go through physical and chemical changes. The total mass of the matter is the same before and after a change. This is known as the law of conservation of mass.

The Water Cycle

Look at a globe or a map. Notice that water surrounds the landmasses. Water covers about 70 percent of Earth's surface.

Most of Earth's water—about 97 percent—is in oceans. Water is also in rivers and streams, lakes, and underground reservoirs. In addition, water is in the atmosphere, icy glaciers, and living things.

Water continually cycles between Earth and its atmosphere. This movement of water is called the water cycle. It involves three processes: evaporation, condensation, and precipitation.

Active Reading **2. Identify** What are three Florida bodies of water near you?

_____

_____

_____

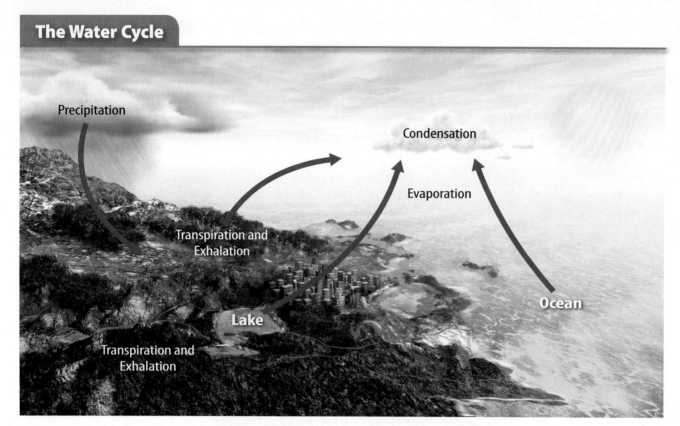

Figure 2 During the water cycle, the processes of evaporation, condensation, and precipitation move water from Earth's surface into the atmosphere and back again.

Evaporation

The Sun supplies the energy for the water cycle, as shown in **Figure 2.** As the Sun heats Earth's surface waters, evaporation occurs. **Evaporation** (ih va puh RAY shun), *is the process during which liquid water changes into a gas called water vapor.* This water vapor rises into the atmosphere. Temperature, humidity, and wind affect how quickly water evaporates.

Water is also released from living things. Transpiration is the release of water vapor from the leaves and stems of plants. Recall that cellular respiration is a process that occurs in many cells. A by-product of cellular respiration is water. This water leaves cells and enters the environment and atmosphere as water vapor.

Condensation

The higher in the atmosphere you are, the cooler the temperature is. As water vapor rises, it cools and condensation occurs.

Condensation (kahn den SAY shun), *is the process during which water vapor changes into liquid water.* Condensation causes clouds. Clouds are made of millions of tiny water droplets or crystals of ice. These form when water vapor condenses on particles of dust and other substances in the atmosphere.

Precipitation

Water that falls from clouds to Earth's surface is called **precipitation** (prih sih puh TAY shun). It enters bodies of water or soaks into soil. Precipitation can be rain, snow, sleet, or hail. It forms as water droplets or ice crystals join together in clouds. Eventually, these droplets or crystals become so large and heavy that they fall to Earth. Over time, living things use this precipitation, and the water cycle continues.

3. NGSSS Check Determine What forms does water take as it moves through ecosystems? SC.8.L.18.4

Inquiry
LAB STATION

Try It!

SC.8.L.18.4,
SC.8.N.1.6

MiniLab *Is your soil rich in nitrogen?* at connectED.mcgraw-hill.com

Apply It! After you complete the lab, answer these questions.

1. How could nitrogen help Florida farmers?

2. How might nitrogen levels differ throughout the state of Florida?

The Nitrogen Cycle

Just as water is necessary for life on Earth, so is the element nitrogen. It is an essential part of proteins, which all organisms need to stay alive. Nitrogen is also an important part of DNA, the molecule that contains genetic information. Nitrogen demonstrates the law of conservation of mass by cycling between Earth and its atmosphere and back again, as shown in **Figure 3.**

Active Reading

4. **Interpret** <u>Underline</u> what living things use nitrogen for.

Figure 3 Different forms of nitrogen are in the atmosphere, soil, and organisms.

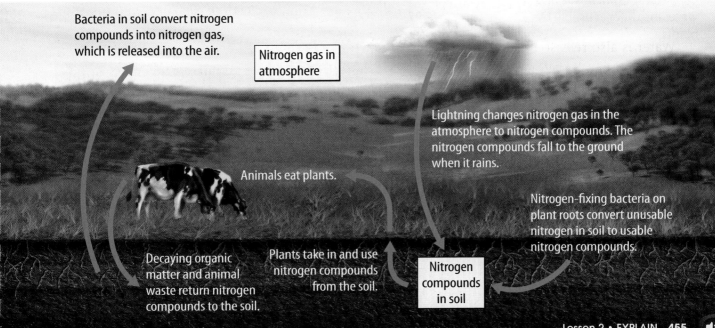

Bacteria in soil convert nitrogen compounds into nitrogen gas, which is released into the air.

Nitrogen gas in atmosphere

Lightning changes nitrogen gas in the atmosphere to nitrogen compounds. The nitrogen compounds fall to the ground when it rains.

Animals eat plants.

Nitrogen-fixing bacteria on plant roots convert unusable nitrogen in soil to usable nitrogen compounds.

Decaying organic matter and animal waste return nitrogen compounds to the soil.

Plants take in and use nitrogen compounds from the soil.

Nitrogen compounds in soil

From the Environment to Organisms

Recall that the atmosphere is mostly nitrogen. However, this nitrogen is in a form that plants and animals cannot use. How do organisms get nitrogen into their bodies? The nitrogen must first be changed into a different form with the help of certain bacteria that live in soil and water. These bacteria take in nitrogen from the atmosphere and change it into nitrogen compounds that other living things can use. *The process that changes atmospheric nitrogen into nitrogen compounds that are usable by living things is called* **nitrogen fixation** (NI truh jun • fihk SAY shun). Nitrogen fixation is shown in **Figure 4.**

Plants and some other organisms take in this changed nitrogen from the soil and water. Then, animals take in nitrogen when they eat the plants or other organisms.

Active Reading

5. Infer Nitrogen fixation occurs in some types of sugar cane, a crop grown in Florida. How does nitrogen fixation benefit the sugar cane plant?

_____

_____

From Organisms to the Environment

Some types of bacteria can break down the tissues of dead organisms. When organisms die, these bacteria help return the nitrogen in the tissues of dead organisms to the environment. This process is shown in **Figure 5.**

Nitrogen also returns to the environment in the waste products of organisms. Farmers often spread animal wastes, called manure, on their fields during the growing season. The manure provides nitrogen to plants for better growth.

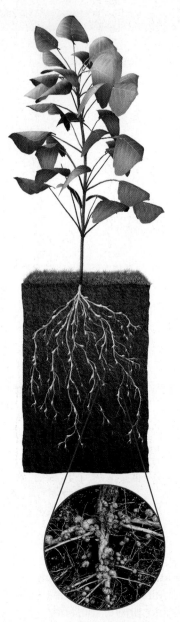

Figure 4 Certain bacteria convert nitrogen in soil and water into a form usable by plants.

Figure 5 Bacteria break down the remains of dead plants and animals.

Figure 6 Most oxygen in the air comes from plants and algae.

 6. NGSSS Check **Identify** Fill in the blanks with the correct term and (circle) your part in the oxygen cycle. SC.8.L.18.4

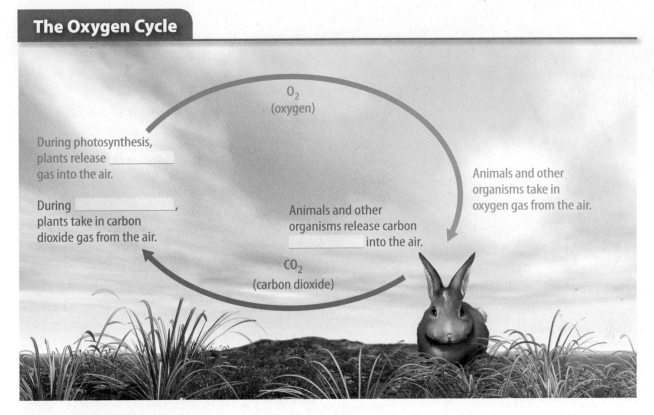

The Oxygen Cycle

O₂ (oxygen)

During photosynthesis, plants release _____ gas into the air.

During _____, plants take in carbon dioxide gas from the air.

Animals and other organisms release carbon _____ into the air.

Animals and other organisms take in oxygen gas from the air.

CO₂ (carbon dioxide)

The Oxygen Cycle

Almost all living things need oxygen for cellular processes that release energy. Oxygen is also part of many substances that are important to life, such as carbon dioxide and water. Oxygen cycles through ecosystems, as shown in **Figure 6.**

Earth's early atmosphere probably did not contain oxygen gas. Oxygen might have entered the atmosphere when certain **bacteria** evolved that could carry out the process of photosynthesis and make their own food. A by-product of photosynthesis is oxygen gas. Over time, other photosynthetic organisms evolved and the amount of oxygen in Earth's atmosphere increased. Today, photosynthesis is the primary source of oxygen in Earth's atmosphere. Some scientists estimate that unicellular organisms in water, called phytoplankton, release more than 50 percent of the oxygen in Earth's atmosphere.

Many living things, including humans, take in the oxygen and release carbon dioxide. The interaction of the carbon and oxygen cycles is one example of a relationship between different types of matter in ecosystems. As the matter cycles through an ecosystem, both the carbon and the oxygen take different forms and play a role in the other element's cycle.

REVIEW VOCABULARY

bacteria
a group of microscopic unicellular organisms without a membrane-bound nucleus

The Carbon Cycle

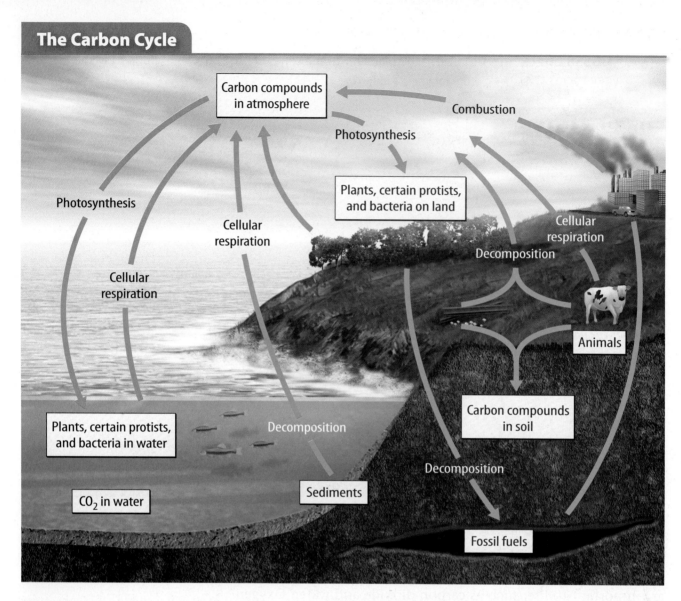

Figure 7 In the carbon cycle, all organisms return carbon to the environment.

The Carbon Cycle

All organisms contain carbon. It is part of proteins, sugars, fats, and DNA. Some organisms, including humans, get carbon from food. Organisms, such as plants, get carbon from the atmosphere or bodies of water. Carbon is conserved as it cycles through ecosystems, as shown in **Figure 7.**

Carbon in Soil

Like nitrogen, carbon can enter the environment when organisms die and decompose. This returns carbon compounds to the soil and releases carbon dioxide (CO_2) into the atmosphere for use by other organisms. Carbon is also found in fossil fuels, which formed when decomposing organisms were exposed to pressure, high temperatures, and bacteria over hundreds of millions of years.

Carbon in Air

Recall that carbon is found in the atmosphere as carbon dioxide. Plants and other photosynthetic organisms take in carbon dioxide and water and produce energy-rich sugars. These sugars are a source of carbon and energy for organisms that eat photosynthetic organisms. When the sugar is broken down by cells and its energy is **released**, carbon dioxide is released as a by-product. This carbon dioxide gas enters the atmosphere, where it can be used again.

The Greenhouse Effect

Carbon dioxide is one of the gases in the atmosphere that absorbs thermal energy from the Sun and keeps Earth warm. This process is called the greenhouse effect. The Sun produces solar radiation, as shown in **Figure 8.** Some of this energy is reflected back into space, and some passes through Earth's atmosphere. Greenhouse gases in Earth's atmosphere absorb thermal energy that reflects off Earth's surface. The more greenhouse gases released, the greater the gas layer becomes and the more thermal energy is absorbed. These gases are one factor that keeps Earth from becoming too hot or too cold.

Active Reading 7. <u>Underline</u> What is the greenhouse effect?

While the greenhouse effect is essential for life, a steady increase in greenhouse gases can harm ecosystems. For example, carbon is stored in fossil fuels such as coal, oil, and natural gas. When people burn fossil fuels to heat homes, for transportation, or to provide electricity, carbon dioxide gas is released into the atmosphere. The amount of carbon dioxide in the air has increased due to both natural and human activities.

ACADEMIC VOCABULARY

release
(**verb**) to set free or let go

Figure 8 Some thermal energy remains close to the Earth due to greenhouse gases.

 Active Reading 8. **Infer** What might Florida be like if heat were not absorbed by greenhouse gases?

_____

_____

_____

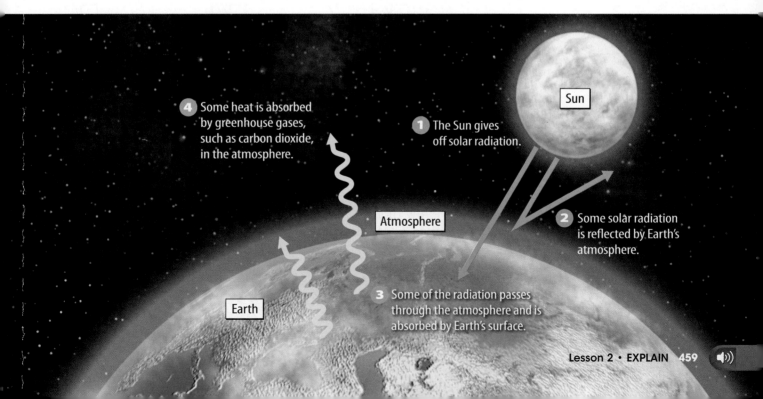

4 Some heat is absorbed by greenhouse gases, such as carbon dioxide, in the atmosphere.

1 The Sun gives off solar radiation.

Sun

Atmosphere

2 Some solar radiation is reflected by Earth's atmosphere.

3 Some of the radiation passes through the atmosphere and is absorbed by Earth's surface.

Earth

Visual Summary

Matter such as water, oxygen, nitrogen, and carbon cycles through ecosystems.

The three stages of the water cycle are evaporation, condensation, and precipitation.

The greenhouse effect helps keep the Earth from getting too hot or too cold.

Inquiry SC.8.N.1.2, SC.8.N.1.3, SC.8.N.1.6, MA.6.S.6.2

LAB STATION Try It!

Skill Lab *How do scientists use variables?* at connectED.mcgraw-hill.com

Use Vocabulary

1. **Distinguish** between evaporation and condensation.

_____

_____

2. **Define** *nitrogen fixation* in your own words.

_____

_____

_____

3. Water that falls from clouds to Earth's surface is called _____.

Understand Key Concepts

4. What is the driving force behind the water cycle?
   - (A) gravity
   - (C) sunlight
   - (B) plants
   - (D) wind

5. **Infer** Farmers add nitrogen to their fields every year to help their crops grow. Why must farmers continually add nitrogen when this element recycles naturally? SC.8.L.18.4

_____

_____

_____

Interpret Graphics

6. **Sequence** Draw a graphic organizer like the one below and sequence the steps in the water cycle. LA.8.2.2.3

```
[     ] ⇒ [     ] ⇒ [     ] ⇒
```

Critical Thinking

7. **Explain** how carbon cycles through the ecosystem in which you live. SC.8.L.18.3, SC.8.L.18.4

_____

_____

8. **Consider** How might ecosystems be affected if levels of atmospheric CO_2 continue to rise?

_____

_____

_____

Florida's Lake Wales Ridge

Cattle egret

Did you know that Florida has one of the most unique ecosystems in the world? Lake Wales Ridge is an ancient beach and sand dune system in central Florida. The approximately 100 miles of sand hills contain many organisms that are found only in the ridge, including 31 species of rare plants. Small shrubs and bushes, called "scrub," cover most of the area. These unique organisms are well-suited for living in a harsh environment like the Lake Wales Ridge.

The soil in the Ridge is very dry and sandy and does not contain many nutrients. Water drains out of the soil very quickly. Most plants would not be able to grow in this kind of environment. However, certain types of trees, such as shrubby oaks and Florida rosemary, grow well in it. These plants provide shelter and food for animals, such as Florida scrub jays and gopher tortoises.

Scientists estimate that about 80,000 acres of scrubland existed in Florida before European settlers arrived. Today, about 85 percent of the ridge has been cleared for use by agriculture, commercial, and residential developments. Government programs have protected the remaining scrubland in a wildlife refuge so generations to come can study this amazing ecosystem.

Sundew plant

It's Your Turn

RESEARCH Investigate how one type of matter moves throughout the Lake Wales Ridge ecosystem, and create a diagram showing the matter cycle.

Energy in ECOSYSTEMS

ESSENTIAL QUESTIONS

 How does energy move in ecosystems?

 How is the movement of energy in an ecosystem modeled?

Vocabulary

photosynthesis p. 464

chemosynthesis p. 464

food chain p. 466

food web p. 467

energy pyramid p. 468

 Florida NGSSS

LA.8.2.2.3 The student will organize information to show understanding or relationships among facts, ideas, and events (e.g., representing key points within text through charting, mapping, paraphrasing, summarizing, or comparing/contrasting);

MA.6.A.3.6 Construct and analyze tables, graphs, and equations to describe linear functions and other simple relations using both common language and algebraic notation.

SC.8.L.18.1 Describe and investigate the process of photosynthesis, such as the roles of light, carbon dioxide, water and chlorophyll; production of food; release of oxygen.

SC.8.L.18.3 Construct a scientific model of the carbon cycle to show how matter and energy are continuously transferred within and between organisms and their physical environment.

SC.8.L.18.4 Cite evidence that living systems follow the Laws of Conservation of Mass and Energy.

SC.8.N.1.6 Understand that scientific investigations involve the collection of relevant empirical evidence, the use of logical reasoning, and the application of imagination in devising hypotheses, predictions, explanations and models to make sense of the collected evidence.

SC.8.N.3.1 Select models useful in relating the results of their own investigations.

 Launch Lab SC.8.N.1.6, SC.8.L.18.4

10 minutes

How does energy change form?

Every day, sunlight travels hundreds of millions of kilometers and brings warmth and light to Earth. Energy from the Sun is necessary for nearly all life on Earth. Without it, most life could not exist.

Procedure

1. Read and complete a lab safety form.

2. Obtain **UV-sensitive beads** from your teacher. Write a description of them below.

3. Place half the beads in a sunny place. Place the other half in a dark place.

4. Wait a few minutes, and then observe both sets of beads. Record your observations.

Data and Observations

Think About This

1. Compare and contrast the two sets of beads after the few minutes. How are they different? How are they the same?

2. Hypothesize why the beads looked different.

3. **Key Concept** How do you think living things use energy?

Inquiry

1. All organisms need energy, and many get it from eating other organisms. How do you think each of the living things in this picture gets the energy it needs?

_____

_____

_____

_____

_____

_____

How does energy move in ecosystems?

When you see a picture of an ecosystem like Lake Okeechobee, Florida, it often looks quiet and peaceful. However, ecosystems are actually full of movement. Birds squawk and beat their wings, plants sway in the breeze, and insects buzz.

Each movement made by a living thing requires energy. All of life's functions, including growth and reproduction, require energy. The main source of energy for most life on Earth is the Sun. Unlike other resources, such as water and carbon, energy does not cycle through ecosystems. Instead, energy flows in one direction, as shown in **Figure 9.** In most cases, energy flow begins with the Sun and moves from one organism to another. Many organisms get energy by eating other organisms. Sometimes organisms change energy into different forms as it moves through an ecosystem. Not all the energy an organism gets is used for life processes. Some is released to the environment as thermal energy. You might have read that energy cannot be created or destroyed, but it can change form. This idea is called the law of conservation of energy.

 2. NGSSS Check Contrast How do the movements of matter and energy differ? SC.8.L.18.4

_____

_____

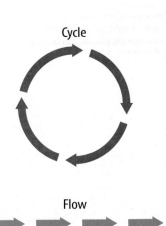

Figure 9 Matter moves in a cycle pattern, and energy moves in a flow pattern.

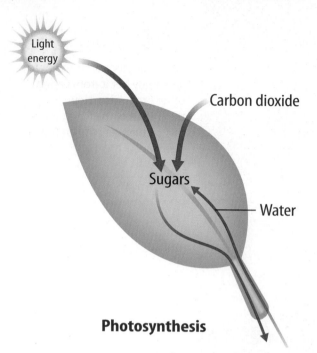

Photosynthesis

Figure 10 Most producers make their food through the process of photosynthesis.

WORD ORIGIN

photosynthesis
from Greek *photo*, meaning "light"; and *synthese*, meaning "synthesis"

Chemosynthesis

Figure 11 The producers at a hydrothermal vent make their food using chemosynthesis.

Producers

People who make things or products are often called producers. In a similar way, living things that make their own food are called producers. Producers make their food from materials found in their environments. Most producers are photosynthetic (foh toh sihn THEH tihk). They use the process of photosynthesis (foh toh SIHN thuh sus), which is described below. Grasses, trees, and other plants are photosynthetic. Algae, some other protists, and certain bacteria are also photosynthetic. Other producers, including some bacteria, are chemosynthetic (kee moh sihn THEH tihk). They make their food using chemosynthesis (kee moh SIHN thuh sus).

Photosynthesis Recall that in the carbon cycle, carbon in the atmosphere cycles through producers, such as plants, into other organisms and then back into the atmosphere. This and other matter cycles involve photosynthesis, as shown in **Figure 10.** **Photosynthesis** *is a series of chemical reactions that convert light energy, water, and carbon dioxide into the food-energy molecule glucose and give off oxygen.*

Chemosynthesis As you read earlier, some producers make food using chemosynthesis. **Chemosynthesis** *is the process during which producers use chemical energy in matter rather than light energy and make food.* One place where chemosynthesis can occur is on the deep ocean floor. There, inorganic compounds that contain hydrogen and sulfur flow out from cracks in the ocean floor. These cracks are called hydrothermal vents. These vents, such as the one shown in **Figure 11,** contain chemosynthetic bacteria. These bacteria use the chemical energy contained in inorganic compounds and produce food.

Active Reading **3. Recall** What materials do producers use to make food during chemosynthesis?

Herbivore

Carnivore

Omnivore

Detritivore

Detritivore—
Decomposer

Figure 12 Organisms can be classified by the type of food that they eat.

Consumers

Some consumers are shown in **Figure 12.** Consumers do not produce their own energy-rich food, as producers do. Instead, they get the energy they need to survive by consuming other organisms.

Consumers can be classified by the type of food they eat. Herbivores feed on only producers. For example, a deer is an herbivore because it eats only plants. Carnivores eat other animals. They are usually predators, such as lions and wolves. Omnivores eat both producers and other consumers. A bird that eats berries and insects is an omnivore.

Another group of consumers is detritivores (dih TRI tuh vorz). They get their energy by eating the remains of other organisms. Some detritivores, such as insects, eat dead organisms. Other detritivores, such as bacteria and mushrooms, feed on dead organisms and help decompose them. For this reason these organisms often are called decomposers. During decomposition, decomposers produce carbon dioxide that enters the atmosphere. Some of the decayed matter enters the soil. In this way detritivores help recycle nutrients through ecosystems. They also help keep ecosystems clean. Without decomposers dead organisms would pile up in an ecosystem.

Active Reading

4. Give Examples List one herbivore, one carnivore, and one detritivore in a Florida ecosystem.

_____

_____

_____

_____

Inquiry

LAB STATION **Try It!**

MiniLab *How can you classify organisms?* at connectED.mcgraw-hill.com

Apply It!

After you complete the lab, answer the question below.

1. Do you think any animals can be classified as more than one type of consumer? Why or why not?

Make a pyramid book from a sheet of paper. Use each side to organize information about one of the ways energy flows in an ecosystem. You can add additional information on the inside of your pyramid book.

Modeling Energy in Ecosystems

Unlike matter, energy does not cycle through ecosystems because it does not return to the Sun. Instead, energy flows through ecosystems. Organisms store energy in their bodies as chemical energy and use some energy for life processes. When consumers eat these organisms, the energy is transferred to the consumer. However, some energy is changed to thermal energy in the process and enters the environment. Decomposers transfer energy back into the environment when organisms die. Scientists use models to study this flow of energy through an ecosystem. They use different models depending on how many organisms they are studying.

Food Chains

A **food chain** *is a model that shows how energy flows in an ecosystem through feeding relationships.* In a food chain, arrows show the transfer of energy. A typical food chain is shown in **Figure 13.** Notice that there are not many links in this food chain. That is because the amount of available energy decreases every time it is transferred from one organism to another.

 5. NGSSS Check Explain How does a food chain model energy flow? SC.8.L.18.4

_____

_____

Figure 13 Energy moves from the Sun to a plant, a mouse, a snake, and a hawk in this food chain.

Food Chain

1 The Sun emits energy.

2 Plants make energy-rich food using sunlight.

3 The mouse obtains energy by eating the plant.

4 The snake obtains energy by eating the mouse.

5 The hawk obtains energy by eating the snake.

Food Webs

Imagine you have a jigsaw puzzle of part of the Everglades. Each piece of the puzzle shows only one small section of a wetland. A food chain is like one piece of an ecosystem jigsaw puzzle. It is helpful when studying certain parts of an ecosystem, but it does not show the whole picture.

In the food chain on the previous page, the mouse might also eat the seeds of several producers, such as corn, berries, or grass. The snake might eat other organisms such as frogs, crickets, lizards, or earthworms. The hawk hunts mice, squirrels, rabbits, and fish, as well as snakes.

Scientists use a model of energy transfer called a **food web** to show how food chains in a community are interconnected, as shown in **Figure 14.** You can think of a food web as many overlapping food chains. Like in a food chain, arrows show how energy flows in a food web. Some organisms in the food web might be part of more than one food chain in that web.

 6. **NGSSS Check Assess** Underline What models show the transfer of energy in an ecosystem? SC.8.L.18.4

Figure 14 A food web shows the complex feeding relationships among organisms in an ecosystem.

Energy Pyramids

Food chains and food webs show how energy moves in an ecosystem. However, they do not show how the amount of energy in an ecosystem changes. *Scientists use a model called an* **energy pyramid** *to show the amount of energy available in each step of a food chain,* as shown in **Figure 15.** The steps of an energy pyramid are also called trophic (TROH fihk) levels.

Producers, such as plants, make up the bottom trophic level. Consumers that eat producers, such as squirrels, make up the next trophic level. Consumers, such as hawks, that eat other consumers make up the highest trophic level. Notice that less energy is available for consumers at each higher trophic level. As you read earlier, organisms use some of the energy they get from food for life processes. During life processes, some energy is changed to thermal energy and is transferred to the environment. Only about 10 percent of the energy available at one trophic level transfers on to the next trophic level.

Figure 15 An energy pyramid shows the amount of energy available at each trophic level.

 8. **Visual Check** **Analyze** How does the amount of available energy change at each trophic level?

 Available energy decreases.

Trophic level 3
(1 percent of energy available)

Trophic level 2
(10 percent of energy available)

Trophic level 1
(100 percent of energy available)

Visual Summary

Energy flows in ecosystems from producers to consumers.

Producers make their own food through the processes of photosynthesis or chemosynthesis.

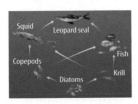

Food chains and food webs model how energy moves in ecosystems.

Inquiry SC.8.N.3.1, SC.8.L.18.3

LAB STATION Try It!

Inquiry Lab *How does soil type affect plant growth?* at connectED.mcgraw-hill.com

Use Vocabulary

1 Scientists use a(n) _____ to show how energy moves in an ecosystem.

2 Distinguish between photosynthesis and chemosynthesis. SC.8.L.18.1

_____

_____

Understand Key Concepts

3 Which organism is a producer?

(A) cow   (C) grass

(B) dog   (D) human

4 Construct a food chain with four links.

_____

_____

Interpret Graphics

5 Assess Which trophic level has the most energy available to living things?

_____

 Trophic level 3

 **Trophic level 2**

 **Trophic level 1**

Critical Thinking

6 Recommend A friend wants to show how energy moves in ecosystems. Which model would you recommend? Explain. SC.8.L.18.4

_____

_____

_____

Math Skills   MA.6.A.3.6

7 The plants in level 1 of a food pyramid obtain 30,000 units of energy from the Sun. How much energy is available for the organisms in level 2? Level 3?

Think About It! Living things interact with and depend on each other and on the nonliving things in an ecosystem.

 Key Concepts Summary

Vocabulary

LESSON 1 Abiotic Factors

- The **abiotic factors** in an environment include sunlight, temperature, climate, air, water, and soil.

ecosystem p. 447
biotic factor p. 447
abiotic factor p. 447
climate p. 448
atmosphere p. 449

LESSON 2 Cycles of Matter

- Matter such as oxygen, nitrogen, water, carbon, and minerals moves in cycles in the ecosystem.

evaporation p. 454
condensation p. 454
precipitation p. 454
nitrogen fixation p. 456

LESSON 3 Energy in Ecosystems

- Energy flows through ecosystems from producers to consumers.
- **Food chains, food webs,** and **energy pyramids** model the flow of energy in ecosystems.

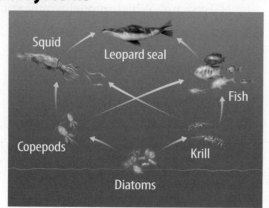

Squid
Leopard seal
Fish
Copepods
Krill
Diatoms

photosynthesis p. 464
chemosynthesis p. 464
food chain p. 466
food web p. 467
energy pyramid p. 468

FOLDABLES® Chapter Project

Assemble your lesson Foldables as shown to make a Chapter Project. Use the project to review what you have learned in this chapter.

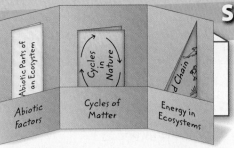

Use Vocabulary

1 Distinguish between climate and atmosphere.

_____

_____

2 The atmosphere is made mainly of the gases

_____ and _____ .

3 Living organisms in an ecosystem are called

_____ , while the nonliving things

are called _____ .

4 The process of converting nitrogen in the air into a form that can be used by living organisms is called

_____ .

5 Use the word *precipitation* in a complete sentence.

_____

_____

6 Define *condensation* in your own words.

_____

_____

7 How does a food chain differ from a food web?

_____

_____

8 The process of _____ uses energy from the Sun.

9 Define *chemosynthesis* in your own words.

_____

_____

Link Vocabulary and Key Concepts

Use vocabulary terms from the previous page and other terms from the chapter to complete the concept map.

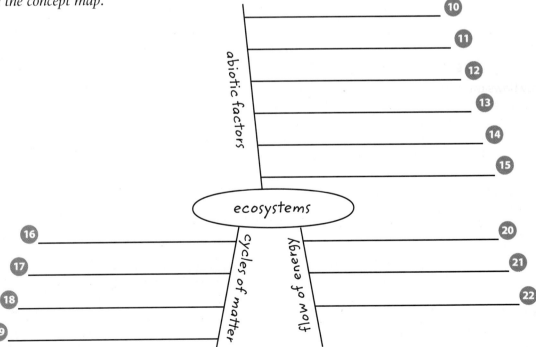

Chapter 12   Review

Fill in the correct answer choice.

🔑 Understand Key Concepts

1. What is the source of most energy on Earth? SC.8.L.18.4
   - Ⓐ air
   - Ⓑ soil
   - Ⓒ the Sun
   - Ⓓ water

2. Which is a biotic factor in an ecosystem? LA.8.2.2.3
   - Ⓐ a plant living near a stream
   - Ⓑ the amount of rainfall
   - Ⓒ the angle of the Sun
   - Ⓓ the types of minerals present in soil

3. Study the energy pyramid shown here.

   Trophic level III
   Trophic level II
   Trophic level I

   Which organism might you expect to find at trophic level I? SC.8.L.18.4
   - Ⓐ fox
   - Ⓑ frog
   - Ⓒ grass
   - Ⓓ grasshopper

4. Which includes both an abiotic and a biotic factor? LA.8.2.2.3
   - Ⓐ a chicken laying an egg
   - Ⓑ a deer drinking from a stream
   - Ⓒ a rock rolling down a hill
   - Ⓓ a squirrel eating an acorn

5. Which process helps keep temperatures on Earth from becoming too hot or too cold? LA.8.2.2.3
   - Ⓐ condensation
   - Ⓑ global warming
   - Ⓒ greenhouse effect
   - Ⓓ nitrogen fixation

Critical Thinking

6. **Compare and contrast** the oxygen cycle and the nitrogen cycle. LA.8.2.2.3

7. **Create** a plan for making an aquatic ecosystem in a jar. Include both abiotic and biotic factors. LA.8.2.2.3

8. **Recommend** a strategy for decreasing the amount of carbon dioxide in the atmosphere. SC.8.L.18.3

9. **Role-Play** Working in a group, perform a skit about organisms living near a hydrothermal vent. Be sure to include information about how the organisms obtain energy. Write your notes below. SC.8.L.18.4

10. **Assess** the usefulness of models as tools for studying ecosystems. SC.8.N.1.6

11 Study the food web below. Classify each organism according to what it eats and write its classification under its name. **LA.8.2.2.3**

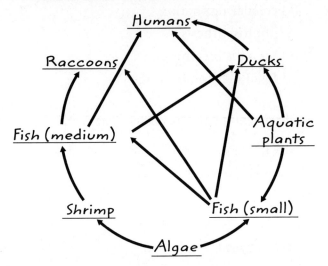

12 **Predict** what would happen if all the nitrogen-fixing bacteria in an ecosystem were removed. **SC.8.N.1.6**

_____

_____

_____

_____

Writing in Science

13 On a separate piece of paper, write an argument for or against the following statement. *The energy humans use in cars originally came from the Sun.* **SC.8.L.18.4**

Big Idea Review

14 Describe an interaction between a living thing and a nonliving thing in an ecosystem. **LA.8.2.2.3**

_____

_____

_____

15 How might the ram interact with nonliving things in its environment? **LA.8.2.2.3**

_____

_____

_____

_____

Math Skills   MA.6.A.3.6

Use Percentages

16 A group of plankton, algae, and other ocean plants absorb 150,000 units of energy.

a. How much energy is available for the third trophic level?

b. How much energy would remain for a fourth trophic level?

17 Some organisms, such as humans, are omnivores. They eat both producers and consumers. How much more energy would an omnivore get from eating the same mass of food at the first trophic level than at the second trophic level?

Fill in the correct answer choice.

Multiple Choice

1 Which process do producers complete to convert light energy, water, and carbon dioxide into glucose? SC.8.L.18.1

Ⓐ chemosynthesis

Ⓑ fermentation

Ⓒ carbon cycle

Ⓓ photosynthesis

Use the image below to answer question 2.

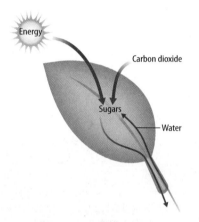

2 What process is shown above? SC.8.L.18.1

Ⓕ chemosynthesis

Ⓖ decomposition

Ⓗ nitrogen fixation

Ⓘ photosynthesis

3 Which process returns carbon compounds to the soil in the carbon cycle? SC.8.L.18.3

Ⓐ decomposition

Ⓑ transpiration

Ⓒ cellular respiration

Ⓓ nitrogen fixation

4 Which activity does NOT release carbon into the atmosphere during the carbon cycle? SC.8.L.18.1, SC.8.L.18.3

Ⓕ cellular respiration

Ⓖ photosynthesis

Ⓗ humans breathing

Ⓘ fossil fuel combustion

Use the diagram below to answer question 5.

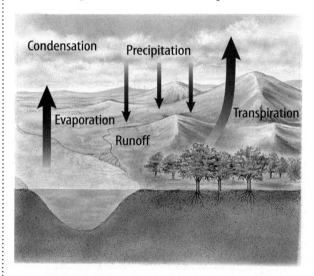

5 To complete the water cycle, which process causes water to fall from clouds to Earth's surface? SC.8.L.18.4

Ⓐ evaporation

Ⓑ condensation

Ⓒ exhalation

Ⓓ precipitation

6 Which is true of energy in ecosystems? SC.8.L.18.4

Ⓕ It never changes form.

Ⓖ It is both created and destroyed.

Ⓗ It flows in one direction.

Ⓘ It follows a cycle pattern.

7 Why is less energy available at each successive trophic level? SC.8.L.18.4

Ⓐ Some energy is given off as thermal energy.

Ⓑ Predators need less energy in higher trophic levels.

Ⓒ Some energy is destroyed in higher trophic levels.

Ⓓ Energy in higher trophic levels is unusable.

Use the diagram below to answer question 8.

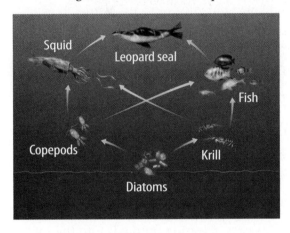

8 How does energy move in the food web pictured above? SC.8.L.18.4

Ⓕ from leopard seal to squid

Ⓖ from diatoms to krill

Ⓗ from fish to krill

Ⓘ from squid to diatoms

9 What three things do producers use during photosynthesis to make sugars? SC.8.L.18.1

Ⓐ oxygen, water, sunlight

Ⓑ oxygen, water, nitrogen

Ⓒ carbon dioxide, water, nitrogen

Ⓓ carbon dioxide, water, light energy

10 During which process is oxygen gas released into the atmosphere? SC.8.L.18.1

Ⓕ chemosynthesis

Ⓖ decomposition

Ⓗ photosynthesis

Ⓘ transpiration

11 Which is a by-product of photosynthesis? SC.8.L.18.1

Ⓐ carbon dioxide

Ⓑ nitrogen

Ⓒ water

Ⓓ oxygen

Use the diagram below to answer question 12.

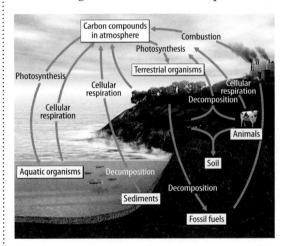

12 In the image of the carbon cycle shown above, which two processes return carbon from the atmosphere to the environment? SC.8.L.18.3

Ⓕ photosynthesis and decomposition

Ⓖ cellular respiration and decomposition

Ⓗ cellular respiration and photosynthesis

Ⓘ photosynthesis and combustion

NEED EXTRA HELP?

If You Missed Question...	1	2	3	4	5	6	7	8	9	10	11	12
Go to Lesson...	3	3	2	2	2	3	3	3	2	2	2	2

Multiple Choice *Bubble the correct answer.*

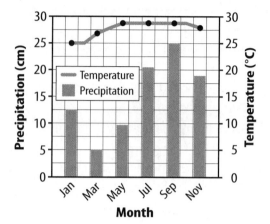

Month

1. Which abiotic factor probably has the greatest effect on the types of plants and animals that live in the environment shown above? **SC.8.L.18.1**

   (A) air

   (B) rocks

   (C) soil

   (D) water

2. Which of these is NOT an abiotic factor in the environment? **SC.8.L.18.1**

   (F) air

   (G) plants

   (H) rocks

   (I) sunlight

3. Which abiotic factors are described in the graph above? **SC.8.L.18.1**

   (A) air and soil

   (B) water and air

   (C) precipitation and soil

   (D) precipitation and temperature

4. In which way does the composition of soil, an abiotic factor, rely on living things? **SC.8.L.18.1**

   (F) Bacteria break down dead matter and wastes, releasing nutrients into the soil.

   (G) Bacteria erode rocks into smaller particles that are added to the soil.

   (H) Insects make small tunnels in the soil. Bacteria use these tunnels to break down dead matter.

   (I) Insects take in air and minerals and release these materials into the soil as wastes.

Multiple Choice *Bubble the correct answer.*

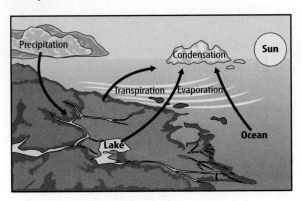

1. Based on the image above, which two processes add water to the atmosphere? **SC.8.L.18.4**

Ⓐ condensation and evaporation

Ⓑ condensation and precipitation

Ⓒ precipitation and evaporation

Ⓓ transpiration and evaporation

2. What might happen to the carbon cycle if dead organisms did not break down? **SC.8.L.18.3**

Ⓕ Bacteria would begin fixing carbon dioxide and adding it to the soil.

Ⓖ Carbon would not be recycled and added to soil, interrupting the cycle.

Ⓗ More carbon dioxide would be available for photosynthesis.

Ⓘ Other sources of carbon would need to be found to keep carbon in the cycle.

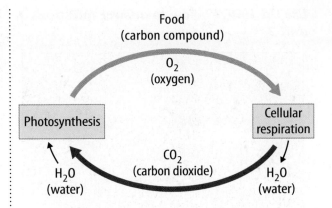

3. According to the image above, what substances are cycled during cellular respiration and photosynthesis? **SC.8.L.18.1**

Ⓐ oxygen and carbon dioxide

Ⓑ oxygen, carbon, and nitrogen

Ⓒ oxygen, carbon, and water

Ⓓ oxygen, nitrogen, and water

4. Without the greenhouse effect, temperatures on Earth would **SC.8.L.18.1**

Ⓕ be too cool or too hot.

Ⓖ become constant.

Ⓗ increase sharply.

Ⓘ vary too much.

Multiple Choice *Bubble the correct answer.*

Use the image below to answer questions 1 and 2.

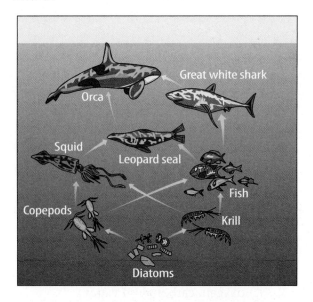

1. A food chain is one part of a food web. Which list is part of the food web shown above? **SC.8.L.18.4**

   Ⓐ diatoms→fish→leopard seal→orca

   Ⓑ diatoms→krill→fish→
      great white shark→orca

   Ⓒ copepods→diatoms→krill→fish→
      leopard seal→orca

   Ⓓ orca→leopard seal→squid→
      copepods→diatoms

2. Which organism forms the base of the food web shown above? **SC.8.L.18.4**

   Ⓕ diatoms

   Ⓖ fish

   Ⓗ krill

   Ⓘ orcas

3. Which organisms are producers? **SC.8.L.18.4**

   Ⓐ algae

   Ⓑ amoebas

   Ⓒ mushrooms

   Ⓓ yeasts

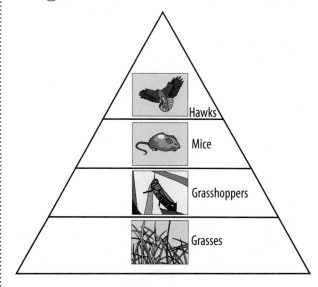

4. Which statement describes the amount of energy passed to the hawk in the energy pyramid above? **SC.8.L.18.4**

   Ⓕ The hawk has 50 percent of the energy available to it.

   Ⓖ The hawk has the least amount of energy available to it.

   Ⓗ The hawk has the most energy available to it.

   Ⓘ The hawk has no energy available to it.

Name _____ Date _____

Glossary/Glosario

Cómo usar el glosario en español:
1. Busca el término en inglés que desees encontrar.
2. El término en español, junto con la definición, se encuentran en la columna de la derecha.

Pronunciation Key
Use the following key to help you sound out words in the glossary.

a	. . . .	back (BAK)	ew	. . . . food (FEWD)
ay	. . . .	day (DAY)	yoo	. . . . pure (PYOOR)
ah	. . . .	father (FAH thur)	yew	. . . . few (FYEW)
ow	. . . .	flower (FLOW ur)	uh	. . . . comma (CAH muh)
ar	. . . .	car (CAR)	u (+ con)	. . . . rub (RUB)
e	. . . .	less (LES)	sh	. . . . shelf (SHELF)
ee	. . . .	leaf (LEEF)	ch	. . . . nature (NAY chur)
ih	. . . .	trip (TRIHP)	g	. . . . gift (GIHFT)
i (i + com + e)	idea (i DEE uh)	j	. . . . gem (JEM)
oh	. . . .	go (GOH)	ing	. . . . sing (SING)
aw	. . . .	soft (SAWFT)	zh	. . . . vision (VIH zhun)
or	. . . .	orbit (OR buht)	k	. . . . cake (KAYK)
oy	. . . .	coin (COYN)	s	. . . . seed, cent (SEED, SENT)
oo	. . . .	foot (FOOT)	z	. . . . zone, raise (ZOHN, RAYZ)

English **A** **Español**

abiotic factor/asteroid   **factor abiótico/asteroide**

abiotic factor (ay bi AH tihk • FAK tuhr): a non-living thing in an ecosystem.

acid: a substance that produces a hydronium ion (H_3O^+) when dissolved in water.

activation energy: the minimum amount of energy needed to start a chemical reaction.

alkali (AL kuh li) metal: an element in group 1 on the periodic table.

alkaline (AL kuh lun) earth metal: an element in group 2 on the periodic table.

apparent magnitude: a measure of how bright an object appears from Earth.

asteroid: a small, rocky object that orbits the Sun.

factor abiótico: componente no vivo de un ecosistema.

ácido: sustancia que produce ión hidronio (H_3O^+) cuando se disuelve en agua.

energía de activación: cantidad mínima de energía necesaria para iniciar una reacción química.

metal alcalino: elemento del grupo 1 de la tabla periódica.

metal alcalinotérreo: elemento del grupo 2 de la tabla periódica.

magnitud aparente: medida del brillo de un objeto visto desde la Tierra.

asteroide: objeto pequeño y rocoso que orbita el Sol.

astrobiology: the study of the origin, development, distribution, and future of life on Earth and in the universe.

astronomical unit (AU): the average distance from Earth to the Sun—about 150 million km.

atmosphere (AT muh sfir): a thin layer of gases surrounding Earth.

atom: a small particle that is the building block of matter.

atomic number: the number of protons in an atom of an element.

astrobiología: estudio del origen, desarrollo, distribución y futuro de la vida en la Tierra y en el universo.

unidad astronómica (UA): distancia media entre la Tierra y el Sol, aproximadamente 150 millones de km.

atmósfera: capa delgada de gases que rodean la Tierra.

átomo: partícula pequeña que es el componente básico de la materia.

número atómico: número de protones en el átomo de un elemento.

B

base: a substance that produces hydroxide ions (OH⁻) when dissolved in water.

Big Bang theory: the scientific theory that states that the universe began from one point and has been expanding and cooling ever since.

biotic factor (bi AH tihk • FAK tuhr): a living or once-living thing in an ecosystem.

black hole: an object whose gravity is so great that no light can escape.

base: sustancia que produce iones hidróxido (OH⁻) cuando se disuelve en agua.

Teoría del Big Bang: teoría científica que establece que el universo se originó de un punto y se ha ido expandiendo y enfriando desde entonces.

factor biótico: vida cosa o anteriormente vida cosa en un ecosistema.

agujero negro: objeto cuya gravedad es tan grande que la luz no puede escapar.)

C

catalyst: a substance that increases reaction rate by lowering the activation energy of a reaction.

chemical bond: a force that holds two or more atoms together.

chemical change: a change in matter in which the substances that make up the matter change into other substances with different chemical and physical properties.

chemical equation: a description of a reaction using element symbols and chemical formulas.

chemical formula: a group of chemical symbols and numbers that represent the elements and the number of atoms of each element that make up a compound.

chemical property: the ability or inability of a substance to combine with or change into one or more new substances.

catalizador: sustancia que aumenta la velocidad de reacción al disminuir la energía de activación de una reacción.

enlace químico: fuerza que mantiene unidos dos o más átomos.

cambio químico: cambio de la materia en el cual las sustancias que componen la materia se transforman en otras sustancias con propiedades químicas y físicas diferentes.

ecuación química: descripción de una reacción con símbolos de los elementos y fórmulas químicas.

fórmula química: grupo de símbolos químicos y números que representan los elementos y el número de átomos de cada elemento que forman un compuesto.

propiedad química: capacidad o incapacidad de una sustancia para combinarse con una o más sustancias o transformarse en una o más sustancias.

chemical reaction: a process in which atoms of one or more substances rearrange to form one or more new substances.

chemosynthesis (kee moh sihn THUH sus): the process during which producers use chemical energy in matter rather than light energy to make food.

chromosphere: the orange-red layer above the photosphere of a star.

climate: the long-term average weather conditions that occur in a particular region.

coefficient: a number placed in front of an element symbol or chemical formula in an equation.

combustion: a chemical reaction in which a substance combines with oxygen and releases energy.

comet: a small, rocky and icy object that orbits the Sun.

compound: a substance containing atoms of two or more different elements chemically bonded together.

concentration: the amount of a particular solute in a given amount of solution.

condensation (kahn den SAY shun): the process by which a gas changes to a liquid.

constants: the factors in an experiment that remain the same.

control group: the part of a controlled experiment that contains the same factors as the experimental group, but the independent variable is not changed.

convection zone: layer of a star where hot gas moves up toward the surface and cooler gas moves deeper into the interior.

corona: the wide, outermost layer of a star's atmosphere.

covalent bond: a chemical bond formed when two atoms share one or more pairs of valence electrons.

critical thinking: comparing what you already know with information you are given in order to decide whether you agree with it.

reacción química: proceso en el cual átomos de una o más sustancias se acomodan para formar una o más sustancias nuevas.

quimiosíntesis: proceso durante el cual los productores usan la energía química en la materia en vez de la energía lumínica, para elaborar alimento.

cromosfera: capa de color rojo anaranjado arriba de la fotosfera de una estrella.

clima: promedio a largo plazo de las condiciones del tiempo atmosférico de una región en particular.

coeficiente: número colocado en frente del símbolo de un elemento o de una fórmula química en una ecuación.

combustión: reacción química en la cual una sustancia se combina con oxígeno y libera energía.

cometa: objeto pequeño, rocoso y helado que orbita el Sol.

compuesto: sustancia que contiene átomos de dos o más elementos diferentes unidos químicamente.

concentración: cantidad de cierto soluto en una cantidad dada de solución.

condensación: proceso mediante el cual un gas cambia a líquido.

constantes: factores que no cambian en un experimento.

grupo de control: parte de un experimento controlado que contiene los mismos factores que el grupo experimental, pero la variable independiente no se cambia.

zona de convección: capa de una estrella donde el gas caliente se mueve hacia arriba de la superficie y el gas más frío se mueve más profundo hacia el interior.

corona: capa extensa más externa de la atmósfera de una estrella.

enlace covalente: enlace químico formado cuando dos átomos comparten uno o más pares de electrones de valencia.

pensamiento crítico: comparación que se hace cuando se sabe algo acerca de información nueva, y se decide si se está o no de acuerdo con ella.

D

dark matter: matter that emits no light at any wavelength.

decomposition: a type of chemical reaction in which one compound breaks down and forms two or more substances.

density: the mass per unit volume of a substance.

dependent variable: the factor a scientist observes or measures during an experiment.

description: a spoken or written summary of an observation.

Doppler shift: the shift to a different wavelength on the electromagnetic spectrum.

double-replacement reaction: a type of chemical reaction in which the negative ions in two compounds switch places, forming two new compounds.

ductility (duk TIH luh tee): the ability to be pulled into thin wires.

materia oscura: materia que no emite luz a ninguna longitud de onda.

descomposición: tipo de reacción química en la que un compuesto se descompone y forma dos o más sustancias.

densidad: cantidad de masa por unidad de volumen de una sustancia.

variable dependiente: factor que el científico observa o mide durante un experimento.

descripción: resumen oral o escrito de una observación.

efecto Doppler: cambio a una longitud de onda diferente en el espectro electromagnético.

reacción de sustitución doble: tipo de reacción química en la que los iones negativos de dos compuestos intercambian lugares, para formar dos compuestos nuevos.

ductilidad: capacidad para formar alambres delgados

E

ecosystem: all the living things and nonliving things in a given area.

electromagnetic (ih lek troh mag NEH tik) spectrum: the entire range of radiant energy carried by electromagnetic waves.

electron: a negatively charged particle that occupies the space in an atom outside the nucleus.

electron cloud: the region surrounding an atom's nucleus where one or more electrons are most likely to be found.

electron dot diagram: a model that represents valence electrons in an atom as dots around the element's chemical symbol.

element: a substance that consists of only one type of atom.

endothermic reaction: a chemical reaction that absorbs thermal energy.

energy pyramid: a model that shows the amount of energy available in each link of a food chain.

ecosistema: todos los seres vivos y los componentes no vivos de un área dada.

espectro electromagnético: gama completa de energía radiante transportada por las ondas electromagnéticas.

electrón: partícula cargada negativamente que ocupa el espacio por fuera del núcleo de un átomo.

nube de electrones: región que rodea el núcleo de un átomo en donde es más probable encontrar uno o más electrones.

diagrama de puntos de Lewis: modelo que representa electrones de valencia en un átomo a manera de puntos alrededor del símbolo químico del elemento.

elemento: sustancia que consiste de un sólo tipo de átomo.

reacción endotérmica: reacción química que absorbe energía térmica.

pirámide energética: modelo que explica la cantidad de energía disponible en cada vínculo de una cadena alimentaria.

enzyme: a catalyst that speeds up chemical reactions in living cells.

equinox: when Earth's rotation axis is tilted neither toward nor away from the Sun.

evaporation (ih va puh RAY shun): the process of a liquid changing to a gas at the surface of the liquid.

exothermic reaction: a chemical reaction that releases thermal energy.

experimental group: the part of the controlled experiment used to study relationships among variables.

explanation: an interpretation of observations.

extraterrestrial (ek struh tuh RES tree ul) life: life that originates outside Earth.

enzima: catalizador que acelera reacciones químicas en las células vivas.

equinoccio: cuando el eje de rotación de la Tierra se inclina sin acercarse ni alejarse del Sol.

evaporación: proceso de cambio de un líquido a un gas en la superficie del líquido.

reacción exotérmica: reacción química que libera energía térmica.

grupo experimental: parte del experimento controlado que se usa para estudiar las relaciones entre las variables.

explicación: interpretación de las observaciones.

vida extraterrestre: vida que se origina fuera de la Tierra.

F

---

food chain: a model that shows how energy flows in an ecosystem through feeding relationships.

food web: a model of energy transfer that can show how the food chains in a community are interconnected.

cadena alimentaria: modelo que explica cómo la energía fluye en un ecosistema a través de relaciones alimentarias.

red alimentaria: modelo de transferencia de energía que explica cómo las cadenas alimentarias están interconectadas en una comunidad.

G

---

galaxy: a huge collection of stars, gas, and dust.

Galilean moons: the four largest of Jupiter's 63 moons discovered by Galileo.

gas: matter that has no definite volume and no definite shape.

greenhouse effect: the natural process that occurs when certain gases in the atmosphere absorb and reradiate thermal energy from the Sun.

group: a column on the periodic table.

galaxia: conjunto enorme de estrellas, gas, y polvo.

lunas de Galileo: las cuatro lunas más grandes de las 63 lunas de Júpiter, descubiertas por Galileo.

gas: materia que no tiene volumen ni forma definidos.

efecto invernadero: proceso natural que ocurre cuando ciertos gases en la atmósfera absorben y vuelven a irradiar la energía térmica del Sol.

grupo: columna en la tabla periódica.

H

---

halogen (HA luh jun): an element in group 17 on the periodic table.

halógeno: elemento del grupo 17 de la tabla periódica.

Hertzsprung-Russell diagram: a graph that plots luminosity v. temperature of stars.

heterogeneous mixture: a mixture in which substances are not evenly mixed.

homogeneous mixture: a mixture in which two or more substances are evenly mixed but not bonded together.

hydronium ion (H_3O^+): a positively charged ion formed when an acid dissolves in water.

hypothesis: a possible explanation for an observation that can be tested by scientific investigations.

diagrama de Hertzsprung-Russell: diagrama que traza la luminosidad frente a la temperatura de las estrellas.

mezcla heterogénea: mezcla en la cual las sustancias no están mezcladas de manera uniforme.

mezcla homogénea: mezcla en la cual dos o más sustancias están mezcladas de manera uniforme, pero no están unidas químicamente.

ión hidronio (H_3O^+): ión cargado positivamente que se forma cuando un ácido se disuelve en agua.

hipótesis: explicación posible para una observación que puede ponerse a prueba en investigaciones científicas.

I

---

impact crater: a round depression formed on the surface of a planet, moon, or other space object by the impact of a meteorite.

independent variable: the factor that is changed by the investigator to observe how it affects a dependent variable.

indicator: a compound that changes color at different pH values when it reacts with acidic or basic solutions.

inference: a logical explanation of an observation that is drawn from prior knowledge or experience.

inhibitor: a substance that slows, or even stops, a chemical reaction.

International System of Units (SI): the internationally accepted system of measurement.

ion (I ahn): an atom that is no longer neutral because it has gained or lost electrons.

ionic bond: the attraction between positively and negatively charged ions in an ionic compound.

isotopes (I suh tohps): atoms of the same element that have different numbers of neutrons.

cráter de impacto: depresión redonda formada en la superficie de un planeta, luna u otro objeto espacial debido al impacto de un meteorito.

variable independiente: factor que el investigador cambia para observar cómo afecta la variable dependiente.

indicador: compuesto que cambia de color a diferentes valores de pH cuando reacciona con soluciones ácidas o básicas.

inferencia: explicación lógica de una observación que se obtiene a partir de conocimiento previo o experiencia.

inhibidor: sustancia que disminuye, o incluso detiene, una reacción química.

Sistema Internacional de Unidades (SI): sistema de medidas aceptado internacionalmente.

ión: átomo que no es neutro porque ha ganado o perdido electrones.

enlace iónico: atracción entre iones cargados positiva y negativamente en un compuesto iónico.

isótopos: átomos del mismo elemento que tienen diferente número de neutrones.

law of conservation of mass: law that states that the total mass of the reactants before a chemical reaction is the same as the total mass of the products after the chemical reaction.

light-year: the distance light travels in one year.

liquid: matter with a definite volume but no definite shape.

luminosity (lew muh NAH sih tee): the true brightness of an object.

lunar: term that refers to anything related to the Moon.

lunar eclipse: an occurrence during which the Moon moves into Earth's shadow.

luster: the way a mineral reflects or absorbs light at its surface.

ley de la conservación de la masa: ley que plantea que la masa total de los reactivos antes de una reacción química es la misma que la masa total de los productos después de la reacción química.

año luz: distancia que recorre la luz en un año.

líquido: materia con volumen definido y forma indefinida.

luminosidad: brillantez real de un objeto.

lunar: término que hace referencia a todo lo relacionado con la luna.

eclipse lunar: ocurrencia durante la cual la Luna se mueve hacia la zona de sombra de la Tierra.

brillo: forma en que un mineral refleja o absorbe la luz en su superficie.

malleability (ma lee uh BIH luh tee): the ability of a substance to be hammered or rolled into sheets.

maria (MAR ee uh): the large, dark, flat areas on the Moon.

mass: the amount of matter in an object.

matter: anything that has mass and takes up space.

metal: an element that is generally shiny, is easily pulled into wires or hammered into thin sheets, and is a good conductor of electricity and thermal energy.

metallic bond: a bond formed when many metal atoms share their pooled valence electrons.

metalloid (MEH tul oyd): an element that has physical and chemical properties of both metals and nonmetals.

meteor: a meteoroid that has entered Earth's atmosphere and produces a streak of light.

meteorite: a meteoroid that strikes a planet or a moon.

maleabilidad: capacidad de una sustancia de martillarse o laminarse para formar hojas.

mares: áreas extensas, oscuras y planas en la Luna.

masa: cantidad de materia en un objeto.

materia: cualquier cosa que tiene masa y ocupa espacio.

metal: elemento que generalmente es brillante, fácilmente puede estirarse para formar alambres o martillarse para formar hojas delgadas y es buen conductor de electricidad y energía térmica.

enlace metálico: enlace formado cuando muchos átomos metálicos comparten su banco de electrones de valencia.

metaloide: elemento que tiene las propiedades físicas y químicas de metales y no metales.

meteoro: meteorito que ha entrado a la atmósfera de la Tierra y produce un haz de luz.

meteorito: meteoroide que impacta un planeta o una luna.

meteoroid: a small rocky particle that moves through space.

mixture: matter that can vary in composition.

molecule (MAH lih kyewl): two or more atoms that are held together by covalent bonds and act as a unit.

meteoroide: partícula rocosa pequeña que se mueve por el espacio.

mezcla: materia cuya composición puede variar.

molécula: dos o más átomos que están unidos mediante enlaces covalentes y actúan como una unidad.

N

nebula: a cloud of gas and dust.

neutron: a neutral particle in the nucleus of an atom.

neutron star: a dense core of neutrons that remains after a supernova.

nitrogen fixation (NI truh jun • fihk SAY shun): the process that changes atmospheric nitrogen into nitrogen compounds that are usable by living things.

noble gas: an element in group 18 on the periodic table.

nonmetal: an element that has no metallic properties.

nuclear fusion: a process that occurs when the nuclei of several atoms combine into one larger nucleus.

nucleus: the region in the center of an atom where most of an atom's mass and positive charge is concentrated.

nebulosa: nube de gas y polvo.

neutrón: partícula neutra en el núcleo de un átomo.

estrella de neutrones: núcleo denso de neutrones que queda después de una supernova.

fijación del nitrógeno: proceso que cambia el nitrógeno atmosférico en componentes de nitrógeno útiles para los seres vivos.

gas noble: elemento del grupo 18 de la tabla periódica.

no metal: elemento que tiene propiedades no metálicas.

fusión nuclear: proceso que ocurre cuando los núcleos de varios átomos se combinan en un núcleo mayor.

núcleo: región en el centro de un átomo donde se concentra la mayor cantidad de masa y las cargas positivas.

O

observation: the act of using one or more of your senses to gather information and take note of what occurs.

orbit: the path an object follows as it moves around another object.

observación: acción de usar uno o más sentidos para reunir información y tomar notar de lo que ocurre.

órbita: trayectoria que un objeto sigue a medida que se mueve alrededor de otro objeto.

P

penumbra: the lighter part of a shadow where light is partially blocked.

percent error: the expression of error as a percentage of the accepted value.

period: a row on the periodic table.

period of revolution: the time it takes an object to travel once around the Sun.

penumbra: parte más clara de una sombra donde la luz se bloquea parcialmente.

error porcentual: expresión del error como porcentaje del valor aceptado.

periodo: hilera en la tabla periódica.

período de revolución: tiempo que gasta un objeto en dar una vuelta alrededor del Sol.

period of rotation: the time it takes an object to complete one rotation.

periodic table: a chart of the elements arranged into rows and columns according to their physical and chemical properties.

pH: an inverse measure of the concentration of hydronium ions (H_3O^+) in a solution.

phase: the lit part of the Moon or a planet that can be seen from Earth.

photoperiodism: a plant's response to the number of hours of darkness in its environment.

photosphere: the apparent surface of a star.

photosynthesis (foh toh SIHN thuh sus): a series of chemical reactions that convert light energy, water, and carbon dioxide into the food-energy molecule glucose and give off oxygen.

physical change: a change in the size, shape, form, or state of matter that does not change the matter's identity.

physical property: a characteristic of matter that you can observe or measure without changing the identity of the matter.

plant hormone: a substance that acts as a chemical messenger within a plant.

polar molecule: a molecule with a slight negative charge in one area and a slight positive charge in another area.

precipitation (prih sih puh TAY shun): water, in liquid or solid form, that falls from the atmosphere.

prediction: a statement of what will happen next in a sequence of events.

product: a substance produced by a chemical reaction.

Project Apollo: a series of space missions designed to send people to the Moon.

proton: a positively charged particle in the nucleus of an atom.

período de rotación: tiempo que gasta un objeto para completar una rotación.

tabla periódica: cuadro en que los elementos están organizados en hileras y columnas según sus propiedades físicas y químicas.

pH: medida inversa de la concentración de iones hidronio (H_3O^+) en una solución.

fase: parte iluminada de la Luna o de un planeta que se ve desde la Tierra.

fotoperiodismo: respuesta de una planta al número de horas de oscuridad en su medioambiente.

fotosfera: superficie luminosa de una estrella.

fotosíntesis: serie de reacciones químicas que convierten la energía lumínica, el agua y el dióxido de carbono en glucosa, una molécula de energía alimentaria.

cambio físico: cambio en el tamaño, la forma o estado de la materia en el que no cambia la identidad de la materia.

propiedad física: característica de la materia que puede observarse o medirse sin cambiar la identidad de la materia.

fitohormona: sustancia que actúa como mensajero químico dentro de una planta.

molécula polar: molécula con carga ligeramente negativa en una parte y ligeramente positiva en otra.

precipitación: agua, en forma líquida o sólida, que cae de la atmósfera.

predicción: afirmación de lo que ocurrirá después en una secuencia de eventos.

producto: sustancia producida por una reacción química.

Proyecto Apolo: serie de misiones espaciales diseñadas para enviar personas a la Luna.

protón: partícula cargada positivamente en el núcleo de un átomo.

Q

qualitative data: the use of words to describe what is observed in an experiment.

datos cualitativos: uso de palabras para describir lo que se observa en un experimento.

quantitative data: the use of numbers to describe what is observed in an experiment.

datos cuantitativos: uso de números para describir lo que se observa en un experimento.

R

radiative zone: a shell of cooler hydrogen above a star's core.

radio telescope: a telescope that collects radio waves and some microwaves using an antenna that looks like a TV satellite dish.

reactant: a starting substance in a chemical reaction.

reflecting telescope: a telescope that uses a curved mirror to concentrate light from distant objects.

refracting telescope: a telescope that uses a convex lens to concentrate light from distant objects.

revolution: the orbit of one object around another object.

rocket: a vehicle propelled by the exhaust made from burning fuel.

rotation: the spin of an object around its axis.

rotation axis: the line on which an object rotates

zona radiativa: capa de hidrógeno más frío por encima del núcleo de una estrella.

radiotelescopio: telescopio que recoge ondas de radio y algunas microondas por medio de una antena parecida a una antena parabólica de TV.

reactivo: sustancia inicial en una reacción química.

telescopio reflector: telescopio que tiene un espejo para reunir y enfocar luz de objetos lejanos.

telescopio refractor: telescopio que usa un lente convexo para enfocar la luz de objetos lejanos.

revolución: movimiento de un objeto alrededor de otro objeto.

cohete: vehículo propulsado por gases de escape producidos por la ignición de combustible.

rotación: movimiento giratorio de un objeto sobre su eje.

eje de rotación: línea sobre la cual un objeto rota.

S

satellite: any small object that orbits a larger object other than a star.

saturated solution: a solution that contains the maximum amount of solute the solution can hold at a given temperature and pressure.

science: the investigation and exploration of natural events and of the new information that results from those investigations.

scientific law: a rule that describes a pattern in nature.

scientific literacy: having knowledge of scientific concepts and being able to use that knowledge in your everyday life.

satélite: cualquier objeto pequeño que orbita un objeto más grande diferente de una estrella.

solución saturada: solución que contiene la cantidad máxima de soluto que la solución puede sostener a cierta temperatura y presión.

ciencia: investigación y exploración de eventos naturales y la información nueva que resulta de dichas investigaciones.

ley científica: regla que describe un patrón en la naturaleza.

saber científico: tener conocimiento de conceptos científicos y ser capaz de usarlo en la vida diaria.

scientific notation: a method of writing or displaying very small or very large numbers.

scientific theory: an explanation of observations or events that is based on knowledge gained from many observations and investigations.

semiconductor: a substance that conducts electricity at high temperatures but not at low temperatures.

single-replacement reaction: a type of chemical reaction in which one element replaces another element in a compound.

solar eclipse: an occurrence during which the Moon's shadow appears on Earth's surface.

solid: matter that has a definite shape and a definite volume.

solstice: when Earth's rotation axis is tilted directly toward or away from the Sun.

solubility (sahl yuh BIH luh tee): the maximum amount of solute that can dissolve in a given amount of solvent at a given temperature and pressure.

solute: any substance in a solution other than the solvent.

solution: another name for a homogeneous mixture.

solvent: the substance that exists in the greatest quantity in a solution.

space probe: an uncrewed spacecraft sent from Earth to explore objects in space.

space shuttles: reusable spacecraft that transport people and materials to and from space.

spectroscope: an instrument that spreads light into different wavelengths.

star: a large sphere of hydrogen gas, held together by gravity, that is hot enough for nuclear reactions to occur in its core.

stimulus: any change in an organism's environment that causes a response.

notación científica: método para escribir o expresar números muy pequeños o muy grandes.

teoría científica: explicación de las observaciones y los eventos basada en conocimiento obtenido en muchas observaciones e investigaciones.

semiconductor: sustancia que conduce electricidad a altas temperaturas, pero no a bajas temperaturas.

reacción de sustitución sencilla: tipo de reacción química en la que un elemento reemplaza a otro en un compuesto.

eclipse solar: acontecimiento durante el cual la sombra de la Luna aparece sobre la superficie de la Tierra.

sólido: materia con forma y volumen definidos.

solsticio: cuando el eje de rotación de la Tierra se inclina acercándose o alejándose del Sol.

solubilidad: cantidad máxima de soluto que puede disolverse en una cantidad dada de solvente a temperatura y presión dadas.

soluto: cualquier sustancia en una solución diferente del solvente.

solución: otro nombre para una mezcla homogénea.

solvente: sustancia que existe en mayor cantidad en una solución.

sonda espacial: nave espacial sin tripulación enviada desde la Tierra para explorar objetos en el espacio.

transbordador espacial: nave espacial reutilizable que transporta personas y materiales hacia y desde el espacio.

espectroscopio: instrumento utilizado para propagar la luz en diferentes longitudes de onda.

estrella: esfera enorme de gas de hidrógeno, que se mantiene unida por la gravedad, lo suficientemente caliente para producir reacciones nucleares en el núcleo.

estímulo: cualquier cambio en el medioambiente de un organismo que causa una respuesta.

substance: matter with a composition that is always the same.

supernova: an enormous explosion that destroys a star.

synthesis (SIHN thuh sus): a type of chemical reaction in which two or more substances combine and form one compound.

sustancia: materia cuya composición es siempre la misma.

supernova: explosión enorme que destruye una estrella.

síntesis: tipo de reacción química en el que dos o más sustancias se combinan y forman un compuesto.

technology: the practical use of scientific knowledge, especially for industrial or commercial use.

terrestrial planets: Earth and the other inner planets that are closest to the Sun including Mercury, Venus, and Mars.

tide: the periodic rise and fall of the ocean's surface caused by the gravitational force between Earth and the Moon, and Earth and the Sun.

transition element: an element in groups 3–12 on the periodic table.

tropism: plant growth toward or away from an external stimulus.

tecnología: uso práctico del conocimiento científico, especialmente para empleo industrial o comercial.

planetas terrestres: la Tierra y otros planetas interiores que están más cerca del Sol, incluidos Mercurio, Venus y Marte.

marea: ascenso y descenso periódico de la superficie del océano causados por la fuerza gravitacional entre la Tierra y la Luna, y entre la Tierra y el Sol.

elemento de transición: elemento de los grupos 3–12 de la tabla periódica.

tropismo: crecimiento de una planta hacia o alejado de un estímulo externo.

umbra: the central, darker part of a shadow where light is totally blocked.

unsaturated solution: a solution that can still dissolve more solute at a given temperature and pressure.

umbra: parte central más oscura de una sombra donde la luz está completamente bloqueada.

solución insaturada: solución que aún puede disolver más soluto a cierta temperatura y presión.

valence electron: the outermost electron of an atom that participates in chemical bonding.

variable: any factor that can have more than one value.

volume: the amount of space a sample of matter occupies.

electrón de valencia: electrón más externo de un átomo que participa en el enlace químico.

variable: cualquier factor que tenga más de un valor.

volumen: cantidad de espacio que ocupa la materia.

waning phases: phases of the Moon during which less of the Moon's near side is lit each night.

fases menguantes: fases de la Luna durante las cuales el lado cercano de la Luna está menos iluminado cada noche.

waxing phases: phases of the Moon during which more of the Moon's near side is lit each night.

white dwarf: a hot, dense, slowly cooling sphere of carbon.

fases crecientes: fases de la Luna durante las cuales el lado cercano de la Luna está más iluminado cada noche.

enana blanca: esfera de carbón caliente y densa que se enfría lentamente.

Index

Photo Credits

COVER Russell Burden/Photolibrary **vii** The McGraw-Hill Companies; **NOS 2-3** photo by John Kaplan, NASA; **NOS 3** (b)The McGraw-Hill Companies; **NOS 4** (inset)SMC Images/Getty Images, (bkgd)Popperfoto/Getty Images; **NOS 5** (t)Maria Stenzel/Getty, (c)Stephen Alvarez/Getty Images, (b)Science Source/Photo Researchers, Inc.; **NOS 6** Martyn Chillmaid/photolibrary.com; **NOS 7** NASA; **NOS 9** (t)Andy Sacks/Getty Images, (c)NASA, H. Ford (JHU), G. Illingworth (UCSC/LO), M.Clampin (STScI), G. Hartig (STScI), the ACS Science Team, and ESA, (b)Brand X Pictures/Punchstock; **NOS 10** (t)Hutchings Photography/Digital Light Source, StockShot/Alamy; **NOS 11** SOHO (ESA & NASA), Michael Newman/PhotoEdit; **NOS 12** Tim Wright/CORBIS; **1 NOS 4** (bkgd) Hutchings Photography/Digital Light Source; **NOS 16** (tl)Matt Meadows, (tr)ASP/YPP/age fotostock, (bl)Blair Seitz/photolibrary.com, (br)The McGraw-Hill Companies, Inc./Louis Rosenstock, photographer; **NOS 17** (t) The McGraw-Hill Comapnies, (c)photostock1/Alamy, (b)Blend Images/Alamy; **NOS 18** The McGraw-Hill Companies; **NOS 20** Hutchings Photography/Digital Light Source, MANDEL NGAN/AFP/Getty Images; **NOS 21** Brian Stevenson/Getty Images; **NOS 22** (t)Hisham Ibrahim/Getty Images, (bl)Bordner Aerials, (br)AP Photo/NTSB; **NOS 24** Larry Downing/Reuters/Landov; **NOS 25** Plus Pix/age fotostock; **NOS 26** ASSOCIATED PRESS; **4** Reuters/CORBIS **5** (l) John W. van de Lindt/Colorado State University, (c) Daniel Cox/O.H. Hinsdale Research Laboratory/Oregon State University, (r) Pacific Marine Environmental Laboratory/NOAA; **8-9** O. Alamany & E. Vicens/CORBIS; **11** NASA Human Spaceflight Collection, **21** NASA; **23** (t)Lunar and Planetary Institute, (c)NASA/JPL/USGS, (bl)ClassicStock/Alamy, (br)Lunar and Planetary Institute, AP Photo/The Minnesota Daily, Stacy Bengs; **25** Eckhard Slawik/Photo Researchers, Inc.; **27** (c)Copyright by Fred Espenak, www.MrEclipse.com; **28** Hutchings Photography/Digital Light Source; **29** (t)Andy Sacks/Getty Images, (l) Copyright by Fred Espenak, www.MrEclipse.com, (r)UPI Photo/NASA/Landov; **30** (l)Eckhard Slawik/Photo Researchers, Inc., (r)Copyright by Fred Espenak www.MrEclipse.com; **31** Copyright by Fred Espenak, www.MrEclipse.com; **33** Robert Estall photo agency/Alamy; **48** (b)Hutchings Photography/Digital Light Source; **48-49** NASA/JPL/Space Science Institute; **50** Hutchings Photography/Digital Light Source; **51** (t) UVImages/amanaimages/Corbis, (b)Diego Barucco/Alamy; **53** NASA/JPL; **56** (t)UVImages/amanaimages/Corbis; **57** (bkgd) NASA and H. Richer (University of British Columbia), (cl)American Museum of Natural History; **58** (t)Hutchings Photography/Digital Light Source; **59** (t)ESA/DLR/FU Berlin (G. Neukum), (bl)NASA/Johns Hopkins University Applied Physics Laboratory/Carnegie Institution of Washington, (bcl)NASA, (bcr)NASA Goddard Space Flight Center, (br)NASA/JPL/Malin Space Science Systems; **60** NASA/Johns Hopkins University Applied Physics Laboratory/Carnegie Institution of Washington; **61** (c)(r)NASA/JPL; **62** (tl)NASA, (r)Image Ideas/PictureQuest; **62** (bl)Comstock/JupiterImages; **63** (tl)NASA/JPL, (b)NASA/JPL/University of Arizona; **64** (t)NASA/JPL/Malin Space Science Systems, (c)Comstock/JupiterImages, (b)NASA; **67** (l)NASA/JPL/USGS, (cl) NASA and The Hubble Heritage Team (STScI/AURA)Acknowledgment: R.G. French (Wellesley College), J. Cuzzi (NASA/Ames), L. Do, (cr)(r)NASA/JPL; **68** (r)NASA/JPL; **70** (t)NASA/JPL/Space Science Institute, (bl)NASA/ESA and Erich Karkoschka, University of Arizona; **71** NASA/JPL; **72** (t)NASA/JPL, (c)NASA/JPL/USGS, (b)NASA/ESA and Erich Karkoschka, University of Arizona; **73** Frederick M. Brown/Getty Images; **74** (t)Hutchings Photography/Digital Light Source; **75** (t)Gordon Garradd/SPL/Photo Researchers, Inc.; **76** (l)Dr. R. Albrecht, ESA/ESO Space Telescope European Coordinating Facility; NASA, (tr)NASA, ESA, and J. Parker (Southwest Research Institute), (br)NASA, ESA, and M. Brown (California Institute of Technology); **77** (l to r, t to b)NASA/JPL/JHUAPL, (2)NASA/Goddard Space Flight Center Scientific Visualization Studio, (3)NASA/JPL/USGS, (4)Ben

Zellner (Georgia Southern University), Peter Thomas (Cornell University), NASA/ESA, (5)Roger Ressmeyer, (6)NASA/JPL-Caltech; **78** (t)Jonathan Blair/CORBIS; **79** (t)NASA/JPL/USGS, (c)Gordon Garradd/SPL/Photo Researchers, Inc., (b)Hutchings Photography/Digital Light Source; **80** (t) UVImages/amanaimages/Corbis, (b)Roger Ressmeyer; **92-93** NASA, ESA, and S. Beckwith (STScI) and the HUDF Team, **94** Hutchings Photography/Digital Light Source; **95** (t)Stephen & Donna O'Meara/Photo Researchers, Inc., (b)Joseph Baylor Roberts/Getty Images; **102** Digital Vision/PunchStock; **103** Science Source/Photo Researchers, Inc; **105** (t)Jerry Lodriguss/Photo Researchers, Inc., (tc)Naval Research Laboratory, (b)Arctic-Images/Getty Images, (bc)SOHO Consortium, ESA, NASA; **106** Photo courtesy of NASA/CORBIS; **109** STEREO Stereoscopic Observations Constraining the Initiation of Polar Coronal Jets J. P. Wuesler/NASA; **111** NASA; **115** (t)X-ray: NASA/CXC/SAO; Optical: NASA/STScI, (b)NASA, The Hubble Heritage Team (STScI/AURA), Y.-H. Chu (UIUC), S. Kulkarni (Caltech) and R. Rothschild (UCSD); **119** NASA/Alamy, NASA/Hubble Heritage Team; **120** (t)CORBIS, (c)Robert Gendler/NASA, (b)NASA, ESA, and The Hubble Heritage Team (STScI/AURA); **137** NASA; **138-139** Stocktrek/age fotostock; **140** Hutchings Photography/Digital Light Source; **141** NASA and The Hubble Heritage Team (AURA/STScI); **144** (inset)Richard Wainscoat/Alamy, (bkgd)Roger Ressmeyer/CORBIS; **145** (t)Images Etc Ltd/Getty Images, (c)Time & Life Pictures/Getty Images, (b)Starfire Optical Range/USAF/Roger Ressmeyer/CORBIS; **146** NASA; **147** (t)NASA/JPL-Caltech/STScI/CXC/SAO, (b)NASA/GSFC; **149** NASA; **150** NASA; **151** NASA/GSFC/METI/ERSDAC/JAROS, and U.S./Japan ASTER Science Team; **152** Hutchings Photography/Digital Light Source; **153** (t)NASA, (b)Stocktrek/age fotostock; **154** (l)NASA, (cl)NASA/SCIENCE PHOTO LIBRARY, (cr)NASA/JPL, (r)Stocktrek/CORBIS; **155** (l)AP Images, (c)NASA/JPL, (r)Atlas Photo Bank/Photo Researchers, Inc; **156** (l)Stocktrek/age fotostock, (r)Stocktrek Images/Getty Images; **157** NASA Marshall Space Flight Center (NASA-MSFC); **159** NASA/Johns Hopkins University Applied Physics Laboratory; **160** Hutchings Photography/Digital Light Source; **161** (t)Brand X Pictures/PunchStock, (b)ESA./NASA/SOHO; **162** (l)NASA/Johns Hopkins University Applied Physics Laboratory/Carnegie Institution of Washington., (r)NASA/epa/CORBIS; **163** (l)Craig Attebery/NASA, (r)NASA/Johns Hopkins University Applied Physics Laboratory/Southwest Research Institute (NASA/JHUAPL/SwRI); **164** Michael Hixenbaugh/National Science Foundation; **165** (t)Arco Images GmbH/Alamy, (bl)Galileo Project/JPL/NASA, (br)NASA/JPL/University of Arizona/University of Colorado; **166** (t)NASA/Ames Wendy Stenzel, (b)NASA; **180** (t)Jochen Tack/photolibrary, (bl)Biophoto Associates/Photo Researchers, Inc., (br)Oleksiy Maksymenko/Alamy; **181** (t) Dennis Kunkel Microscopy, Inc./Visuals Unlimited, Inc.,(c) PhotoAlto/PunchStock, (b)Amanda Hall/Robert Harding World Imagery/Getty Images; **184-185** age fotostock/SuperStock; **186** Hutchings Photography/Digital Light Source; **187** Stephen Frink/Getty Images; **188** (tl)(tr)(cl)(cr) Hutchings Photography/Digital Light Source, (bl)(br)Richard Megna, Fundamental Photographs, NYC; **189** David Madison/CORBIS; **190** Hutchings Photography/Digital Light Source; **191** Hutchings Photography/Digital Light Source; **192** (l)Royalty-Free/CORBIS, (c)Getty Images, (r) ULTRA.F/Getty Images; **193** (l to r)Brand X/CORBIS, (2)Studio/age fotostock, (3)Steve Shott/Steve Shott/Getty Images, (4)Dorling Kindersley/Getty Images, (5)Crawford/Dorling Kindersley/Getty Images; **194** Hutchings Photography/Digital Light Source; **195** (l)Underwood Photo Archives/SuperStock, Nick Koudis/Stockbyte; **196** (c)John A. Rizzo/Getty Images, (b)Hutchings Photography/Digital Light Source; **198** Hutchings Photography/Digital Light Source; **199** (t)BRUCE DALE/National Geographic Image Collection, (b)Brand X Pictures/PunchStock; **200, 201** Hutchings Photography/Digital Light Source; **202** (l)Creatas/PunchStock, (r)Stockbyte/Getty Images; **203** (t)Charles D. Winters/Photo Researchers, Inc., (c)Richard Megna, Fundamental Photographs, NYC, (b)Phil Degginger/Alamy; **204** (t)Brand X Pictures/PunchStock, (c)AP Photo/Courier Post,

Science Benchmark Practice Test

Multiple Choice *Bubble the correct answer.*

1 Amil hypothesized that plants exposed to rock music would grow faster than plants exposed to classical music. He used identical cuttings from a houseplant to grow 10 plants. He placed five of the plants in one room and five in another room. All conditions for the plants were identical, except he played rock music 24 hours per day in one room and classical 24 hours per day in the other room. He measured the starting height of each plant and then the final height of each plant after 30 days. The table below lists his data.

Rock Music	Plant Growth after 30 Days	Classical Music	Plant Growth after 30 Days
Plant 1R	1.8 cm	Plant 1C	2.0 cm
Plant 2R	2.5 m	Plant 2C	1.9 cm
Plant 3R	2.1 cm	Plant 3C	1.8 cm
Plant 4R	2.0 cm	Plant 4C	1.6 cm
Plant 5R	2.4 cm	Plant 5C	1.9 cm

Based on the data in this table, which conclusion can Emil make? **SC.8.N.1.3**

(A) The hypothesis is true.

(B) The results support the hypothesis.

(C) The results prove that the hypothesis is correct.

(D) The hypothesis must be revised to match the results.

2 Ivan studies the movement of plant cell components around a cell's central vacuole, or storage compartment. He is devising a model to show this movement. He wants the model to reflect accurately his observations of real plant cells. What does Ivan need to do to make his model? **SC.8.N.1.6**

(F) collect empirical evidence for at least one year

(G) conduct experiments with controlled variables

(H) form a conclusion based on similar results

(I) use logical thinking and apply his imagination

3 A city resident posted a letter on the city's Web site. He said that he saw a spaceship land in a parking lot. The letter did not contain any photographs of the event. He said that although this happened once before, it can never happen again. What is one reason that this person's claim is pseudoscientific? **SC.8.N.2.1**

(A) He posted it on a city Web site.

(B) The city resident is not a scientist.

(C) The alien landing is not reproducible.

(D) There were no photographs of the event.

NGSSS for Science Benchmark Practice continued

4 While conducting research, Suzanne made observations, formed a hypothesis, designed an experiment, and collected data. She repeated her experiment several times. Each time, the results of her experiment did not support her hypothesis. What should she do next? **SC.8.N.2.2**

(F) revise her data

(G) revise her conclusion

(H) revise her hypothesis

(I) revise her observations

5 A growing community needs more electrical power to meet increased demand. The power supplier has proposed three different routes for new electrical wires. The community hires a research group to survey wildlife abundance and diversity along each route. The community wants to choose a route that will cause the least harm to local wildlife. This story is an example of which concept? **SC.8.N.4.1**

(A) an idea of pseudoscientific science

(B) new evidence that changes scientific knowledge

(C) the role of opinion in scientific investigation

(D) the use of science to inform decision making

6 This organelle is involved in a process that cells of different organisms use to maintain homeostasis.

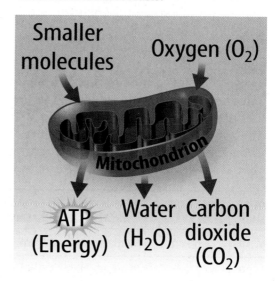

How does this organelle help maintain homeostasis? **SC.8.L.18.2**

(F) by forming sugars

(G) by obtaining water

(H) by getting rid of waste

(I) by releasing energy from food

NGSSS for Science Benchmark Practice *continued*

7 Most of the structures in plant cells also are in animal cells.

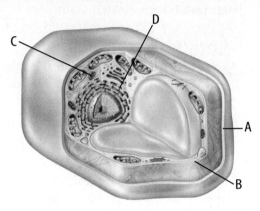

Which structure shown in this plant cell is not in animal cells? **SC.8.N.1.1**

(A) A

(B) B

(C) C

(D) D

8 This pedigree chart shows a family's phenotypes for earlobes.

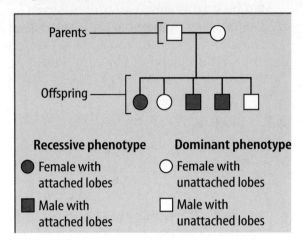

Based on this pedigree chart, what is the probability that a sixth child will have the recessive trait, attached earlobes? **SC.8.N.1.1**

(F) 10 percent

(G) 25 percent

(H) 50 percent

(I) 75 percent

9 Astronomers describe distances in space using a unit called a light-year. A light-year is equal to about 10 trillion km. The center of our galaxy, the Milky Way galaxy, is about 26,000 light-years away from Earth. About how far away is the center of our galaxy? **SC.8.E.5.1**

(A) 2,600 trillion km

(B) 260,000 trillion km

(C) 2,600,000 trillion km

(D) 26,000,000,000,000 trillion km

10 Steve observes through a telescope two stars that appear to have the same color and brightness. However, he knows that one of the stars is farther away from Earth than the other. Which conclusion could he draw about these two stars? **SC.8.E.5.5**

(F) They have the same apparent magnitude but different luminosities.

(G) They have the same apparent magnitude but different temperatures.

(H) They have the same luminosity but different absolute magnitudes.

(I) They have the same luminosity but different apparent magnitudes.

11 The outer planets—Jupiter, Saturn, Uranus, and Neptune—have larger orbits than Earth does. Which statement best compares the periods of revolution of the planets? **SC.8.E.5.7**

(A) All the outer planets have periods of revolution that are greater than Earth has.

(B) All the outer planets have periods of revolution that are smaller than Earth has.

(C) Some of the outer planets have the same period of revolution as the Earth has.

(D) Some of the outer planets have periods of revolution that are greater than Earth has.

NGSSS for Science Benchmark Practice continued

12 LaShawn finds this diagram in a science book.

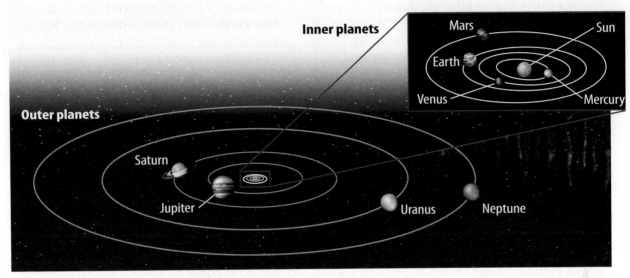

What can he conclude about scientists' understanding of the solar system at the time the illustration was made? **SC.8.E.5.8**

(F) Astronomers classified the inner planets as dwarf planets.

(G) Astronomers thought that the planets revolved around Earth.

(H) Astronomers thought that the planets revolved around the Sun.

(I) Astronomers thought that the outer planets were a new solar system.

13 Cassandra starts a soccer team in her hometown on Merritt Island, Florida, near the Kennedy Space Center. Because the *Apollo 11* space mission was launched from this center, Cassandra wants to name the soccer team "The Apollos." Which picture would best represent a mascot named after the *Apollo 11* space mission? **SC.8.E.5.12**

(A) *Sputnik 1*

(B) space shuttle

(C) astronaut on the moon

(D) *International Space Station*

NGSSS for Science Benchmark Practice continued

14 These diagrams model the atoms of a substance in three different states.

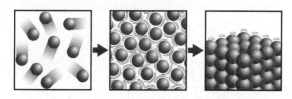

Which set of labels correctly identifies each state in the diagrams? **SC.8.P.8.1**

- (F) gas → liquid → solid
- (G) gas → solid → liquid
- (H) liquid → gas → solid
- (I) liquid → solid → gas

15 Imagine that an astronaut takes his lucky horseshoe to the Moon. What characteristic of the lucky horseshoe will be different on the Moon than on Earth? **SC.8.P.8.2**

- (A) density
- (B) mass
- (C) volume
- (D) weight

16 Antonio measures the masses and volumes of samples of four different liquids and lists them in the table below.

Liquid	Volume (mL)	Mass (g)
corn syrup	20	27.6
rubbing alcohol	30	23.5
vegetable oil	15	13.8
water	25	25

Which liquid has the greatest density? **SC.8.P.8.3**

- (F) corn syrup
- (G) cold water
- (H) rubbing alcohol
- (I) vegetable oil

NGSSS for Science Benchmark Practice continued

17 Dr. Phung is trying to identify a substance. She knows that is a white powder at room temperature (20°C), a liquid at 750°C, and a gas at 1400°C. She knows that this means the substance melts (turns from solid to liquid) at a temperature below 750°C and boils (turns from liquid to gas) at a temperature above 1400°C. She looks up the melting and boiling points of four different white powders in a science handbook and lists them in the table below.

Substance	Melting Point (°C)	Boiling Point (°C)
Potassium bromide	734	1435
Potassium chloride	790	1420
Sodium bromide	747	1396
Sodium chloride	801	1413

What most likely is the substance based on the information in the table? **SC.8.P.8.4**

(A) potassium bromide

(B) potassium chloride

(C) sodium bromide

(D) sodium chloride

18 Janice's candy recipe states that she must stir the hot mixture constantly for 10 minutes over medium heat. She decides to stir with a wooden spoon rather than a metal spoon to avoid burning her hand. Which physical property did Janice use to choose a spoon for stirring the hot candy mixture? **SC.8.P.8.4**

(F) boiling point

(G) melting point

(H) electrical conductivity

(I) thermal conductivity

19 Hayden uses a pH meter to test an unknown liquid. He finds that it has a pH of 4. What can Hayden conclude about the unknown liquid? **SC.8.P.8.8**

(A) It is an acid.

(B) It is a base.

(C) It is a mixture.

(D) It is a solution.

NGSSS for Science Benchmark Practice continued

20 A scientist examines a sample of a liquid under a high-powered microscope. He observes that the liquid has the same color and texture throughout. He cannot find different particles. Which term or terms could he use to describe the liquid? **SC.8.P.8.9**

(F) "substance" only

(G) "heterogeneous mixture" only

(H) "substance" or "homogeneous mixture"

(I) "substance" or "heterogeneous mixture"

21 Rhonda is a park ranger at Three Rivers State Park in northwest Florida. She prepares a yearly report on how much the park staff spends on damage repairs. She classifies each type of damage as the result of a physical change or a chemical change. Which would she describe as a physical change? **SC.8.P.9.2**

(A) burning of wooden trail markers during the spread of forest fires

(B) crumbling of roads due to water that freezes and expands in cracks

(C) rotting of wooden railings along lookouts and fishing docks

(D) rusting of steel parts in bridges and water tanks that have no paint

22 Javier saw a sign at a store that said, "Be prepared for this week's changing climate: Buy an umbrella and snow shoes!" What is wrong with the sign's use of the term "climate"? **LA.8.2.2.3**

(F) Climate describes a permanent thing that does not ever change.

(G) Climate describes conditions over a period of time longer than a week.

(H) Climate describes rainfall patterns, but it does not describe temperature patterns.

(I) Climate describes temperature patterns, but it does not describe rainfall patterns.

23 Emily is riding in a roller coaster car that slows down as it climbs each hill and speeds up as it rolls down the hill. How can she describe this pattern in terms of potential energy and kinetic energy? **SC.8.N.1.1**

(A) Potential energy transforms into kinetic energy as the car climbs each hill.

(B) Potential energy transforms into kinetic energy as the car rolls down each hill.

(C) Potential energy is created while kinetic energy is destroyed as the car rolls down each hill.

(D) Potential energy is destroyed while kinetic energy is created as the car rolls down each hill.

NGSSS for Science Benchmark Practice continued

24 A scientist tests a fossil containing a certain radioactive element whose rate of decay is described by the graph below.

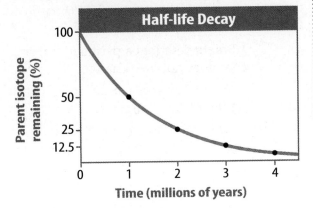

She determines that the fossil contains only about 25% of its original amount of the radioactive element. About how old is the fossil? **SC.8.N.1.1**

(F) 1 million years old

(G) 2 million years old

(H) 3 million years old

(I) 4 million years old

25 Babcock Ranch is a planned city that will be built in southwestern Florida. If the project is successful, it will be the first U. S. city to use only solar power during the day. Its power will come from a large photovoltaic solar plant. What kind of energy transformation occurs at a solar power plant? **SC.8.N.1.1**

(A) Chemical energy transforms to electric energy.

(B) Electric energy transforms to light energy.

(C) Light energy transforms to electric energy.

(D) Nuclear energy transforms to light energy.

Science Benchmark Mastery Test

This test covers benchmarks from grades 6, 7, and 8.

Multiple Choice *Bubble the correct answer.*

1 Bianca conducts an experiment. She fills three 1-L glass beakers with the same liquid. Her set-up is shown here.

Based on the illustration, what is the independent variable in Bianca's experiment? **SC.8.N.1.1**

(A) air pressure

(B) type of liquid

(C) volume of the beaker

(D) mass of the rectangular object

2 Ben claims that he conducted careful investigations and concluded that his pet fish could read basic English words. After a long and healthy life, the fish died. Ben says that the investigations cannot be replicated with other fish. Ben's big sister says that his investigation is not scientific. Why is that so? **SC.8.N.1.2**

(F) Other researchers cannot change more than one variable.

(G) Other researchers cannot observe his original experiments.

(H) Other researchers cannot confirm the results of the investigation.

(I) Other researchers cannot practice his techniques to improve their skills.

3 Which question could be answered through modeling but not through experimentation? **SC.8.N.1.3**

(A) At what rate is global warming occurring?

(B) How does UV light affect protein structure?

(C) Which material that is used in running shoe soles is most likely shock-absorbent?

(D) What is the minimum level of carbon monoxide detected by a carbon monoxide detector?

4 Which human body system includes organs that are also part of the digestive and respiratory systems? **SC.6.L.14.5**

(F) excretory system

(G) nervous system

(H) reproductive system

(I) skeletal system

5 A class conducted an investigation to see if the freshness of an unfertilized chicken egg relates to its mass. Unfertilized eggs decrease in mass over time. Three groups used the same research methods. Each group measured and recorded the mass of three eggs on the day the eggs were laid and then again 30 days later.

Group 1	
Mass of Fresh Egg	**Mass of Egg after 30 Days**
51 g	48 g
48 g	46 g
47 g	45 g

Group 2	
Mass of Fresh Egg	**Mass of Egg after 30 Days**
49 g	49 g
52 g	53 g
50 g	50 g

Group 3	
Mass of Fresh Egg	**Mass of Egg after 30 Days**
47 g	45 g
49 g	47 g
51 g	48 g

Groups 1 and 3 concluded that the freshness related to mass. Group 2 concluded that the freshness of the eggs was not related to mass. Based on the data, what could explain why Group 2 came to a different conclusion? **SC.8.N.1.4**

(A) They made an error in measuring the mass of the eggs.

(B) They measured mass in standard rather than metric units.

(C) Their eggs decreased in mass more than the other groups' eggs.

(D) Their eggs were smaller, on average, than the other groups' eggs.

6 The table shows data for the speed, mass, and kinetic energy of three vehicles.

Vehicle	Speed	Mass	Relative Kinetic Energy of Vehicle
Vehicle 1	15 m/s	1,500 kg	least
Vehicle 2	25 m/s	1,500 kg	middle
Vehicle 3	15 m/s	8,000 kg	most

Based on the data, which is the best conclusion about the relationship between speed, mass, and kinetic energy? **SC.8.N.1.1**

(F) All objects with the same mass have equal amounts of kinetic energy.

(G) All objects moving at the same speed have equal amounts of kinetic energy.

(H) The kinetic energy of an object depends on both its speed and its mass.

(I) If two objects have the same mass, the one moving faster has less kinetic energy.

NGSSS for Science Benchmark Mastery continued

7 All living things can be classified into one of three domains then one of six kingdoms, based on shared characteristics.

To which Domain and Kingdom does this organism belong? **SC.6.L.15.1**

- (A) Animalia, Protista
- (B) Archaea, Eukarya
- (C) Eukarya, Animalia
- (D) Protista, Archaea

8 The theory of seafloor spreading explains how continents move apart as the seafloor spreads along a mid-ocean ridge. Evidence for this theory includes parallel magnetic signatures in rocks on either side of a mid-ocean ridge. What makes this a scientific theory and not a scientific law? **SC.8.N.3.1**

- (F) It is supported by most of the scientists but not by all of the scientists.
- (G) It contains one supported hypothesis rather than many supported hypotheses.
- (H) It explains why seafloor spreading occurs rather than predicting it will happen.
- (I) It's based on enough evidence to make it a theory but not enough to make it a law.

9 Over many generations, neck length changed in a turtle population on the Galápagos Islands. The ancestral population had shorter necks than the current population. What explains this change in neck length? **SC.7.L.15.2**

- (A) Evolution favored larger body structures in organisms.
- (B) Environmental factors caused the increase in neck length.
- (C) Generations of turtles stretched their necks to reach higher plants and passed on the trait.
- (D) The turtles with longer necks could reach higher plants and produced more offspring.

10 Biological adaptations enable species to survive within a constantly changing environment. Which condition could result in extinction due to a species' inability to adapt? **SC.7.L.15.3**

- (F) rapid evolution
- (G) short life cycle
- (H) a lack of genetic variation
- (I) to much genetic variation

11 Mustafa wants to make a model of the relationship between genes and chromosomes. Which model most accurately represents this relationship? **SC.7.L.16.1**

(A) Batteries that represent genes power a robot that represents a chromosome.

(B) Paragraphs that represent genes make up a book that represents a chromosome.

(C) Potatoes representing genes are blended into a soup representing a chromosome.

(D) Pulleys that represent genes lift a metal weight that represents a chromosome.

12 Biotechnology includes techniques such as cloning, genetic engineering, and artificial selection. In general, what type of issue represents a main concern about the impact of biotechnology on society? **SC.7.L.16.4**

(F) criminal issues, especially those related to DNA fingerprinting

(G) ethical issues, especially related to the manipulation of human DNA

(H) environmental issues, especially related to an increase in global warming

(I) overpopulation issues, especially related to the creation of a new species

13 The illustration shows a desert food web.

How would an increase in the population of the large black birds directly affect the food web? **SC.7.L.17.1**

(A) The cactus population would increase.

(B) The coyote population would increase.

(C) The lizard population would decrease.

(D) The rattlesnake population would decrease.

NGSSS for Science Benchmark Mastery *continued*

14 Green sea turtles nest in coastal areas of the southeast United States. On a particular Florida beach, there is plenty of space for the turtles to nest. The nearby waters contain large amounts of seagrass and algae that the turtles eat. Raccoons, birds, and humans prey upon hatchlings. Based on this information, which limiting factor has the greatest impact on the turtle population on this beach? **SC.7.L.17.3**

(F) food

(G) nesting

(H) predation

(I) shelter

15 The graph shows the relationship between the luminosities of four types of stars and their surface temperatures.

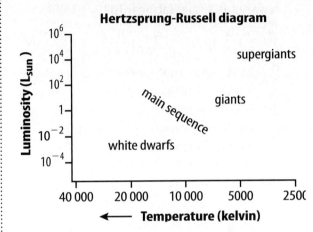

Which conclusion can you come to from the information in this graph? **SC.8.N.1.1**

(A) In general, the supergiants have a higher surface temperature than giants.

(B) In general, the greater the surface temperature, the greater the luminosity.

(C) In general, main sequence stars have a higher surface temperature than giants.

(D) In general, luminosity of white dwarfs is higher than that of main sequence stars.

NGSS for Science Benchmark Mastery continued

16 Leif is conducting research on a small freshwater flatworm. He hypothesized that the flatworm would move more quickly toward egg yolk than toward lettuce. He timed how long it took one flatworm to travel 10 cm down a channel toward the lettuce at its end. Then he timed the flatworm in the same way as it moved toward egg yolk. It took the flatworm longer to reach the lettuce. Leif concluded that flatworms prefer egg yolk. Based on the information given, what is one problem with Leif's experiment? **SC.8.N.1.2**

- (F) He changed the wrong variable.
- (G) He did not use repeated trials.
- (H) He had an incorrect hypothesis.
- (I) He should have used egg white.

17 A hypothesis is a possible explanation for an observation. A hypothesis can be tested with scientific investigations. Which generalization can be made about hypotheses that are not supported by data? **SC.8.N.1.4**

- (A) They are still valuable if they lead to further investigations.
- (B) They are probably true but the experiment was done incorrectly.
- (C) They are a waste of resources because they did not turn out to be true.
- (D) They are showing researchers were not properly prepared for the experiment.

18 Which two fields of science are most likely to use observation and models instead of controlled experiments? **SC.8.N.1.5**

- (F) astronomy and geology
- (G) biology and astronomy
- (H) chemistry and biology
- (I) geology and chemistry

19 Sadie and Joe learn that the properties of water, including adhesion and cohesion, are due to the shape of the water molecule. Which molecular model should Sadie and Joe use to represent this information? **SC.8.N.3.1**

- (A) CO_2
- (B) $:O::C::O:$
- (C) $O=C=O$
- (D) ●●●

NGSSS for Science Benchmark Mastery continued

20 Scientific theories are based on large quantities of accumulated evidence. Sometimes, new evidence contradicts a scientific theory. If the scientific community agrees that the new evidence contradicts a theory, what is most likely to happen? **SC.8.N.3.2**

(F) The theory will be modified to incorporate the new evidence.

(G) The theory will stay the same, because most evidence supports it.

(H) The theory will have a new theory being written based on new evidence.

(I) The theory will be discarded, because the new evidence does not support it.

21 Wind energy is renewable and does not create greenhouse gases. Energy released from burning coal is less expensive than wind energy but is nonrenewable and causes pollution. Some town governments help residents pay for the use of wind energy and other renewable energy. This story is an example of which concept? **SC.8.N.4.2**

(A) how science can affect economic concerns

(B) how science can affect legal concerns

(C) how political concerns can affect science

(D) how social concerns can affect science

22 Photosynthesis involves a series of chemical processes that result in energy transformations. Which of these represents the energy transformation that occurs in photosynthesis? **SC.8.L.18.1**

(F) chemical energy $\rightarrow$ kinetic energy

(G) chemical energy $\rightarrow$ solar energy

(H) solar energy $\rightarrow$ chemical energy

(I) solar energy $\rightarrow$ thermal energy

23 Cellular respiration provides energy that an organism can use. Which of these represents the process that occurs in cellular respiration? **SC.8.L.18.2**

(A) ATP + oxygen $\rightarrow$ carbon dioxide + water

(B) glucose + oxygen $\rightarrow$ carbon dioxide + water + ATP

(C) glucose + water $\rightarrow$ oxygen + water + ATP

(D) light energy + water $\rightarrow$ ATP + oxygen

24 Kia uses this model to study the carbon cycle.

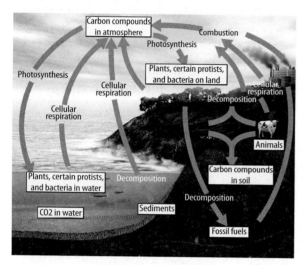

Which two processes release carbon compounds into the atmosphere? **SC.8.L.18.3**

- (F) cellular respiration and combustion
- (G) cellular respiration and photosynthesis
- (H) photosynthesis and combustion
- (I) photosynthesis and decomposition

25 Energy pyramids show that less energy is available at each trophic level in an ecosystem. If energy is conserved—that is, it cannot be created or destroyed—how is it possible that less energy is available at each trophic level? **SC.8.L.18.4**

- (A) Living systems do not conserve energy.
- (B) Living systems do not create energy.
- (C) Some energy changes to thermal energy.
- (D) Some energy will return to the Sun.

26 Imagine that you could follow the path of one particle in the air within a global wind belt. The diagram shows the global wind belts.

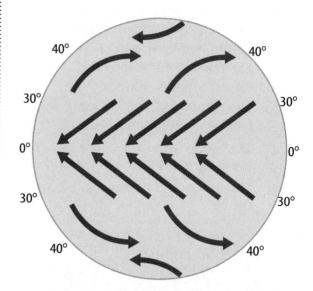

The particle travels from the 30° latitude in the Northern Hemisphere toward the equator. Where does the particle in the air go after it reaches the equator? **SC.6.E.7.3**

- (F) Its air mass cools, causing the particle to cross the equator and head south.
- (G) Its air mass cools, causing the particle to rise in altitude and head north again.
- (H) Its air mass warms up, causing the particle to cross the equator and head south.
- (I) Its air mass warms up, causing the particle to rise in altitude and head north again.

NGSSS for Science Benchmark Mastery continued

27 Earth's geosphere is the rocky part of the planet—its crust, mantle, and core. Its hydrosphere is made up of all of the water on Earth. This includes water in oceans, in streams and lakes, in the atmosphere, and underground. Which describes an interaction between Earth's geosphere and hydrosphere? **SC.6.E.7.4**

(A) Groundwater is heated by nearby underground pockets of magma, forming steam that rises out of a geyser.

(B) Ocean water warmed by sunlight evaporates, enters the atmosphere, and condenses to form clouds.

(C) Plants take up carbon dioxide from air and produce oxygen, which is released into the atmosphere.

(D) Snow that falls during the winter eventually melts in the spring and flows through rivers to the ocean.

28 Some of the Sun's radiation is absorbed or reflected back to space by clouds. How might this reflection and absorption affect the temperature on Earth's surface on a cloudy day? **SC.6.E.7.5**

(F) Radiation absorbed by clouds causes rain to form, making it feel warmer on a cloudy day.

(G) Radiation reflected by clouds bounces toward Earth's surface, making it feel cooler on a cloudy day.

(H) Radiation absorbed or reflected by clouds cannot heat Earth's surface, making it feel cooler on a cloudy day.

(I) Radiation absorbed or reflected by clouds brings the heat near Earth's surface, making it feel warmer on a cloudy day.

29 In August 1992, hurricane Andrew struck the coast of southern Florida. Andrew is estimated to have cost more than $25 billion dollars. Until Katrina hit Louisiana in 2005, Andrew was the costliest hurricane in U.S. history. Which was a positive effect of hurricane Andrew? **SC.6.E.7.7**

(A) Andrew knocked down one quarter of the trees in nearby protected wetlands.

(B) Lessons learned from Andrew prevented future storms from costing so much.

(C) Many insurance agencies that covered the costs of damage went bankrupt.

(D) Reconstruction after the storm created many jobs for builders and developers.

30 The graph shows the distance that an object moved over time.

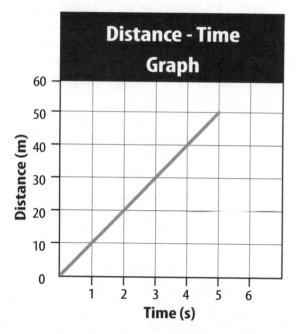

The object moved at a constant speed of 10 m/s. How would the graph be different if the object had moved at a constant speed of 20 m/s? **SC.6.P.12.1**

- (F) The line would be horizontal (flat).
- (G) The line would be curved instead of straight.
- (H) The line would have a greater slope (be steeper).
- (I) The line would have a smaller slope (be less steep).

31 The diagram shows the rock cycle. The numbered squares stand for types of rocks and rocky materials.

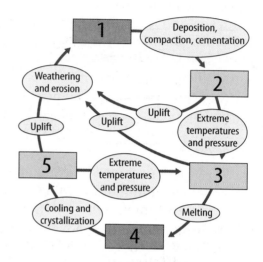

A mountain can form when lava erupts from a volcano and cools, causing a buildup of rock around the mouth of a volcano. Which number represents the kind of rock that forms this kind of mountain? **SC.7.E.6.2**

- (A) 2
- (B) 3
- (C) 4
- (D) 5

NGSSS for Science Benchmark Mastery continued

32 A scientist uses radioactive decay to compare the ages of fossils in the different rock layers, A through E, of the canyon wall shown here.

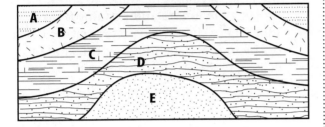

She compares the amount of radioactive parent element to the amount of stable daughter element in each fossil sample. Which layer has fossils that contain the smallest percentage of radioactive parent element still left in the sample and why? **SC.7.E.6.4**

(F) A, because it makes up the smallest part of the canyon wall

(G) C, because it is the oldest layer that has been eroded

(H) D, because it is the youngest layer that has not been eroded

(I) E, because it is the oldest layer in the canyon wall

33 Rainey wants to create an exciting light display to accompany her band's rock show. She plans to shine beams of colored light on objects that will either focus the light or cause it to bounce off in a different direction. Which could she use to bounce light in a different direction? **SC.7.P.10.2**

(A) a rough surface that absorbs most light

(B) a clear material that transmits most light

(C) a smooth surface that does not reflect much light

(D) a shiny material that does not transmit or absorb light

34 When Abe drops a basketball, its potential energy transforms into kinetic energy. When the ball hits the ground, its kinetic energy is great, but it has no potential energy. The ball bounces upward and its kinetic energy transforms into potential energy as it rises higher and higher. If Abe doesn't touch the ball, the ball will stop bouncing eventually. What causes the ball to stop bouncing eventually? **SC.7.P.11.3**

(F) Each time the ball touches the ground, some of the ball's kinetic energy will be destroyed.

(G) Each time the ball reaches the height of its bounce, some of the ball's potential energy is destroyed.

(H) With each bounce, some of the ball's kinetic energy is transformed into thermal energy, which is lost to the surroundings.

(I) With each bounce, some of the ball's potential energy is transformed into chemical energy, which cannot be transformed into kinetic energy.

35 Terrance's model compares different types of astronomical bodies. He has four objects: the school, a classroom, a table, and a ball. Which list shows the best way to use each object if he wants to model the relative size of the astronomical bodies? **SC.8.E.5.3**

(A) school = galaxy, classroom = universe, table = solar system, ball = star

(B) school = star, classroom = solar system, table = galaxy, ball = universe

(C) school = universe, classroom = galaxy, table = solar system, ball = star

(D) school = universe, classroom = galaxy, table = star, ball = solar system

36 The diagram shows Earth's motion around the Sun.

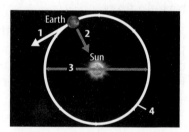

Which properties does the force represented by arrow 2 depend on?
SC.8.E.5.4

(F) the mass of the Earth and the speed at which the Earth travels

(G) the masses of Earth and the Sun and the distance between them

(H) the shape of the Earth's orbit and the speed that the Earth travels

(I) the speed that Earth travels and the distance of Earth from the Sun

37 Dan wants to show his little brother how eclipses work. He uses a flashlight to model the Sun. He uses a ball to model either Earth or the moon, depending on the kind of eclipse he wants to show. He projects the umbra and penumbra on a wall. These shadows represent either Earth or the moon, depending on the kind of eclipse, as shown in the diagram.

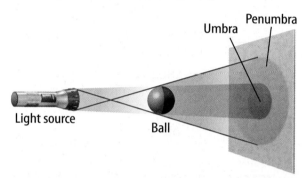

He forms a small piece of clay into a ball, which represents the observer of an eclipse. Where should he place the lump of clay if he wants to show the location of the observer of a total lunar eclipse?
SC.8.E.5.9

(A) on the lit side of the ball

(B) on the dark side of the ball

(C) within the umbra on the wall

(D) within the penumbra on the wall

38 Improvements to telescopes in the early 1600s enabled Galileo to observe Jupiter's moons. Before this discovery, no planet other than Earth was known to have moons. This discovery influenced people's rejection of the Earth-centered model of the solar system. How does technology improve scientific investigation? **SC.8.E.5.10**

(F) New technologies can be very costly and very dangerous.

(G) New technologies can lead to moral and ethical controversies.

(H) New technologies can offer opportunities for new discoveries.

(I) New technologies can slow down the process of observation.

39 The diagram represents many different wavelengths of light in the electromagnetic spectrum.

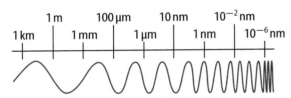

Which characteristic of light increases as you go from left to right in the diagram? **SC.8.E.5.11**

(A) amplitude

(B) energy

(C) visibility

(D) wavelength

40 A scientist carried out a chemical reaction between reactants A and B, which generated products C and D. She recorded the mass of each solid or liquid reactant and product in a table.

Product or Reactant	Mass (g)
A	25.6
B	14.3
C	?
D	21.8

Product C was a gas, so she was unable to capture it and find its mass. Given the law of conservation of mass, what can she conclude about the mass of product C? **SC.8.P.9.1**

(F) Its mass must be 18.1 g.

(G) Its mass must be 20.6 g.

(H) Its mass must be greater than 39.9.

(I) Its mass cannot be determined.

41 Jaime carries out a chemical reaction between a solid and a liquid in a laboratory flask. However, the reaction is happening too quickly for him to observe it. How might he slow down the chemical reaction? **SC.8.P.9.3**

(A) He could add a catalyst to the flask.

(B) He could shine a bright light on the flask.

(C) He could place the flask in the refrigerator.

(D) He could break the solid reactant into smaller bits.

42 In an area where wind blows mostly in one direction, a mountain can cast a rain shadow—an area on one side of the mountain where there is little rain compared with the other side. Suppose the wind blows from east to west and carries warm moist air toward the east side of a mountain. When this air reaches the east side of the mountain, it rises. On the west side of the mountain, the air sinks again. What precipitation pattern results and causes a rain shadow? **SC.6.E.7.2**

(F) The air cools as it rises, causing the moisture it holds to condense and form rain on the east side of the mountain.

(G) The air cools as it sinks, causing the moisture it holds to condense and form rain on the west side of the mountain.

(H) The air gets warmer as it rises, causing the moisture it holds to condense and form rain on the east side of the mountain.

(I) The air gets warmer as it sinks, causing the moisture it holds to condense and form rain on the west side of the mountain.

PERIODIC TABLE OF THE ELEMENTS

Legend

- 🎈 Gas
- 💧 Liquid
- ▢ Solid
- ⊙ Synthetic

Element — Hydrogen
Atomic number — 1
Symbol — H
Atomic mass — 1.01
State of matter

A column in the periodic table is called a **group**.

A row in the periodic table is called a **period**.

Group	1	2	3	4	5	6	7	8	9
1	Hydrogen 1 H 1.01								
2	Lithium 3 Li 6.94	Beryllium 4 Be 9.01							
3	Sodium 11 Na 22.99	Magnesium 12 Mg 24.31							
4	Potassium 19 K 39.10	Calcium 20 Ca 40.08	Scandium 21 Sc 44.96	Titanium 22 Ti 47.87	Vanadium 23 V 50.94	Chromium 24 Cr 52.00	Manganese 25 Mn 54.94	Iron 26 Fe 55.85	Cobalt 27 Co 58.93
5	Rubidium 37 Rb 85.47	Strontium 38 Sr 87.62	Yttrium 39 Y 88.91	Zirconium 40 Zr 91.22	Niobium 41 Nb 92.91	Molybdenum 42 Mo 95.96	Technetium 43 Tc (98)	Ruthenium 44 Ru 101.07	Rhodium 45 Rh 102.91
6	Cesium 55 Cs 132.91	Barium 56 Ba 137.33	Lanthanum 57 La 138.91	Hafnium 72 Hf 178.49	Tantalum 73 Ta 180.95	Tungsten 74 W 183.84	Rhenium 75 Re 186.21	Osmium 76 Os 190.23	Iridium 77 Ir 192.22
7	Francium 87 Fr (223)	Radium 88 Ra (226)	Actinium 89 Ac (227)	Rutherfordium 104 Rf (267)	Dubnium 105 Db (268)	Seaborgium 106 Sg (271)	Bohrium 107 Bh (272)	Hassium 108 Hs (270)	Meitnerium 109 Mt (276)

The number in parentheses is the mass number of the longest lived isotope for that element.

Lanthanide series	Cerium 58 Ce 140.12	Praseodymium 59 Pr 140.91	Neodymium 60 Nd 144.24	Promethium 61 Pm (145)	Samarium 62 Sm 150.36	Europium 63 Eu 151.96
Actinide series	Thorium 90 Th 232.04	Protactinium 91 Pa 231.04	Uranium 92 U 238.03	Neptunium 93 Np (237)	Plutonium 94 Pu (244)	Americium 95 Am (243)

Metal
Metalloid
Nonmetal
Recently discovered

			13	**14**	**15**	**16**	**17**	**18**
								Helium 2 **He** 4.00
			Boron 5 **B** 10.81	Carbon 6 **C** 12.01	Nitrogen 7 **N** 14.01	Oxygen 8 **O** 16.00	Fluorine 9 **F** 19.00	Neon 10 **Ne** 20.18
10	**11**	**12**	Aluminum 13 **Al** 26.98	Silicon 14 **Si** 28.09	Phosphorus 15 **P** 30.97	Sulfur 16 **S** 32.07	Chlorine 17 **Cl** 35.45	Argon 18 **Ar** 39.95
Nickel 28 **Ni** 58.69	Copper 29 **Cu** 63.55	Zinc 30 **Zn** 65.38	Gallium 31 **Ga** 69.72	Germanium 32 **Ge** 72.64	Arsenic 33 **As** 74.92	Selenium 34 **Se** 78.96	Bromine 35 **Br** 79.90	Krypton 36 **Kr** 83.80
Palladium 46 **Pd** 106.42	Silver 47 **Ag** 107.87	Cadmium 48 **Cd** 112.41	Indium 49 **In** 114.82	Tin 50 **Sn** 118.71	Antimony 51 **Sb** 121.76	Tellurium 52 **Te** 127.60	Iodine 53 **I** 126.90	Xenon 54 **Xe** 131.29
Platinum 78 **Pt** 195.08	Gold 79 **Au** 196.97	Mercury 80 **Hg** 200.59	Thallium 81 **Tl** 204.38	Lead 82 **Pb** 207.20	Bismuth 83 **Bi** 208.98	Polonium 84 **Po** (209)	Astatine 85 **At** (210)	Radon 86 **Rn** (222)
Darmstadtium 110 **Ds** (281)	Roentgenium 111 **Rg** (280)	Copernicium 112 **Cn** (285)	Ununtrium * 113 **Uut** (284)	Ununquadium * 114 **Uuq** (289)	Ununpentium * 115 **Uup** (288)	Ununhexium * 116 **Uuh** (293)		Ununoctium * 118 **Uuo** (294)

* The names and symbols for elements 113-116 and 118 are temporary. Final names will be selected when the elements' discoveries are verified.

Gadolinium 64 **Gd** 157.25	Terbium 65 **Tb** 158.93	Dysprosium 66 **Dy** 162.50	Holmium 67 **Ho** 164.93	Erbium 68 **Er** 167.26	Thulium 69 **Tm** 168.93	Ytterbium 70 **Yb** 173.05	Lutetium 71 **Lu** 174.97
Curium 96 **Cm** (247)	Berkelium 97 **Bk** (247)	Californium 98 **Cf** (251)	Einsteinium 99 **Es** (252)	Fermium 100 **Fm** (257)	Mendelevium 101 **Md** (258)	Nobelium 102 **No** (259)	Lawrencium 103 **Lr** (262)